4/5/93

Integral equations

Cambridge texts in applied mathematics

Maximum and minimum principles: a unified approach with applications
M. J. SEWELL

Introduction to numerical linear algebra and optimization
P. G. CIARLET

Solitons: an introduction
P. G. DRAZIN AND R. S. JOHNSON

The kinematics of mixing: stretching, chaos and transport
J. M. OTTINO

Integral equations: a practical treatment, from spectral theory to applications
D. PORTER AND D. S. G. STIRLING

Integral equations
A practical treatment,
from spectral theory to applications

DAVID PORTER
Senior Lecturer in Applied Mathematics,
University of Reading

DAVID S. G. STIRLING
Lecturer in Pure Mathematics,
University of Reading

CAMBRIDGE
UNIVERSITY PRESS

Published by the Press Syndicate of the University of Cambridge
The Pitt Building, Trumpington Street, Cambridge CB2 1RP
40 West 20th Street, New York NY 10011-4211, USA
10 Stamford Road, Oakleigh, Victoria 3166, Australia

© Cambridge University Press 1990

First published 1990
Reprinted 1993

Printed in Canada

British Library cataloguing in publication data

Porter, David
Integral equations.
1. Integral equations
I. Title   II. Stirling, David S. G.
515.4′5

Library of Congress cataloguing in publication data

Porter, David
Integral equations: a rigorous and practical treatment/
David Porter and David S. G. Stirling.
p.   cm.
ISBN 0 521 33151 X. – ISBN 0 521 33742 9 (paperback)
1. Integral equations.   I. Stirling, David S. G.
II. Title.
QA431.P67 1990
515′45—dc20   89-37313   CIP

ISBN 0-521-33151-X hardback
ISBN 0-521-33742-9 paperback

Contents

Preface   ix

1   Classification and examples of integral equations

1.1   Introduction   1
1.2   Classification of integral equations   2
1.3   A collection of examples   7
       1.3.1   Initial value problems for ordinary differential equations   7
       1.3.2   Boundary value problems for ordinary differential equations   13
       1.3.3   Boundary value problems for elliptic partial differential
               equations   16
       1.3.4   Abel's problem   21
       Problems   24

2   Second order ordinary differential equations and integral equations

2.1   Introduction   31
2.2   Differential equation theory   31
2.3   Initial value problems   33
2.4   Boundary value problems   34
2.5   Singular boundary value problems   44
       Problems   48

3   Integral equations of the second kind

3.1   Introduction   55
3.2   Degenerate kernels   57
3.3   A different approach   62
3.4   Operators   72
3.5   The Neumann series   78
       Problems   88

4   Compact operators

4.1   Introduction   95
4.2   General properties   95
4.3   Adjoint operators   101
4.4   The Spectral Theorem   107

4.5	Applications to integral equations	120
	Problems	133

5 The spectrum of a compact self-adjoint operator

5.1	Introduction	139
5.2	The Rayleigh quotient	139
5.3	Eigenvalue inequalities	146
	Problems	159

6 Positive operators

6.1	Introduction	163
6.2	General properties	163
6.3	The Sturm-Liouville problem revisited	182
	6.3.1 A simple example	182
	6.3.2 The general case	183
	6.3.3 The eigenvalues	185
	6.3.4 The eigenvectors	191
	6.3.5 The inhomogeneous problem	194
	Problems	195

7 Approximation methods for eigenvalues and eigenvectors of self-adjoint operators

7.1	Introduction	203
7.2	Variational principles	203
	7.2.1 The basic ideas	204
	7.2.2 The Rayleigh–Ritz and Galerkin methods	207
	7.2.3 Upper and lower bounds	214
7.3	Kellogg's method	217
7.4	The trace method	223
7.5	Comparison and related methods	227
	Problems	231

8 Approximation methods for inhomogeneous integral equations

8.1	Introduction	241
8.2	Well-posed problems	243
8.3	Methods based on variational principles	246
	8.3.1 A maximum principle	248
	8.3.2 Second kind integral equations	249
	8.3.3 Further extremum principles and bounds	254
	8.3.4 First kind integral equations	259
	8.3.5 Bounds on other quantities	259
	8.3.6 Pointwise bounds	264
	8.3.7 Non-self-adjoint operator equations	266

	8.3.8 The Rayleigh–Ritz method	267
8.4	Galerkin's method and related topics	269
	Problems	282

9   Some singular integral equations
9.1	Introduction	291
9.2	Volterra operators with weakly singular kernels	291
9.3	First kind Fredholm equations	295
9.4	Fourier transforms and the Hilbert transform	301
9.5	Equations with Cauchy singular kernels	308
	9.5.1 The second kind equation	308
	9.5.2 The first kind equation	319
9.6	Fourier transform methods	330
9.7	An example	334
	Problems	339

Appendix A: Functional analysis   351

Appendix B: Measure theory and integration   358

Appendix C: Miscellaneous results   362

Notation Index   367

Index   369

Preface

It often happens that the most concise and illuminating method of solving even the most practical problem in mathematics involves the use of abstract ideas and techniques. This is particularly true of integral equations, where much progress can be made by using both direct and abstract techniques side by side.

The advantage of reformulating an equation, such as an integral equation, as an 'abstract' problem in a Hilbert space is that many of the important issues become clearer. In the abstract setting, a function is regarded as a 'point' in some suitable space and an integral operator as a transformation of one 'point' into another. Since a point is conceptually simpler than a function this view has the merit of removing some of the mathematical clutter from the problem, making it possible to see the salient issues more clearly. It is thus easy to visualise elegant general structures which can be translated into results about the original concrete problem. To obtain these results in a useful form, however, a second step is needed, for elegant general results tend to produce only elegant generalities and a further process is required to recover hard specific facts about the solutions sought. We use the abstract framework of functional analysis to derive the general structures and more *ad hoc* techniques for the recovery.

There is all too often a gap between the approaches of a pure and an applied mathematician to the same problem, to the extent that they may have little in common. We offer, in this book, a middle road where we develop, rigorously, the general structures associated with the problems which arise in application areas and also pay attention to the recovery of information of practical interest. We do not, however, pursue the structural results to the ultimate generality where this would not have any bearing on the issues which motivate our study, integral equations. Conversely, we do not avoid substantial matters of calculation where these are necessary to adapt the general methods to cope with classes of integral equations which arise in application areas. Our approach is, we hope, both rigorous and satisfying to the pure mathematician and accessible and useful to those dealing with concrete problems.

The book deals with linear integral equations, that is, equations involving an unknown function which appears under an integral sign (and where the dependence on this function is linear). Such equations occur widely in diverse areas of applied mathematics and physics. They offer a powerful (sometimes the only) technique for solving a variety of practical problems. One reason for this utility is that all of the conditions specifying an initial value or boundary value problem for a differential equation can often be condensed into a single integral equation. In the case of partial differential equations the dimension of the problem is reduced in this process so that, for example, a boundary value problem for a partial differential equation in two independent variables transforms into an integral equation involving an unknown function of only one variable. This reduction of what may represent a complicated mathematical model of a physical situation into a single equation is itself a significant step, but there are other advantages to be gained by replacing differentiation with integration. Some of these advantages arise because integration is a 'smoothing' process, a feature which has significant implications when approximate solutions are sought. Whether one is looking for an exact solution to a given problem or having to settle for an approximation to it, an integral equation formulation can often prove to be a useful way forward. For this reason integral equations have attracted attention for most of this century and their theory is well-developed.

This book has its origins in a course of lectures given for a number of years at the University of Reading to final year Honours Mathematics and M.Sc. students. One of the aims of this course has been to bring together strands from pure mathematics and applied mathematics, often regarded by students as totally unconnected. To do this we have, of necessity, had to remain within areas that one student might be expected to know, so that we have not presumed, for example, a familiarity with specific applications, but have taken the mathematical modelling of these as given. For similar reasons we have not presumed knowledge of more sophisticated analysis, such as distribution theory, even if there are places where a little extra progress might have resulted.

We have followed the same guidelines in the book, which is also aimed at final year undergraduate and first year postgraduate students. We assume that the reader is familiar with classical real analysis, basic linear algebra and the rudiments of ordinary differential equation theory. In addition, some acquaintance with functional analysis and Hilbert spaces is necessary, roughly at the level of a first course in the subject, although we have found that even a limited familiarity with these topics is easily consolidated as a by-product of using them in the setting of integral equations. Because of the scope of the text and the emphasis on practical issues, we hope that the book will prove useful to those working in application areas who find that they need to know about, or more about, integral equations. Functional analysts, too, may find it useful in providing concrete examples of their art.

Although the treatment is our own, our ideas have of course been influenced by others. We have both felt for many years that integral equations should be treated in the fashion of this book and we derived much benefit from reading H. Hochstadt's text *Integral Equations*, Wiley–Interscience, 1973, and F. Smithies' now classic *Integral Equations*, Cambridge University Press, 1958. Other influences, in some cases acting more in spirit by making us aware of the sort of results we might seek, have been M. J. Sewell's *Maximum and Minimum Principles*, Cambridge University Press, 1987, papers by J. B. Reade (for certain eigenvalue approximations), by A. S. Peters (for the concept of a simplifying operator) and by I. H. Sloan (for the iterated Galerkin method). Most of the material in the book has been known for many years, although not necessarily in the form in which we have presented it, but the later chapters do contain some results we believe to be new.

Finally we should like to thank Rosemary Pellew for her skill, patience and enthusiasm in typing the manuscript at the same time as coping with the mysteries of a new word processor.

David Porter
David S. G. Stirling
Reading

1
Classification and examples of integral equations

1.1 Introduction

The mechanics problem of calculating the time a particle takes to slide under gravity down a given smooth curve, from any point on the curve to its lower end, leads to an exercise in integration. The time, $f(Y)$ say, for the particle to descend from the height Y is given by an expression of the form

$$f(Y) = \int_0^Y \frac{\phi(y)\,dy}{(Y-y)^{\frac{1}{2}}} \quad (0 \leqslant Y \leqslant b), \tag{1.1}$$

where $\phi(y)$ embodies the shape of the given curve.

The converse problem, in which the time of descent from height Y is given and the particular curve which produces this time has to be found, is less straightforward, as it entails the determination of the function ϕ from (1.1), $f(Y)$ now being assigned for $0 \leqslant Y \leqslant b$. From this point of view, (1.1) is called an integral equation, this description expressing the fact that the function to be determined appears under an integral sign. The equation (1.1) is one of historical importance, attributed to Abel.

Many readers will have already encountered integral equations, but perhaps only in a context where this terminology is not used. For example, the pair of equations

$$\frac{1}{\sqrt{2\pi}} \int_{-\infty}^{\infty} e^{ixt}\phi(t)\,dt = f(x) \quad (-\infty < x < \infty),$$
$$\left. \right\} \tag{1.2}$$
$$\frac{1}{\sqrt{2\pi}} \int_{-\infty}^{\infty} e^{-ixt}f(t)\,dt = \phi(x) \quad (-\infty < x < \infty),$$

which defines the Fourier transform and its inverse, may be viewed as an integral equation and its solution. If $f(x)$ is regarded as known for $-\infty < x < \infty$ in the first equation, then the solution for $\phi(x)$ is provided by the second equation, also for $-\infty < x < \infty$.

Our illustration (1.1) from particle mechanics is one of many integral equations which result directly from a physical problem. Other problems, in elasticity theory and fluid mechanics for example, whose natural formulations are in terms of differential equations also provide a plentiful supply of integral equations. This is because initial and boundary value problems for differential equations can often be converted into integral equations and there are usually significant advantages to be gained from making use of this conversion. The intimate relationship between differential and integral equations forms the basis of a theme running through the book and is used to provide many of the illustrations. It allows us to present neutral examples, removed from the particular application areas where they originated and accessible to all readers, although we shall outline the context from which a problem has been drawn where this is appropriate.

Before deriving some examples of integral equations, it is convenient to introduce the principal features of their classification, so that the different types can be identified as they arise. We do this in the next section and follow it in §1.3 by providing illustrations of those varieties of integral equation with which we shall be concerned in subsequent chapters. We also take the opportunity to introduce at an early stage some ideas which can be used in the development of the general theory. The reader not familiar with partial differential equations can omit the details given in §1.3.3. This is the only section in which a knowledge of partial differential equations is needed.

1.2 Classification of integral equations

We are concerned, for the most part, with integral equations in which integration is with respect to a single real variable. The extension of the terminology and methods to higher order integral equations, where these do appear, is straightforward.

The notation adopted in this section, and throughout much of the text, is as follows. The unknown function will be denoted by ϕ or $\phi(x)$. Every integral equation contains a function obtained from ϕ by integration and of the form $\int_a^b k(x, t)\phi(t) \, dt$, where k is called the *kernel* and is assumed known. For example, in the integral equation

$$\phi(x) = \int_0^1 |x - t|\phi(t) \, dt + f(x) \quad (0 \leqslant x \leqslant 1)$$

the kernel is given by $k(x, t) = |x - t|$, and the function f, called the *free term*, is also assumed known. In general the kernel and free term will be complex-valued functions of real variables. A condition such as $(0 \leqslant x \leqslant 1)$

following an equation indicates that the equation holds for all values of x in the given interval. Thus for the integral equation given above, we seek a solution $\phi(x)$ satisfying the equation for all x in $[0, 1]$.

For definiteness we shall assume for the present that each integral equation holds in a closed interval. In fact some equations encountered later may hold only in open, or half-open, intervals. Analogous use of conditions like $(0 \leqslant x \leqslant 1)$ will be made in other circumstances.

The classification of integral equations centres on three basic characteristics which together describe their overall structure, and it is useful to set these down briefly before entering into greater detail.

(i) The *kind* of an equation refers to the location of the unknown function. *First* kind equations have the unknown function present under the integral sign only; *second* and *third* kind equations also have the unknown function outside the integral.

(ii) The historical descriptions *Fredholm* and *Volterra* are concerned with the integration interval. In a Fredholm equation the integral is over a finite interval with fixed end-points; in a Volterra equation the integral is indefinite.

(iii) The adjective *singular* is sometimes used when the integration is improper, either because the interval is infinite, or because the integrand is unbounded within the given interval. Obviously an integral equation can be singular on both counts.

Fredholm equations are therefore distinguished by having fixed, finite limits of integration. We denote these limits by a and b here, but we shall usually take $a = 0$ and $b = 1$ later, noting that the interval $[0, 1]$ can be transformed to a general finite interval $[a, b]$ by a simple change of variable. The Fredholm equation of the first kind is

$$f(x) = \int_a^b k(x, t)\phi(t) \, dt \quad (a \leqslant x \leqslant b), \tag{1.3}$$

and the Fredholm equation of the second kind is

$$\phi(x) = f(x) + \lambda \int_a^b k(x, t)\phi(t) \, dt \quad (a \leqslant x \leqslant b). \tag{1.4}$$

The quantity appearing in (1.4) which we have not mentioned so far, λ, is a numerical parameter, generally complex. It plays a crucial part in the theory of (1.4); in practical applications λ is usually composed of physical quantities and our later examples will show why it is not simply absorbed into the kernel.

Equation (1.4) is the standard form of the second kind Fredholm equation, but an alternative version which sometimes proves to be more convenient to deal with is

$$\mu\phi(x)=f(x)+\int_a^b k(x,t)\phi(t)\,dt \quad (a\leqslant x\leqslant b). \tag{1.5}$$

In this representation, the parameter μ takes over the role of λ, the two being related simply by

$$\lambda\mu=1. \tag{1.6}$$

We are merely observing here that λ can be divided through (1.4) (the case $\lambda=0$ can be disregarded as it is trivial) and (1.6) used to replace $1/\lambda$ by μ; μ can be absorbed into the given free term f without any loss of generality. There are advantages in each of the versions (1.4) and (1.5) and for this reason we retain both λ and μ, using whichever is the more convenient at any stage; the two symbols will have no other uses and will always be related by (1.6). One obvious advantage of the form (1.5) is that setting $\mu=0$ reduces it to the first kind equation (1.3), apart from a sign change which can be easily accommodated in the free term f.

The Fredholm equation of the third kind,

$$\mu\psi(x)\phi(x)=f(x)+\int_a^b k(x,t)\phi(t)\,dt \quad (a\leqslant x\leqslant b), \tag{1.7}$$

where ψ is an assigned function, may be regarded as a generalisation of (1.5). If ψ does not vanish in $[a,b]$, it can be divided out and absorbed into f and k. For this reason (1.7) is of limited practical significance.

We note that the specification of the interval in which the integral equation holds is an inherent part of the statement of the problem. For the first kind equation (1.3) it is not necessary that this interval should coincide with the integration interval, although it usually does.

Volterra equations differ from Fredholm equations, as we have already noted, in that the integration is indefinite. Retaining the understanding that a and b are fixed and finite, the classical form of the first kind Volterra equation is

$$f(x)=\int_a^x k(x,t)\phi(t)\,dt \quad (a\leqslant x\leqslant b), \tag{1.8}$$

and the corresponding second kind Volterra equation can be written as

$$\phi(x) = f(x) + \lambda \int_a^x k(x, t)\phi(t)\, dt \quad (a \leqslant x \leqslant b), \tag{1.9}$$

or in the alternative form

$$\mu\phi(x) = f(x) + \int_a^x k(x, t)\phi(t)\, dt \quad (a \leqslant x \leqslant b). \tag{1.10}$$

The third kind equation has the given multiplier $\psi(x)$ present on the left of (1.9) or (1.10).

There is obviously ample scope for variations on these standard forms. The integration interval can be altered to give, for example, the first kind equation

$$f(x) = \int_x^b k(x, t)\phi(t)\, dt \quad (a \leqslant x \leqslant b), \tag{1.11}$$

which will prove to be of some importance in due course. Other generalisations of (1.8) and (1.9) or (1.10) can be considered.

One further item of terminology associated with both Fredholm and Volterra equations is that they are said to be *homogeneous* if $f(x) = 0$ in $[a, b]$, and *inhomogeneous* otherwise. In effect we have previously set down the inhomogeneous versions of the various equations and, for example, either of the forms

$$\left.\begin{array}{l} \phi(x) = \lambda \displaystyle\int_a^b k(x, t)\phi(t)\, dt \\[4mm] \mu\phi(x) = \displaystyle\int_a^b k(x, t)\phi(t)\, dt \end{array}\right\} \quad (a \leqslant x \leqslant b), \tag{1.12}$$

is referred to as the homogeneous Fredholm equation of the second kind.

It must be remarked here that one property of the integral equations we have introduced is that the principle of superposition applies to their homogeneous versions. That is, if ϕ_1 and ϕ_2 are both solutions of (1.12) for given values of λ and μ, then the function $c_1\phi_1 + c_2\phi_2$ is also a solution for any constants c_1 and c_2. Thus the condition defining a *linear* problem is met, Fredholm and Volterra equations being examples of linear integral equations.

We should also note that Fredholm equations reduce to those of Volterra type if their kernels are defined to have the property that

$$k(x, t) = 0 \quad (a \leqslant x \leqslant t \leqslant b). \tag{1.13}$$

This relationship between the two varieties of equation is a useful one, but it is wrong to infer that the differences between them are minimal.

The third feature of the principal classification given earlier relates to improper integrals, which occur widely in application areas. Strictly speaking, an integral equation should be described as singular if

(a) at least one of the integration limits, or the interval in which the equation holds, is infinite,

or (b) the kernel is unbounded in the given interval.

It turns out, however, that rigid adherence to this terminology is both unnecessary and misleading.

As the definitions of Fredholm and Volterra equations specifically excluded infinite integrals, case (a) does provide us with a new type of equation which includes the Fourier transform pair (1.2). On the other hand we made no stipulations regarding the kernels of Fredholm and Volterra equations. Whilst it is premature to attempt to do this here, we can anticipate the analysis to the extent of remarking that certain types of unboundedness in kernels are permitted in our general theory of Fredholm and Volterra equations. In such cases we do not need to distinguish the equations by the addition of the adjective 'singular' to the usual Fredholm or Volterra description, although we may choose to do so for emphasis.

The so-called *weakly singular* kernel

$$k(x, t) = \frac{\hat{k}(x, t)}{|x - t|^\alpha},\tag{1.14}$$

where $\alpha \in (0, 1)$ is given and $\hat{k}$ is a bounded function, is an example of an unbounded kernel which needs no special treatment. However, we shall frequently attach the description 'weakly singular' to an integral equation with a kernel of the form (1.14), because of the prominent place occupied by such equations in application areas. It should be noted in particular that the equally important logarithmically singular kernel

$$k(x, t) = \hat{k}(x, t) \log|x - t|,\tag{1.15}$$

where $\hat{k}$ is bounded, may be regarded as weakly singular, as it can be written in the form

$$k(x, t) = \frac{\hat{k}(x, t)|x - t|^\epsilon \log|x - t|}{|x - t|^\epsilon},$$

the numerator of which is bounded for any $\epsilon > 0$.

Our first example, (1.1), of an integral equation has a kernel of the form

(1.14) with $\alpha = \frac{1}{2}$ and its full description is as a weakly singular Volterra equation of the first kind.

Special consideration does have to be given to what are often termed *strongly singular* or *Cauchy singular* kernels. These are typified by

$$k(x, t) = \frac{\hat{k}(x, t)}{x - t}, \tag{1.16}$$

where $\hat{k}$ is bounded. Within the framework we develop these are the only truly singular kernels of practical interest. We shall, however, allow a less restrictive condition on $\hat{k}$ than that given here, as regards (1.14) and (1.16).

We have now covered the principal elements arising in the classification of integral equations. There are of course items of subsidiary terminology which we do not introduce until the need arises.

1.3 A collection of examples

Here we give particular illustrations of the integral equation types introduced in the previous section, drawing on some straightforward problems which can be dealt with in an *ad hoc* manner. We also use these examples to expose some features of integral equations which can be pursued more fully in later chapters.

1.3.1 Initial value problems for ordinary differential equations

Suppose that ϕ satisfies

$$\left. \begin{array}{l} \phi'(x) = F(x, \phi(x)) \quad (0 < x < 1), \\ \phi(0) = \phi_0, \end{array} \right\} \tag{1.17}$$

where the function F and the number ϕ_0 are given. We assume that ϕ is continuous in the closed interval $[0, 1]$ which, in particular, allows the initial condition to be sensibly interpreted. Integration then gives

$$\phi(x) = \int_0^x F(t, \phi(t)) \, dt + \phi_0 \quad (0 \leqslant x \leqslant 1). \tag{1.18}$$

Conversely if ϕ is a continuous function satisfying (1.18) then $\phi(0) = \phi_0$ and the integral may be differentiated to give (1.17). Therefore, provided that all the functions are sufficiently well-behaved that the integration and differentiation may be performed, (1.17) and (1.18) have the same solutions and are in this sense equivalent. (Although we do not wish to become involved with the analysis at this stage, it is worth noting that if F is continuous then a continuous function ϕ which satisfies (1.17) also satisfies (1.18) and a solution of (1.18) is continuous and satisfies (1.17).)

We can proceed in the same way for the second order initial value problem

$$
\left.\begin{array}{l}
\phi''(x) = F(x, \phi(x)) \quad (0 < x < 1), \\
\phi(0) = \phi_0, \quad \phi'(0) = \phi'_0,
\end{array}\right\} \tag{1.19}
$$

where the number ϕ'_0 is additionally assigned. Again we must make a stipulation regarding continuity, and to avoid the need to raise this issue repeatedly we adopt the convention that, unless otherwise indicated, in a problem such as (1.19) ϕ and its derivatives up to the highest order specified at end points of the interval are extended to continuous functions on the closed interval.

One integration gives

$$
\phi'(x) = \int_0^x F(t, \phi(t))\, dt + \phi'_0 \quad (0 \leqslant x \leqslant 1),
$$

satisfying $\phi'(0) = \phi'_0$ and a second produces

$$
\phi(x) = \int_0^x ds \int_0^s F(t, \phi(t))\, dt + \phi'_0 x + \phi_0 \quad (0 \leqslant x \leqslant 1), \tag{1.20}
$$

the further constant of integration having been chosen so that $\phi(0) = \phi_0$.

Simplification of the repeated integral in (1.20) follows on using the result

$$
\int_0^x ds \int_0^s G(s, t)\, dt = \int_0^x dt \int_t^x G(s, t)\, ds, \tag{1.21}
$$

for which it is sufficient that G be a continuous function of both variables. To establish (1.21) note that the repeated integral on the left hand side is evaluated over the shaded triangular region of the t–s plane shown in Figure 1.1. The inner integral is evaluated at a fixed s from $t=0$ to $t=s$ and the outer integral then runs from $s=0$ to $s=x$. On reversing the integration order the same triangular region must be covered. This is achieved by integrating from $s=t$ to $s=x$ at a fixed t, followed by integration with respect to t from $t=0$ to $t=x$.

If we assume that F is a continuous function of both variables, then (1.21) gives

$$
\int_0^x ds \int_0^s F(t, \phi(t))\, dt = \int_0^x (x-t)F(t, \phi(t))\, dt,
$$

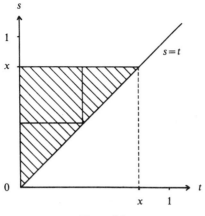

Figure 1.1

and the integral equation corresponding to (1.19) in its simplest form is

$$\phi(x) = \int_0^x (x - t)F(t, \phi(t))\, dt + \phi_0' x + \phi_0, \quad (0 \leqslant x \leqslant 1). \tag{1.22}$$

If ϕ is a continuous solution of (1.22) then differentiation under the integral sign shows that ϕ also satisfies (1.19) and so the two problems for ϕ correspond.

So far the formulation has allowed for non-linear equations. Clearly, if $F(x, y)$ is linear in y, then the integral equation (1.22) is also linear and in this case we can describe it as an inhomogeneous second kind Volterra equation.

Example 1.1

Let $\phi(x)$ satisfy Airy's equation

$$\phi''(x) = x\phi(x) \quad (0 < x < 1), \tag{1.23}$$

with

$$\phi(0) = 1, \quad \phi'(0) = 0.$$

Putting $F(x, y) = xy$, $\phi_0 = 1$ and $\phi_0' = 0$ in (1.22) gives the integral equation

$$\phi(x) = \int_0^x (x - t)t\phi(t)\, dt + 1 \quad (0 \leqslant x \leqslant 1), \tag{1.24}$$

the solution of which is also that solution of (1.23) which satisfies the prescribed initial conditions. ☐

We can conduct a preliminary investigation of second kind Volterra equations at this point, using (1.24) for illustration. In the interests of algebraic simplicity, suppose that we introduce the operator K by

$$(K\phi)(x) = \int_0^x (x-t)t\phi(t)\,dt \quad (0 \leqslant x \leqslant 1) \tag{1.25}$$

and let $f(x) = 1$. Then, for example,

$$(Kf)(x) = \int_0^x (x-t)t\,dt = [\tfrac{1}{2}xt^2 - \tfrac{1}{3}t^3]_{t=0}^{t=x} = \tfrac{1}{6}x^3 \quad (0 \leqslant x \leqslant 1). \tag{1.26}$$

Although our use of K at this stage is merely as a shorthand, we can anticipate our later central viewpoint by noting that K defines a linear mapping of functions into functions.

The integral equation (1.24) can now be written in the abbreviated form

$$\phi(x) = f(x) + (K\phi)(x) \quad (0 \leqslant x \leqslant 1),$$

or even more briefly as

$$\phi = f + K\phi, \tag{1.27}$$

where the notation assumes that the interval in which the equation holds is prescribed; in this case the interval is, of course, $[0, 1]$. If we substitute the expression for ϕ given by the right hand side of (1.27) into the term $K\phi$ of (1.27), we obtain

$$\phi = f + K(f + K\phi),$$

that is

$$\phi = f + Kf + K^2\phi, \tag{1.28}$$

where $K^2\phi = K(K\phi)$ denotes that the operator K has to be applied to the function ϕ twice in succession. Referring to (1.26) we have, for example,

$$(K^2f)(x) = \int_0^x (x-t)t(Kf)(t)\,dt$$

$$= \int_0^x (x-t)t(\tfrac{1}{6}t^3)\,dt = \tfrac{1}{180}x^6 \quad (0 \leqslant x \leqslant 1). \tag{1.29}$$

Substituting (1.27) into the right hand side of (1.28) results in

$$\phi = f + Kf + K^2f + K^3\phi, \tag{1.30}$$

which we can make more explicit by putting $f = 1$ and using (1.26) and

(1.29) to give

$$\phi(x) = 1 + \tfrac{1}{6}x^3 + \tfrac{1}{180}x^6 + (K^3\phi)(x) \quad (0 \leqslant x \leqslant 1). \tag{1.31}$$

We have therefore replaced (1.27) by two alternative integral equations for ϕ, (1.28) and (1.30), and we can obviously continue the substitution procedure to generate further equations, each being associated with a different integral operator. In (1.28) for example, the operator is given by

$$(K^2\phi)(x) = \int_0^x (x-s)s(K\phi)(s)\,\mathrm{d}s$$

$$= \int_0^x (x-s)s\,\mathrm{d}s \int_0^s (s-t)t\phi(t)\,\mathrm{d}t$$

$$= \int_0^x t\phi(t)\,\mathrm{d}t \int_t^x (x-s)s(s-t)\,\mathrm{d}s$$

$$= \int_0^x \tfrac{1}{12}t(x-t)^3(x+t)\phi(t)\,\mathrm{d}t \quad (0 \leqslant x \leqslant 1),$$

where (1.21) has been used to reverse the order of integration. It is now clear that there is no immediate advantage in replacing (1.27) by (1.28), as the latter is a good deal more complicated than the former.

The reader acquainted with the method of obtaining series solutions of differential equations will easily be able to verify that the first three terms on the right of (1.31) are in fact the leading terms in that series solution of (1.23) which satisfies the given initial conditions. This suggests that although, in this example, the sequence of integral equations generated by successive substitutions from (1.27) is increasingly complicated, the successive substitution process could actually provide the solution of the integral equation if carried out indefinitely.

In order to pursue this idea further without being drawn prematurely into detailed analysis, we need to consider an operator K rather simpler than that defined in (1.25).

Example 1.2

Let ϕ satisfy the integral equation $\phi = f + K\phi$ where

$$(K\phi)(x) = \int_0^x xt\phi(t)\,\mathrm{d}t \quad (0 \leqslant x \leqslant 1).$$

For this K we have, using (1.21) to reverse the integration order,

$$(K^2\phi)(x) = \int_0^x xs\,ds \int_0^s st\phi(t)\,dt = \int_0^x xt\phi(t)\,dt \int_t^x s^2\,ds$$

$$= \int_0^x \tfrac{1}{3}(x^3 - t^3)xt\phi(t)\,dt \quad (0 \leqslant x \leqslant 1),$$

and

$$(K^3\phi)(x) = \int_0^x xs\,ds \int_0^s \tfrac{1}{3}(s^3 - t^3)st\phi(t)\,dt$$

$$= \int_0^x \frac{1}{3^2 2!}(x^3 - t^3)^2 xt\phi(t)\,dt \quad (0 \leqslant x \leqslant 1).$$

At this point the effect of applying each additional operator K is clear and an induction argument can be used to verify that

$$(K^n\phi)(x) = \int_0^x \frac{1}{3^{n-1}(n-1)!}(x^3 - t^3)^{n-1}xt\phi(t)\,dt \quad (0 \leqslant x \leqslant 1).$$

After n successive substitutions, $\phi = f + K\phi$ gives the integral equation

$$\phi = f + Kf + K^2 f + \cdots + K^{n+1}\phi.$$

We notice the kernel of $K + K^2 + \cdots + K^n$ consists of the first n terms of the Taylor expansion of $xt\exp\{\tfrac{1}{3}(x^3 - t^3)\}$, and therefore consider the function

$$\phi(x) = f(x) + \int_0^x xt\exp\{\tfrac{1}{3}(x^3 - t^3)\}f(t)\,dt \quad (0 \leqslant x \leqslant 1). \tag{1.32}$$

It is an easy matter to verify that this ϕ does indeed satisfy the integral equation

$$\phi(x) = f(x) + \int_0^x xt\phi(t)\,dt \quad (0 \leqslant x \leqslant 1). \tag{1.33}$$

$\square$

Of course, we shall eventually address the issues raised by formally taking the limit $n \to \infty$ in the successive substitution process, which we have deliberately avoided here. The question of whether (1.32) provides the only solution of (1.33) is also open at this stage. Despite the shortcomings of our present speculative method we have established the useful fact that the solution of $\phi = f + K\phi$ can be expressed in the form $\phi = f + Rf$, where R, the *resolvent operator* for K, is also an integral operator and is defined by

$$(Rf)(x) = \int_0^x xt \exp\{\tfrac{1}{3}(x^3 - t^3)\} f(t)\, dt \quad (0 \leqslant x \leqslant 1).$$

The kernel of this operator is called the *resolvent kernel*.

It is also worth noting that if the roles of f and ϕ are reversed then $f = \phi - K\phi$ is a solution of $f = \phi - Rf$, with ϕ now regarded as the free term.

1.3.2 Boundary value problems for ordinary differential equations

Consider the determination of ϕ from

$$\left.\begin{array}{ll} \phi''(x) = F(x, \phi(x)), & (0 < x < 1), \\ \phi(0) = \phi_0, & \phi(1) = \phi_1, \end{array}\right\} \tag{1.34}$$

where we have again allowed for the possibility of a non-linear problem at this stage.

We can follow the procedure used for (1.19), to obtain

$$\phi'(x) = \int_0^x F(t, \phi(t))\, dt + A \quad (0 \leqslant x \leqslant 1),$$

and

$$\phi(x) = \int_0^x (x - t) F(t, \phi(t))\, dt + Ax + \phi_0 \quad (0 \leqslant x \leqslant 1), \tag{1.35}$$

the only difference between this and our earlier calculation being that the value $\phi'(0) = A$ is not now given and A has to be determined by imposing the condition $\phi(1) = \phi_1$. This implies that

$$A = \phi_1 - \phi_0 - \int_0^1 (1 - t) F(t, \phi(t))\, dt,$$

and therefore (1.35) can be written as

$$\phi(x) = \int_0^x (x - t) F(t, \phi(t))\, dt - x \int_0^1 (1 - t) F(t, \phi(t))\, dt + (\phi_1 - \phi_0)x + \phi_0$$

$$= -\int_0^x t(1 - x) F(t, \phi(t))\, dt - \int_x^1 x(1 - t) F(t, \phi(t))\, dt + (\phi_1 - \phi_0)x + \phi_0$$

$$(0 \leqslant x \leqslant 1).$$

The advantage of this rearrangement is that it leads to the form

$$\phi(x) = -\int_0^1 k(x, t)F(t, \phi(t))\,dt + (\phi_1 - \phi_0)x + \phi_0 \quad (0 \leqslant x \leqslant 1), \quad (1.36)$$

in which

$$k(x, t) = \begin{cases} t(1-x) & (t \leqslant x) \\ x(1-t) & (x \leqslant t) \end{cases}. \quad (1.37)$$

Once again we can reverse the process and deduce that the function ϕ which satisfies the integral equation (1.36) also satisfies the boundary value problem (1.34).

If we now specialise (1.34) to the simple, linear, boundary value problem

$$\left.\begin{array}{ll} \phi''(x) = -\lambda\phi(x) & (0 < x < 1), \\ \phi(0) = \phi_0, \qquad \phi(1) = \phi_1, \end{array}\right\} \quad (1.38)$$

then (1.36) reduces to the second kind Fredholm equation

$$\phi(x) = \lambda \int_0^1 k(x, t)\phi(t)\,dt + f(x) \quad (0 \leqslant x \leqslant 1), \quad (1.39)$$

where

$$f(x) = (\phi_1 - \phi_0)x + \phi_0.$$

In this case $k(x, t)$ is the kernel (our use of the symbol k in (1.37) anticipated this specialisation) and we note that

$$k(x, t) = k(t, x) \quad (0 \leqslant x, t \leqslant 1).$$

Any real-valued kernel with this property is said to be *symmetric*.

Both the boundary value problem (1.38) and the integral equation (1.39) are homogeneous if $\phi_0 = \phi_1 = 0$ and inhomogeneous otherwise. In the homogeneous case a non-trivial solution of the boundary value problem exists only if λ is equal to an eigenvalue, of which there are infinitely many. Because of the correspondence between (1.38) and (1.39) we deduce that the homogeneous version of the latter also has non-trivial solutions only for certain values of λ; in other words a homogeneous Fredholm equation is an example of an eigenvalue problem. The eigenvalue parameter λ is often composed of physical quantities, as Example 1.3 illustrates.

We can introduce a shorthand operator notation for Fredholm equations just as we did for Volterra equations in the previous section. Here our definition of K is

$$(K\phi)(x) = \int_0^1 k(x, t)\phi(t)\, dt \quad (0 \leqslant x \leqslant 1), \tag{1.40}$$

with k given by (1.37), and it abbreviates (1.39) to

$$\phi = \lambda K \phi + f.$$

We build on this illustration by considering the following application of integral equations to a fourth order boundary value problem.

Example 1.3

The normal modes of free flexural vibration of a thin, uniform rod of unit length are governed approximately by the differential equation

$$\frac{d^4\phi}{dx^4} = \lambda\phi(x) \quad (0 < x < 1). \tag{1.41}$$

Here $\phi(x)$ represents the transverse displacement of the centroid of the cross-section of the rod, at position x, from its equilibrium position, and λ is proportional to σ^2, where σ, the frequency of vibration, is not known in advance.

Various boundary conditions may be applied at the ends $x = 0$ and $x = 1$ of the rod, according to the way in which it is supported there. Here we take

$$\phi(0) = \phi''(0) = 0, \qquad \phi(1) = \phi''(1) = 0, \tag{1.42}$$

which corresponds to two simply supported (i.e. hinged) ends.

The problem may be decomposed by setting $\psi(x) = -\phi''(x)$, so that (1.41) and (1.42) are replaced by

$$\left.\begin{array}{ll} \psi''(x) = -\lambda\phi(x) & (0 < x < 1) \quad \text{with} \quad \psi(0) = \psi(1) = 0, \text{ (a)} \\ \phi''(x) = -\psi(x) & (0 < x < 1) \quad \text{with} \quad \phi(0) = \phi(1) = 0. \text{ (b)} \end{array}\right\} \tag{1.43}$$

The effect of integrating each of these differential equations twice subject to the appropriate end conditions can be deduced at once by reference to (1.34) and (1.36). The result is

$$\left.\begin{array}{ll} \psi(x) = \lambda \displaystyle\int_0^1 k(x, t)\phi(t)\, dt & \text{(a)} \\[3mm] \phi(x) = \displaystyle\int_0^1 k(x, t)\psi(t)\, dt & \text{(b)} \end{array}\right\} \quad (0 \leqslant x \leqslant 1), \tag{1.44}$$

where k is given by (1.37), and these constitute two *coupled*, homogeneous Fredholm equations of the second kind. Elimination of ψ produces the integral equation for ϕ equivalent to (1.41) and (1.42); replacing t by s as the integration variable in (1.44b) and substituting for $\psi(s)$ from (1.44a), we obtain

$$\phi(x) = \int_0^1 k(x, s)\, ds\, \lambda \int_0^1 k(s, t)\phi(t)\, dt \quad (0 \leqslant x \leqslant 1), \qquad (1.45)$$

that is,

$$\phi(x) = \lambda \int_0^1 \left\{ \int_0^1 k(x, s)k(s, t)\, ds \right\} \phi(t)\, dt \quad (0 \leqslant x \leqslant 1). \qquad (1.46)$$

Alternatively, if we express (1.44) in terms of the operator K defined by (1.40), we have

$$\psi = \lambda K\phi \quad \text{(a)},$$

$$\phi = K\psi \quad \text{(b)}.$$

The elimination of ψ is immediate, giving

$$\phi = \lambda K^2 \phi, \qquad (1.47)$$

which is just the shorthand operator form of (1.46). The kernel of the operator K^2, which is equal to the inner integral in (1.46), is given in Problem 1.9; it is a good deal more complicated than the kernel $k(x, t)$ from which it is derived. In the previous section we examined the effect of replacing a given integral equation with an operator K by one involving K^n, for some $n \geqslant 2$. Here the opposite viewpoint presents itself; the current example produces the operator K^2 directly, and the complexity of its kernel provokes the question: is it possible to deduce the solution of (1.47) from that of the much simpler equation $\phi = \lambda K\phi$? This is another issue which we shall resolve in more generality at an appropriate point in the theory. $\square$

1.3.3 Boundary value problems for elliptic partial differential equations

Let the function $u(x, y)$ satisfy the differential equation

$$y^s u_{xx} + u_{yy} = 0, \qquad (1.48)$$

where s is a parameter satisfying $s \geqslant 0$.

In the particular case $s = 0$, (1.48) reduces to Laplace's equation, which arises in numerous application areas. With $s > 0$, (1.48) is often referred to as the generalised Tricomi equation, the value $s=1$ corresponding to the prototype Tricomi equation occurring in compressible, irrotational fluid flow. The variable transformation $y=\{\frac{1}{2}(s+2)\eta\}^{2/(s+2)}$ reduces (1.48) to its canonical form

$$\psi_{xx}+\psi_{\eta\eta}+\frac{s}{s+2}\cdot\frac{1}{\eta}\psi_\eta=0.$$

To illustrate the relationship between partial differential equations and integral equations, we consider the following boundary value problem for u. Let

$$
\begin{aligned}
y^s u_{xx}+u_{yy}=0 &\quad (x>0, y>0), &\text{(a)}\\
u(0, y)=0 &\quad (y>0), &\text{(b)}\\
u_y(x, 0)=0 &\quad (x>1), &\text{(c)}\\
u(x, 0)=f(x) &\quad (0<x<1), &\text{(d)}\\
u\to0 \text{ as } x^2+y^2\to\infty, & &\text{(e)}
\end{aligned}
\qquad (1.49)
$$

where f denotes a given function satisfying $f(0)=0$. A graphical representation of this problem is given in Figure 1.2. It is the mixed boundary condition on $y=0$ which leads to the involvement of an integral equation. Our derivation is governed by the straightforward condition on $x=0$, which suggests use of the Fourier sine transform

$$U(\alpha, y)=\left(\frac{2}{\pi}\right)^{\frac{1}{2}}\int_0^\infty u(x, y)\sin(\alpha x)\,dx \quad (\alpha>0), \qquad (1.50)$$

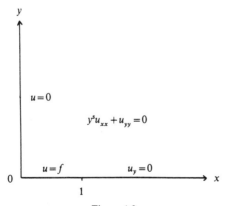

Figure 1.2

for which the inverse (or solution, in integral equation terms) is

$$u(x, y) = \left(\frac{2}{\pi}\right)^{\frac{1}{2}} \int_0^\infty U(\alpha, y) \sin(\alpha x) \, d\alpha \quad (x > 0). \tag{1.51}$$

As is usual in transform theory, we shall assume that the transforms exist and converge sufficiently well to permit the necessary manipulations to be carried out.

Laplace's equation

First consider $s = 0$ in (1.49). Taking the sine transform, Laplace's equation gives

$$U_{yy} - \alpha^2 U = 0,$$

the appropriate solution of which is

$$U(\alpha, y) = A(\alpha) e^{-\alpha y}, \tag{1.52}$$

since we must have $U \to 0$ as $y \to \infty$ by (1.49e) and (1.50). Therefore

$$U_y(\alpha, 0) = -\alpha A(\alpha) = \left(\frac{2}{\pi}\right)^{\frac{1}{2}} \int_0^\infty u_y(x, 0) \sin(\alpha x) \, dx$$

so that

$$U(\alpha, y) = -\left(\frac{2}{\pi}\right)^{\frac{1}{2}} \frac{e^{-\alpha y}}{\alpha} \int_0^\infty u_y(x, 0) \sin(\alpha x) \, dx.$$

This allows us to write (1.51) as

$$u(x, y) = -\frac{2}{\pi} \int_0^\infty \sin(\alpha x) \frac{e^{-\alpha y}}{\alpha} \, d\alpha \int_0^\infty u_y(t, 0) \sin(\alpha t) \, dt$$

$$= -\frac{2}{\pi} \int_0^\infty u_y(t, 0) \, dt \int_0^\infty \sin(\alpha x) \sin(\alpha t) \frac{e^{-\alpha y}}{\alpha} \, d\alpha.$$

The inner integral in this last expression is easily evaluated (for example, one may differentiate under the integral sign with respect to x, noting that the integral vanishes at $x = 0$), and we obtain

$$u(x, y) = \frac{1}{2\pi} \int_0^\infty u_y(t, 0) \log \left\{ \frac{(x-t)^2 + y^2}{(x+t)^2 + y^2} \right\} dt.$$

It can be verified that this function $u(x, y)$ does satisfy Laplace's equation for $x, y > 0$ together with the conditions (1.49b) and (1.49e) and it remains to impose the two elements of the mixed boundary condition. Using (1.49c) gives

$$u(x, y) = \frac{1}{2\pi} \int_0^1 u_y(t, 0) \log\left\{\frac{(x-t)^2 + y^2}{(x+t)^2 + y^2}\right\} dt, \tag{1.53}$$

and enforcing (1.49d) implies that

$$f(x) = \frac{1}{\pi} \int_0^1 u_y(t, 0) \log\left|\frac{x-t}{x+t}\right| dt \quad (0 < x < 1). \tag{1.54}$$

The solution of the boundary value problem is therefore given by (1.53) for $x > 0$, $y > 0$, where the function $\phi(t) = u_y(t, 0)$ ($0 \leqslant t \leqslant 1$) is to be found from the logarithmically (i.e. weakly) singular first kind Fredholm integral equation (1.54).

The reader familiar with Green's function techniques will be able to derive (1.53) without recourse to integral transforms. One reason for adopting the present formulation is that it allows us to produce an alternative integral equation from the boundary value problem, quite different in structure to (1.54). If we use (1.52) in (1.51) we have

$$u(x, y) = \left(\frac{2}{\pi}\right)^{\frac{1}{2}} \int_0^\infty A(\alpha) \sin(\alpha x) \, e^{-\alpha y} \, d\alpha, \tag{1.55}$$

which satisfies (1.49a) and (b) by construction. It also satisfies (1.49c) and (d) if

$$\left. \begin{array}{l} \left(\dfrac{2}{\pi}\right)^{\frac{1}{2}} \displaystyle\int_0^\infty \alpha A(\alpha) \sin(\alpha x) \, d\alpha = 0 \quad (x > 1), \\[3mm] \left(\dfrac{2}{\pi}\right)^{\frac{1}{2}} \displaystyle\int_0^\infty A(\alpha) \sin(\alpha x) \, d\alpha = f(x) \quad (0 < x < 1), \end{array} \right\} \tag{1.56}$$

which constitute a pair of *dual* integral equations for $A(\alpha)$, whose solution gives $u(x, y)$ via (1.55). At first glance, the problem posed by (1.56) may appear to be overspecified, but it is merely an example of a singular (in the sense of an infinite integral) equation of the first kind,

$$\int_0^\infty k(x, \alpha) A(\alpha) \, d\alpha = F(x) \quad (x > 0),$$

in which the kernel

$$k(x, \alpha) = \left(\frac{2}{\pi}\right)^{\frac{1}{2}} \alpha \sin(\alpha x) \quad (x > 1, \alpha > 0),$$

$$= \left(\frac{2}{\pi}\right)^{\frac{1}{2}} \sin(\alpha x) \quad (0 < x < 1, \alpha > 0),$$

is discontinuous, as may also be the free term

$$F(x)=0 \qquad (x>1),$$
$$= f(x) \quad (0<x<1).$$

Two further observations can be made at this point. First, if we note that

$$\frac{\partial}{\partial x}\log|x+t|=(x+t)^{-1} \quad (x+t>0),$$

$$\frac{\partial}{\partial x}\log|x-t|=(x-t)^{-1} \quad (x-t\neq 0)$$

and assume that f is differentiable and that differentiation under the integral sign is permissible, we can transform (1.54) into the integral equation

$$f'(x)=\frac{1}{\pi}\int_0^1 u_y(t,0)\left\{\frac{1}{x-t}-\frac{1}{x+t}\right\}dt \quad (0<x<1),$$

whose kernel is Cauchy singular. In other cases, Cauchy singular integral equations arise directly, not merely as alternative versions of other equations with less singular kernels.

In linearised water wave theory and other areas boundary conditions of the form

$$u_y(x,0)=\gamma u(x,0) \quad (0<x<1), \tag{1.57}$$

occur. Replacing the boundary condition (1.49d) by (1.57) results in the second kind equation

$$u_y(x,0)=\frac{\gamma}{\pi}\int_0^1 u_y(t,0)\log\left|\frac{x-t}{x+t}\right|dt \quad (0<x<1). \tag{1.58}$$

If γ is a constant, γ/π plays the part of the parameter λ used in §1.2.

Tricomi's equation

The boundary value problem (1.49) with $s>0$ can be converted into an integral equation using the method we have described for the case $s=0$. The details are necessarily more complicated; for instance (1.52) is replaced by

$$U(\alpha, y)= A(\alpha)y^{\frac{1}{2}}K_{\frac{1}{2}(1-\beta)}\{\alpha(1-\beta)y^{1/(1-\beta)}\},$$

where K_ν denotes the modified Bessel function of order ν and $\beta=s/(s+2)$, so that $0<\beta<1$. For the present purpose it is sufficient to give only the integral equation which emerges in place of (1.54), and this is

$$f(x) = C \int_0^1 u_y(t, 0) \left\{ \frac{1}{(x+t)^\beta} - \frac{1}{|x-t|^\beta} \right\} dt \quad (0 < x < 1), \tag{1.59}$$

where $C = (1 - \beta)^\beta \Gamma(\tfrac{1}{2}\beta)/2\pi^{\frac{1}{2}} \Gamma(\tfrac{1}{2} + \tfrac{1}{2}\beta)$, Γ denoting the gamma function (see Appendix C).

We infer that (1.49) can be resolved for $s > 0$ by solving the weakly singular first kind Fredholm equation (1.59) and we do this in Chapter 9 (see Example 9.3). If (1.57) replaces (1.49d), then the second kind equation corresponding to (1.59) must be solved. Note that, as in the case $s = 0$, the kernel consists of a singular part and a part which is bounded (for $x > 0$). This is a familiar situation in practical examples; it is the singular part of the kernel which determines the type and the character of the equation.

The most significant feature of this example is that the whole of the boundary value problem (1.49) is condensed to a single integral equation, which, as we have seen explicitly in the case $s = 0$, can appear in a variety of forms. The reduction of the differential equation and the boundary conditions into a single problem also occurs for ordinary differential equations, as we have noted, but with the example above there is the additional feature that the two-dimensional boundary value problem is transformed into a one-dimensional integral equation. This reduction in the dimension of the problem is one of the advantages of using integral equations in connection with partial differential equations. The kernels obtained are unbounded functions; the reader acquainted with Green's functions will be aware that this feature is not confined to the particular example we have used for illustration.

1.3.4 Abel's problem

We now consider the particle mechanics problem described at the beginning of the chapter.

Suppose that a particle of mass m is free to slide from rest, under gravity, down a smooth wire held fixed in a vertical plane. We can arrange the coordinates as shown in Figure 1.3, supposing that the shape of the wire is given by $x = \psi(y)$ for $0 \leqslant y \leqslant b$, where $\psi(0) = 0$ and ψ' is continuous.

In addition to the cartesian coordinates we shall need to use the variable s which measures arc length along the wire from the origin, so that

$$\frac{ds}{dy} = \{1 + (\psi'(y))^2\}^{\frac{1}{2}}. \tag{1.60}$$

By conservation of energy, if the particle which starts from rest at the point

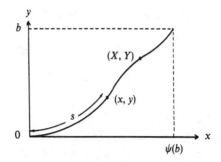

Figure 1.3

(X, Y) on the wire has speed v at the general point (x, y) on its descent, then $\frac{1}{2}mv^2 + mgy = mgY$. Therefore $v = \{2g(Y-y)\}^{\frac{1}{2}}$ and so

$$\frac{ds}{dt} = -\{2g(Y-y)\}^{\frac{1}{2}},$$

the minus sign appearing because s is a decreasing function of the time t. Using (1.60) we also have

$$\frac{ds}{dt} = \frac{ds}{dy}\frac{dy}{dt} = \{1 + (\psi'(y))^2\}^{\frac{1}{2}}\frac{dy}{dt},$$

so that

$$\{1 + (\psi'(y))^2\}^{\frac{1}{2}}\frac{dy}{dt} = -\{2g(Y-y)\}^{\frac{1}{2}}.$$

Separating the variables in this last expression and integrating with respect to y from Y to 0, and correspondingly with respect to t from 0 to $f(Y)$, the time which the particle takes to descend from height Y, we obtain

$$\int_Y^0 \left\{\frac{1 + (\psi'(y))^2}{2g(Y-y)}\right\}^{\frac{1}{2}} dy = -\int_0^{f(Y)} dt \quad (0 \leqslant Y \leqslant b).$$

Thus

$$\int_0^Y \frac{\phi(y)\,dy}{(Y-y)^{\frac{1}{2}}} = f(Y) \quad (0 \leqslant Y \leqslant b), \tag{1.1}$$

where

$$\phi(y) = \left\{\frac{1 + (\psi'(y))^2}{2g}\right\}^{\frac{1}{2}}.$$

For a given shape of the wire $x = \psi(y)$, we can determine $\phi(y)$ and integration then gives the time of descent $f(Y)$ for any $Y \in [0, b]$. The converse problem, in which $f(Y)$ is prescribed for $Y \in [0, b]$, and the corresponding shape of the wire has to be found, requires the solution of Abel's equation (1.1), as we asserted earlier. Having determined ϕ, we can find the required curve $x = \psi(y)$ by solving the separable differential equation

$$2g(\phi(y))^2 = 1 + (\psi'(y))^2 \quad (0 < y < b)$$

for ψ, subject to $\psi(0) = 0$. Notice that the time of descent for a straight, vertical wire is given by (1.1) with $\phi(y) = 1/(2g)^{\frac{1}{2}}$ $(0 \leq y \leq b)$ and is $(2Y/g)^{\frac{1}{2}}$ $(0 \leq Y \leq b)$. The assigned function f must obviously satisfy $f(0) = 0$ and it must be such that the function $\hat{f}$ defined by $\hat{f}(Y) = f(Y) - (2Y/g)^{\frac{1}{2}}$ $(0 \leq Y \leq b)$ is non-decreasing.

To examine (1.1) we set $b = 1$, return to our standard variables and introduce the operator

$$(K\phi)(x) = \int_0^x \frac{\phi(t)\,\mathrm{d}t}{(x-t)^{\frac{1}{2}}} \quad (0 \leq x \leq 1), \tag{1.61}$$

in terms of which Abel's equation is $K\phi = f$, where f is a given function, continuous in $[0, 1]$.

As K is a Volterra operator it is natural to revive the ideas of §1.3.1. The method of successive substitutions cannot be applied to a first kind equation of course, but we can replace $K\phi = f$ by $K^{n+1}\phi = K^n f$ in the hope that, for some n, this new equation is more manageable than Abel's.

Now, assuming that the integration order may be changed,

$$(K^2\phi)(x) = \int_0^x \frac{\mathrm{d}s}{(x-s)^{\frac{1}{2}}} \int_0^s \frac{\phi(t)\,\mathrm{d}t}{(s-t)^{\frac{1}{2}}} = \int_0^x \phi(t)\,\mathrm{d}t \int_t^x \frac{\mathrm{d}s}{(x-s)^{\frac{1}{2}}(s-t)^{\frac{1}{2}}}.$$

To evaluate the kernel of K^2 we set $s = x \sin^2\theta + t \cos^2\theta$, so that

$$\int_t^x \frac{\mathrm{d}s}{(x-s)^{\frac{1}{2}}(s-t)^{\frac{1}{2}}} = \int_0^{\pi/2} 2\,\mathrm{d}\theta = \pi.$$

Therefore, since $K^2\phi = Kf$, ϕ satisfies

$$\int_0^x \pi\phi(t)\,\mathrm{d}t = (Kf)(x) \quad (0 \leq x \leq 1),$$

an integral equation which is solved merely by differentiating to give

$$\phi(x) = \frac{1}{\pi}\frac{\mathrm{d}}{\mathrm{d}x}\int_0^x \frac{f(t)\,\mathrm{d}t}{(x-t)^{\frac{1}{2}}} \quad (0 \leq x \leq 1). \tag{1.62}$$

Abel's equation requires that $f(0)=0$ and if we assume that f is differentiable in $[0, 1]$, an integration by parts followed by differentiation under the integral sign reduces (1.62) to

$$\phi(x)=\frac{1}{\pi}\int_0^x \frac{f'(t)\,dt}{(x-t)^{\frac{1}{2}}}\quad(0\leqslant x\leqslant 1). \tag{1.63}$$

It is entirely straightforward to show that this function satisfies Abel's equation. In fact since $\phi=\pi^{-1}Kf'$ and K^2 represents indefinite integration multiplied by π, we can see at once that $K\phi=f$. It is rather more difficult to show that ϕ in the form (1.62) satisfies $K\phi=f$.

The application to Abel's problem of the solution of Abel's equation is covered in Problem 1.15.

It is worth noting that since $K^2=L$, say, where

$$(L\phi)(x)=\pi\int_0^x \phi(t)\,dt\quad(0\leqslant x\leqslant 1),$$

then we can regard the operators $\pm K$ as square roots of the operator L. This provides us with a simple first illustration of fractional powers of operators.

Problems

1.1   Convert the initial value problem

$$\phi''(x)=x\phi(x)\quad(0<x<1),\quad\phi(0)=0,\quad\phi'(0)=1$$

into an integral equation. Write the integral equation in the form $\phi=f+K\phi$, where K is an appropriately defined operator, and evaluate Kf and K^2f.

(This problem and Example 1.1 together define two linearly independent solutions of Airy's equation.)

1.2   Let ϕ satisfy the initial value problem

$$\begin{cases}\phi''(x)+a(x)\phi'(x)+b(x)\phi(x)=0 & (0<x<1),\\ \phi(0)=\phi_0,\quad\phi'(0)=\phi_0',\end{cases}$$

where a and b are continuous in $[0, 1]$. Verify that

$$\phi'(x)=\int_0^x \phi''(t)\,dt+\phi_0',\quad\phi(x)=\int_0^x (x-t)\phi''(t)\,dt+\phi_0'x+\phi_0\quad(0\leqslant x\leqslant 1)$$

and deduce that

(i) ϕ'' satisfies the integral equation

$$\phi''(x) = -\int_0^x \{a(x) + (x-t)b(x)\}\phi''(t)\,dt - \phi_0'\{a(x) + xb(x)\} - \phi_0 b(x)$$

$$(0 \leqslant x \leqslant 1);$$

(ii) ϕ' satisfies the integral equation

$$\phi'(x) = -\int_0^x \{a(x) + c(x) - c(t)\}\phi'(t)\,dt + \phi_0' - \phi_0 c(x)$$

$$(0 \leqslant x \leqslant 1),$$

where $c(x) = \int_0^x b(t)\,dt \quad (0 \leqslant x \leqslant 1)$.

Determine integral equations for ϕ'' and ϕ' where $\phi''(x) = x\phi(x) \quad (0 < x < 1)$, $\phi(0) = 1$ and $\phi'(0) = 0$. Write each of these equations in the form $\psi = f + K\psi$ and evaluate Kf and $K^2 f$.

(Example 1.1 gives a third integral equation associated with the same initial value problem.)

1.3   Find a solution of the Volterra integral equation

$$\phi(x) = f(x) + \lambda \int_0^x (\cos x \cos t)^{\frac{1}{2}} \phi(t)\,dt \quad (0 \leqslant x \leqslant 1)$$

by following the procedure used in Example 1.2.

1.4   Let ϕ satisfy the integral equation

$$\int_0^x \phi(t)\,dt = \mu x \phi(x) \quad (0 \leqslant x \leqslant 1)$$

where μ is a real constant. Show that there is an integrable solution ϕ for any $\mu > 0$ but that there is no continuous solution for $\mu > 1$.

(This simple example shows the importance of specifying the type of solution of an integral equation which is sought.)

1.5   (i)   Let ϕ be continuous and have a continuous derivative in $[0, 1]$. By deducing an equivalent initial value problem, show that there is no value of λ for which the integral equation

$$\phi(x) = \lambda \int_0^x \phi(t)\,dt \quad (0 < x < 1)$$

has a non-trivial solution with the assumed continuity properties.

(ii)   Let ϕ be continuous and have continuous first and second derivatives in $[0, 1]$ and let ϕ satisfy the integral equation

$$\phi(x) = \lambda \int_0^{1-x} \phi(t)\,dt \quad (0 \leqslant x \leqslant 1).$$

Show that $\phi''(x) + \lambda^2 \phi(x) = 0 \ (0 \leqslant x \leqslant 1)$ and by deducing two boundary conditions, verify that there are infinitely many discrete values of λ for which the integral equation has a non-trivial solution with the assumed continuity properties.

(The integral equation in (i) is a simple example of a Volterra equation. Although the integral equation in (ii) would seem to be of the same general type as that in (i), this problem shows that the two equations are in fact fundamentally different in character.)

1.6 Show that the function ϕ which satisfies the boundary value problem

$$\begin{cases} \phi''(x) + \lambda\phi(x) = h(x) & (0 < x < 1), \\ \phi'(0) = \phi(0), \quad \phi(1) = 0, \end{cases}$$

also satisfies an integral equation in which the kernel is

$$\tfrac{1}{2}(1 - \max(x, t))(1 + \min(x, t)) \quad (0 \leqslant x, t \leqslant 1).$$

1.7 *Let f and g be continuous functions of both variables on $[0, 1] \times [0, 1]$. Show that*

$$\int_0^x g(x, s)\, ds \int_s^1 f(s, t)\, dt = \int_0^1 \left\{ \int_0^{\min(x, t)} g(x, s) f(s, t)\, ds \right\} dt,$$

$$\int_x^1 g(x, s)\, ds \int_0^s f(s, t)\, ds = \int_0^1 \left\{ \int_{\max(x, t)}^1 g(x, s) f(s, t)\, ds \right\} dt.$$

(Diagrams like Figure 1.1 may help in carrying out the necessary changes in integration orders. The results given are true under less stringent conditions on f and g.)

1.8 Show that the boundary value problem

$$\begin{cases} \phi''(x) + \lambda r(x)\phi(x) = 0 & (0 < x < 1), \\ \phi(0) = 0, \quad \phi'(1) = 1, \end{cases}$$

can be integrated to give

$$\phi(x) = \lambda \int_0^x ds \int_s^1 r(t)\phi(t)\, dt + x \quad (0 \leqslant x \leqslant 1)$$

and by using the appropriate result of Problem 1.7 to reverse the integration order, deduce that ϕ satisfies an integral equation whose kernel is $\min(x, t)r(t)$ $(0 \leqslant x, t \leqslant 1)$. Show similarly that the boundary value problem

$$\begin{cases} \phi''(x) + \lambda r(x)\phi(x) = 0 & (0 < x < 1), \\ \phi'(0) = 1, \quad \phi(1) = 0, \end{cases}$$

can be converted into an integral equation whose kernel is $(1 - \max(x, t))r(t)$ $(0 \leqslant x, t \leqslant 1)$.

1.9 Show by induction that if $n \in \mathbb{N}$ and if ϕ satisfies

$$\frac{d^n \phi(x)}{dx^n} = F(x)(x > 0), \quad \frac{d^{n-1}\phi(0)}{dx^{n-1}} = \cdots = \frac{d\phi(0)}{dx} = \phi(0) = 0$$

then

$$\phi(x) = \int_0^x \frac{(x-t)^{n-1}}{(n-1)!} F(t)\, dt \quad (x \geqslant 0).$$

Hence show that if

$$\frac{d^4\phi(x)}{dx^4} = \lambda\phi(x) \quad (x > 0)$$

then

$$\phi(x) = \lambda \int_0^x \frac{(x-t)^3}{3!} \phi(t)\, dt + Ax^3 + Bx^2 + Cx + D \quad (x \geqslant 0).$$

By choosing the constants A, B, C and D appropriately, deduce that the function ϕ which satisfies the boundary value problem

$$\frac{d^4\phi(x)}{dx^4} = \lambda\phi(x) \quad (0 < x < 1), \quad \phi(0) = \phi(1) = \phi''(0) = \phi''(1) = 0$$

also satisfies the integral equation

$$\phi(x) = \lambda \int_0^1 k_2(x,t)\phi(t)\, dt \quad (0 \leqslant x \leqslant 1)$$

where

$$k_2(x,t) = \begin{cases} t(1-x)(2x - t^2 - x^2)/6 & (0 \leqslant t \leqslant x \leqslant 1), \\ x(1-t)(2t - x^2 - t^2)/6 & (0 \leqslant x \leqslant t \leqslant 1). \end{cases}$$

This provides an integral equation conversion method different from that given, for the same boundary value problem, in Example 1.3. Confirm that the integral equations given by the two methods are the same by evaluating the kernel of equation (1.46), that is,

$$\int_0^1 k(x,s)k(s,t)\, ds \quad (0 \leqslant x, t \leqslant 1)$$

where $k(x,t) = \min(x,t) - xt \quad (0 \leqslant x, t \leqslant 1)$.

1.10 Let ϕ satisfy the boundary value problem

$$\frac{d^4\phi(x)}{dx^4} = \lambda\phi(x) \, (0 < x < 1), \qquad \phi(0) = \phi'(0) = \phi''(1) = \phi'''(1) = 0.$$

By writing the differential equation as the coupled pair $\phi'' = -\psi$, $\psi'' = -\lambda\phi$ and integrating these two component equations separately, show that

$$\phi(x) = \lambda \int_0^x (x-s)\, ds \int_s^1 (t-s)\phi(t)\, dt \quad (0 \leqslant x \leqslant 1).$$

Using the appropriate formula in Problem 1.7 to reverse the integration order, deduce that ϕ satisfies the integral equation

$$\phi(x) = \lambda \int_0^1 k(x, t)\phi(t)\, dt \quad (0 \leqslant x \leqslant 1),$$

where $k(x, t) = t^2 x/2 - t^3/6 \quad (0 \leqslant t \leqslant x \leqslant 1)$ and $k(x, t) = k(t, x) \quad (0 \leqslant x, t \leqslant 1)$.

(The boundary value problem can be interpreted in relation to the flexural oscillations of a rod, as described in Example 1.3. Here the boundary conditions correspond to the rod being clamped at $x = 0$ and having a free end at $x = 1$.)

1.11   Show that the integral equation

$$\phi(x) = \lambda \int_0^1 k(x, t)\phi(t)\, dt \quad (0 \leqslant x \leqslant 1),$$

where $k(x, t) = t^2(1 - x)^2(3t - x - 2xt)/6 \quad (0 \leqslant t \leqslant x \leqslant 1)$ and $k(x, t) = k(t, x)$ $(0 \leqslant x, t \leqslant 1)$, can be converted into the boundary value problem

$$\frac{d^4\phi(x)}{dx^4} = \lambda\phi(x) \ (0 < x < 1), \quad \phi(0) = \phi'(0) = \phi(1) = \phi'(1) = 0.$$

(Note that, if $f(x, t)$ and $f_x(x, t)$ are continuous in both variables for $0 \leqslant x \leqslant 1$ and $0 \leqslant t \leqslant 1$ then

$$\left. \begin{array}{l} \dfrac{d}{dx}\displaystyle\int_0^x f(x, t)\, dt = \int_0^x f_x(x, t)\, dt + f(x, x) \\[3mm] \dfrac{d}{dx}\displaystyle\int_x^1 f(x, t)\, dt = \int_x^1 f_x(x, t)\, dt - f(x, x) \end{array} \right\} \quad (0 \leqslant x \leqslant 1).$$

The boundary value problem can be related to the flexural oscillations of a rod which is clamped at both ends. Example 1.3, Problem 1.10 and the present problem together show how strongly the boundary conditions influence the kernels of the associated integral equations.)

1.12   Let ϕ satisfy

$$(p(x)\phi'(x))' + \lambda r(x)\phi(x) = h(x) \quad (0 < |x| < 1)$$

where r and h are continuous in $[-1, 1]$ and

$$p(x) = \begin{cases} p_1 \ (-1 \leqslant x < 0) \\ p_2 \ (0 < x \leqslant 1) \end{cases},$$

p_1 and p_2 being distinct positive constants. Let $\phi(-1) = 0 = \phi'(1)$ and let ϕ and $p\phi'$ be continuous at $x = 0$. By integrating the equation $(p\phi')' = F$ separately for $-1 < x < 0$ and $0 < x < 1$ and applying the continuity conditions at $x = 0$, show that

$$\phi(x) = \lambda \int_{-1}^1 g(x, t)r(t)\phi(t)\, dt - \int_{-1}^1 g(x, t)h(t)\, dt \quad (-1 \leqslant x \leqslant 1),$$

where g is the continuous function defined on $[-1,1] \times [-1,1]$ by

$$g(x,t) = \frac{1}{p_1}(1 + \min(x,t)) \quad (-1 \leqslant x \leqslant 0, \ -1 \leqslant t \leqslant 1)$$

$$\text{or } (-1 \leqslant t \leqslant 0, \ 0 \leqslant x \leqslant 1)$$

$$= \frac{1}{p_1}\left(1 + \frac{p_1}{p_2}\min(x,t)\right) \quad (0 \leqslant t \leqslant 1, \ 0 \leqslant x \leqslant 1).$$

(Boundary value problems of this type arise in situations where a certain physical quantity, represented here by p, changes abruptly from one constant value to another within the region of interest. The continuity conditions at the interface are determined by the underlying physical principles. In a heat flow problem, for example, where p is the thermal conductivity, the temperature ϕ is sought in a composite material made up of two substances of different conductivities joined at $x=0$.)

1.13 Let ϕ satisfy $\phi = f + K\phi$ in $[0,1]$ where

$$(K\phi)(x) = \frac{1}{\pi^{\frac{1}{2}}} \int_0^x \frac{\phi(t)\,dt}{(x-t)^{\frac{1}{2}}} \quad (0 \leqslant x \leqslant 1).$$

Show that $\phi - K_1\phi = f + Kf$ in $[0,1]$ where

$$(K_1\phi)(x) = \int_0^x \phi(t)\,dt \quad (0 \leqslant x \leqslant 1)$$

and, by putting $\psi = K_1\phi$ and solving a differential equation for ψ, deduce that

$$\phi(x) = f(x) + (Kf)(x) + \int_0^x e^{x-t}(f(t) + (Kf)(t))\,dt \quad (0 \leqslant x \leqslant 1).$$

1.14 Let

$$(K\phi)(x) = \frac{1}{\pi^{\frac{1}{2}}} \int_x^1 \frac{\phi(t)\,dt}{(t-x)^{\frac{1}{2}}} \quad (0 \leqslant x \leqslant 1).$$

Show that

$$(K^2\phi)(x) = \int_x^1 \phi(t)\,dt \quad (0 \leqslant x \leqslant 1)$$

and deduce the solutions of the integral equations $K\phi = f$ and $\phi = f + \lambda K\phi$ in $[0,1]$.

1.15 In Abel's problem (see §1.3.4) let $b=1$ and let the time of descent be given by $f(Y) = (8/g)^{\frac{1}{2}} Y^{\frac{1}{2}}$ $(0 \leqslant Y \leqslant 1)$. Show, by using the solution (1.62) of Abel's equation, that the corresponding shape of wire is given by

$$x = \int_0^y \left(\frac{c^2 - s^{\frac{1}{2}}}{s^{\frac{1}{2}}}\right)^{\frac{1}{2}} ds \quad (0 \leqslant y \leqslant 1),$$

where $c = \dfrac{3}{\pi} \displaystyle\int_0^1 s^{\frac{3}{2}} (1-s)^{-\frac{1}{2}} \, ds = 1.669 \ldots$.

(The integral defining the curve can be evaluated by putting $s^{\frac{1}{2}} = c^2 \sin^2\theta$; the integral occurring in c is an example of the beta function, which is referred to in Appendix C.)

1.16   Let $u(x, y)$ be defined for $x \geqslant 0$ and $0 \leqslant y \leqslant b$ by the boundary value problem

$$u_{xx} + u_{yy} = 0 \quad (x > 0, \, 0 < y < b),$$
$$u(x, 0) = u(x, b) = 0 \quad (x > 0),$$
$$u_x(0, y) = F(y) \quad (a < y < b),$$
$$u(0, y) = G(y) \quad (0 < y < a),$$
$$u \to 0 \text{ as } x \to \infty \text{ with } y \in [0, b],$$

F and G being given functions.

By using the separation solutions $e^{-nx} \sin(n\pi y/b)$ ($n \in \mathbb{N}$) and Fourier series, show that

$$u(x, y) = -\frac{2}{b} \sum_{n=1}^{\infty} \frac{e^{-nx} \sin(n\pi y/b)}{n} \int_0^b u_x(0, \eta) \sin(n\pi \eta/b) \, d\eta$$

for $x \geqslant 0$ and $0 \leqslant y \leqslant b$, where $\phi(y) = u_x(0, y)$ $(0 < y < a)$, is obtained by solving an integral equation of the form

$$f(y) = \sum_{n=1}^{\infty} \frac{\sin(n\pi y/b)}{n} \int_0^a \phi(\eta) \sin(n\pi \eta/b) \, d\eta \quad (0 < y < a).$$

(Note that

$$\sum_{n=1}^{\infty} \frac{1}{n} \sin(n\pi y/b) \sin(n\pi \eta/b) = \tfrac{1}{2} \log \left| \frac{\sin\{\pi(y+\eta)/(2b)\}}{\sin\{\pi(y-\eta)/(2b)\}} \right|$$

for $0 < y, \eta < b$ and $y \neq \eta$; see Appendix C.)

Second order ordinary differential equations and integral equations

2.1 Introduction

In the first chapter we referred to the general relationship between differential and integral equations and gave some particular examples of how initial and boundary value problems for the former can be converted into the latter. To broaden the scope of our applications we now consider the transformation of a class of differential equations into integral equations. We limit the technical requirements of differential equation theory by dealing solely with ordinary equations and we are therefore concerned with physical problems which are only one-dimensional in character or which result from two- or three-dimensional problems after a separation of variables.

By using only an elementary knowledge of differential equations we can develop routines which allow the transformations from initial and boundary value problems to integral equations to be readily carried out. In addition to replacing the *ad hoc* manipulations of §1.3.1 and §1.3.2 by techniques which are more widely applicable, our analysis also reveals the distinctive form of the integral equations associated with second order ordinary differential equations and we shall take advantage of this structure to illustrate subsequent parts of the theory.

2.2 Differential equation theory

We consider the general second order equation in the form

$$\{p(x)\phi'(x)\}' - q(x)\phi(x) + \lambda r(x)\phi(x) = h(x) \quad (0 < x < 1). \qquad (2.1)$$

For brevity, the differential operator L defined by

$$(L\phi)(x) = -\{p(x)\phi'(x)\}' + q(x)\phi(x) \qquad (2.2)$$

is used, and it allows us to write (2.1) as

$$(L\phi)(x) = F(x) \quad (0 < x < 1), \quad \text{(a)} \left.\begin{array}{c} \\ \end{array}\right\}$$

where $\qquad\qquad F(x) = \lambda r(x)\phi(x) - h(x). \qquad \text{(b)} \qquad (2.3)$

Our objective is to solve (2.3a) subject to given initial or boundary conditions, thereby expressing ϕ in terms of an integral involving F, just as we did in §1.3.1 and §1.3.2 in the simple case with $(L\phi)(x) = -\phi''(x)$. Substituting $F = \lambda r\phi - h$ then gives a second kind integral equation for ϕ. The main advantage of this exercise from a practical point of view is that, while equation (2.1) may be intractable, it may be possible to deduce information about its solution from the corresponding integral equation, which is obtained by solving the simpler equation $L\phi = F$ with F regarded as given.

We assume that all quantities are real and that

$$p, p', q, r \text{ and } h \text{ are continuous in } [0, 1], \left.\begin{array}{c}\\\\\end{array}\right\} \qquad (2.4)$$
$$p > 0, \quad r > 0 \text{ in } [0, 1],$$

although these conditions will later be relaxed slightly. The main steps in constructing the solution of (2.3a) are as follows: let ψ_1 and ψ_2 be linearly independent solutions of $(L\phi)(x) = 0$ in $[0, 1]$. It follows from our assumptions and standard differential equation theory that ψ_1 and ψ_2 are continuous and have two continuous derivatives in $[0, 1]$, that the Wronskian

$$W(\psi_1, \psi_2) = \psi_1(x)\psi_2'(x) - \psi_1'(x)\psi_2(x) \neq 0 \quad (0 \leqslant x \leqslant 1), \qquad (2.5)$$

and that

$$p(\psi_1\psi_2' - \psi_1'\psi_2) = \text{constant} \neq 0 \text{ in } [0, 1].$$

This last result follows from (2.5), the assumption that $p > 0$ and the identity

$$\psi_2 L\psi_1 - \psi_1 L\psi_2 = p(\psi_1\psi_2'' - \psi_1''\psi_2) + p'(\psi_1\psi_2' - \psi_1'\psi_2) = 0.$$

As ψ_1 and ψ_2 are arbitrary to the extent of constant multipliers we can choose them so that

$$p(\psi_1\psi_2' - \psi_1'\psi_2) = -1 \text{ in } [0, 1]. \qquad (2.6)$$

The particular integral of (2.3a) is obtained by the method of variation of parameters. Setting

$$\phi = \psi_1\chi_1 + \psi_2\chi_2, \quad 0 = \psi_1\chi_1' + \psi_2\chi_2',$$

it is not difficult to show that $L\phi = F$ is satisfied by taking

$$\chi_1'(x) = -\psi_2(x)F(x), \quad \chi_2'(x) = \psi_1(x)F(x) \quad (0 \leqslant x \leqslant 1). \qquad (2.7)$$

From this point we need to vary the development according to whether we are dealing with an initial or a boundary value problem for (2.1).

2.3 Initial value problems

Suppose that we are interested in the initial value problem

$$-(L\phi)(x)+\lambda r(x)\phi(x)=h(x) \quad (0<x<1), \\ \phi(0)=\phi_0, \qquad\qquad\qquad\qquad \phi'(0)=\phi'_0, \tag{2.8}$$

where ϕ_0 and ϕ'_0 are given real numbers. From (2.7), χ_1 and χ_2 can be taken in the forms

$$\chi_1(x)=-\int_0^x \psi_2(t)F(t)\,dt, \quad \chi_2(x)=\int_0^x \psi_1(t)F(t)\,dt,$$

giving the general solution of $L\phi=F$ as

$$\phi(x)=C_1\psi_1(x)+C_2\psi_2(x)-\psi_1(x)\int_0^x \psi_2(t)F(t)\,dt+\psi_2(x)\int_0^x \psi_1(t)F(t)\,dt,$$

valid for $0\leqslant x\leqslant 1$.

Applying the initial conditions produces

$$\phi_0=C_1\psi_1(0)+C_2\psi_2(0), \quad \phi'_0=C_1\psi'_1(0)+C_2\psi'_2(0), \tag{2.9}$$

which have a unique solution for C_1 and C_2 for any ϕ_0 and ϕ'_0 because $\psi_1(0)\psi'_2(0)-\psi'_1(0)\psi_2(0)\neq 0$ by (2.5).

Referring to (2.3) and substituting $F=\lambda r\phi-h$, we conclude that ϕ satisfies the second kind Volterra equation

$$\phi(x)=f(x)+\lambda\int_0^x \{\psi_1(t)\psi_2(x)-\psi_1(x)\psi_2(t)\}r(t)\phi(t)\,dt \quad (0\leqslant x\leqslant 1), \\ f(x)=C_1\psi_1(x)+C_2\psi_2(x)-\int_0^x \{\psi_1(t)\psi_2(x)-\psi_1(x)\psi_2(t)\}h(t)\,dt, \tag{2.10}$$

where C_1 and C_2 are given by (2.9). It is not difficult to deduce (2.8) from (2.9) and (2.10), establishing that ϕ is a solution of (2.8) if and only if it satisfies (2.9) and (2.10).

For illustration consider the initial value problem

$$\phi''(x)=x\phi(x) \quad (0<x<1), \quad \phi(0)=1, \quad \phi'(0)=0$$

which corresponds to

$$(L\phi)(x)=-\phi''(x), \quad p(x)=1, \quad q(x)=0, \\ r(x)=-x, \quad h(x)=0 \quad (0\leqslant x\leqslant 1),$$

and $\lambda=1$, $\phi_0=1$, $\phi'_0=0$. Two linearly independent solutions of $L\phi=0$ in this case are $\psi_1(x) = A$ and $\psi_2(x) = Bx$. Noting that (2.6) reduces to $AB = -1$, we take

$$\psi_1(x) = 1, \quad \psi_2(x) = -x \quad (0 \leqslant x \leqslant 1)$$

and then (2.9) gives $C_1 = 1$ and $C_2 = 0$. It now follows that the version of (2.10) corresponding to this case is

$$\phi(x) = 1 + \int_0^x (x-t)t\phi(t)\,dt \quad (0 \leqslant x \leqslant 1).$$

2.4 Boundary value problems

Let ϕ satisfy the Sturm–Liouville boundary value problem

$$\left.\begin{aligned}
&-(L\phi)(x) + \lambda r(x)\phi(x) = h(x) \quad (0 \leqslant x \leqslant 1), \quad \text{(a)}\\
&(L\phi)(x) = -\{p(x)\phi'(x)\}' + q(x)\phi(x),\\
&a_0\phi(0) + b_0\phi'(0) = 0,\ a_1\phi(1) + b_1\phi'(1) = 0, \quad \text{(b)}
\end{aligned}\right\} \quad (2.11)$$

where a_0, b_0, a_1 and b_1 are real constants such that $a_0^2 + b_0^2 \neq 0$ and $a_1^2 + b_1^2 \neq 0$. A problem of this form often arises following separation of the variables in a linear partial differential equation. In such a case $h = 0$ and λ, which is proportional to the separation constant, is to be determined.

The transformation of the boundary value problem (2.11) into an integral equation differs from that for an initial value problem only in the determination of the constants occurring in the general solution of $L\phi = F$. However we can reduce the algebraic manipulation and expose the structure more clearly by modifying the procedure of the last section slightly.

We saw in (2.7) that if we seek a particular solution of $L\phi = F$ in the form $\phi = \psi_1\chi_1 + \psi_2\chi_2$ with $\psi_1\chi_1' + \psi_2\chi_2' = 0$ then $\chi_1' = -\psi_2 F$ and $\chi_2' = \psi_1 F$. We now choose χ_1 and χ_2 so that $\chi_1(1) = 0$ and $\chi_2(0) = 0$ and therefore

$$\chi_1(x) = \int_x^1 \psi_2(t)F(t)\,dt, \quad \chi_2(x) = \int_0^x \psi_1(t)F(t)\,dt \quad (0 \leqslant x \leqslant 1).$$

This adjustment to the approach used for initial value problems (in which the choice of χ_1 and χ_2 is biased towards the application of conditions only at $x = 0$) makes it easier to handle a condition at each end of the interval and gives the general solution of $L\phi = F$ in the form

$$\phi(x) = D_1\psi_1(x) + D_2\psi_2(x)$$
$$+ \psi_1(x)\int_x^1 \psi_2(t)F(t)\,dt + \psi_2(x)\int_0^x \psi_1(t)F(t)\,dt \quad (0 \leqslant x \leqslant 1). \quad (2.12)$$

Since

$$\phi'(x) = D_1\psi'_1(x) + D_2\psi'_2(x)$$
$$+ \psi'_1(x)\int_x^1 \psi_2(t)F(t)\,dt + \psi'_2(x)\int_0^x \psi_1(t)F(t)\,dt \quad (0 \leqslant x \leqslant 1),$$

the boundary conditions (2.11b) are satisfied if

$$\left.\begin{aligned}
0 &= (a_0\psi_1(0) + b_0\psi'_1(0))\left\{D_1 + \int_0^1 \psi_2(t)F(t)\,dt\right\} \\
&\quad + (a_0\psi_2(0) + b_0\psi'_2(0))D_2, \\
0 &= (a_1\psi_1(1) + b_1\psi'_1(1))D_1 \\
&\quad + (a_1\psi_2(1) + b_1\psi'_2(1))\left\{D_2 + \int_0^1 \psi_1(t)F(t)\,dt\right\}.
\end{aligned}\right\} \quad (2.13)$$

The structure of these equations for D_1 and D_2 suggests that we should now attempt to specify ψ_1 and ψ_2 more closely than we have needed to so far. It transpires that two cases arise, the distinction between them being of some importance in the theory of the resulting integral equations.

Case 1

Let ψ_1 and ψ_2 satisfy

$$\left.\begin{aligned}
(L\psi_1)(x) &= 0 \quad (0 < x < 1), \quad a_0\psi_1(0) + b_0\psi'_1(0) = 0, \\
(L\psi_2)(x) &= 0 \quad (0 < x < 1), \quad a_1\psi_2(1) + b_1\psi'_2(1) = 0.
\end{aligned}\right\} \quad (2.14)$$

Previously we have chosen ψ_1 and ψ_2 to be linearly independent solutions of $L\phi = 0$. Now that each is required to satisfy one boundary condition ψ_1 and ψ_2 may still be linearly independent, but they may be linearly dependent. In the latter circumstance a different strategy is required to the one we are about to describe and this constitutes our Case 2.

In Case 1 therefore we assume that the functions ψ_1 and ψ_2 defined by (2.14) are linearly independent and then $\psi_1(x)\psi'_2(x) - \psi'_1(x)\psi_2(x) \neq 0$ in $[0, 1]$. In particular it follows that

$$a_0\psi_2(0) + b_0\psi'_2(0) \neq 0, \quad a_1\psi_1(1) + b_1\psi'_1(1) \neq 0, \quad (2.15)$$

since at least one of a_0, b_0 and at least one of a_1, b_1 must be non-zero. Because ψ_1 and ψ_2 are only determined to within arbitrary constant multipliers we can still impose the Wronskian normalisation condition

$$p(\psi_1\psi'_2 - \psi'_1\psi_2) = -1 \quad (2.6)$$

as before.

The simplification resulting from our choice of ψ_1 and ψ_2 is evident when (2.14) and (2.15) are used in (2.13), since it follows that $D_1 = D_2 = 0$. Therefore (2.12) reduces to

$$\phi(x) = \int_0^1 g(x,t)F(t)\,dt \quad (0 \leqslant x \leqslant 1), \tag{2.16}$$

in which

$$g(x,t) = \begin{cases} \psi_1(t)\psi_2(x) & (t \leqslant x), \\ \psi_1(x)\psi_2(t) & (x \leqslant t). \end{cases} \tag{2.17}$$

Recalling that $F = \lambda r\phi - h$ by (2.3b), we conclude that the boundary value problem (2.11) can be written as the second kind Fredholm equation

$$\phi(x) = \lambda \int_0^1 g(x,t)r(t)\phi(t)\,dt + f(x) \quad (0 \leqslant x \leqslant 1),$$

where

$$f(x) = -\int_0^1 g(x,t)h(t)\,dt \qquad (0 \leqslant x \leqslant 1),$$

$$\left. \right\} \tag{2.18}$$

provided, of course, that ψ_1 and ψ_2 are linearly independent.

The converse result, that solutions of (2.18) also satisfy (2.11), is easy to establish.

Example 2.1

Let ϕ satisfy

$$\phi''(x) + \lambda r(x)\phi(x) = 0 \quad (0 < x < 1),$$
$$\phi(0) = 0, \qquad\qquad \phi(1) = 0.$$

Here $(L\phi)(x) = -\phi''(x)$ and the functions ψ_1 and ψ_2 satisfying

$$-\psi_1''(x) = 0 \ (0 < x < 1), \quad \psi_1(0) = 0,$$
$$-\psi_2''(x) = 0 \ (0 < x < 1), \quad \psi_2(1) = 0,$$

are $\psi_1(x) = Ax$ and $\psi_2(x) = B(1-x)$. Imposing $\psi_1\psi_2' - \psi_1'\psi_2 = -1$ gives $AB = 1$ and therefore

$$\phi(x) = \lambda \int_0^1 g(x,t)r(t)\phi(t)\,dt \quad (0 \leqslant x \leqslant 1), \tag{2.19}$$

where

$$g(x,t) = \begin{cases} t(1-x) & (t \leqslant x), \\ x(1-t) & (x \leqslant t). \end{cases}$$

Conversely, given (2.19), we write it explicitly as

$$\phi(x) = \lambda \int_0^x t(1-x)r(t)\phi(t)\,dt + \lambda \int_x^1 x(1-t)r(t)\phi(t)\,dt \quad (0 \leqslant x \leqslant 1),$$

showing at once that $\phi(0) = 0 = \phi(1)$. Two successive differentiations produce

$$\phi'(x) = -\lambda \int_0^x tr(t)\phi(t)\,dt + \lambda \int_x^1 (1-t)r(t)\phi(t)\,dt$$

and

$$\phi''(x) = -\lambda x r(x)\phi(x) - \lambda(1-x)r(x)\phi(x) = -\lambda r(x)\phi(x),$$

showing that the function ϕ defined by (2.19) satisfies the original boundary value problem. □

We note that the function g given by (2.17) is symmetric, that is, $g(x,t) = g(t,x)$ $(0 \leqslant x, t \leqslant 1)$. The kernel of (2.18), $g(x,t)r(t)$, is not symmetric, however, unless r is constant. It will become apparent later that there is a considerable advantage to be gained by transforming a non-symmetric kernel into symmetric form, if possible, and this can be achieved for (2.18) if $r(x) > 0$ in $[0,1]$, which is one of our assumptions (2.4). We can therefore rewrite (2.18) as

$$(r(x))^{\frac{1}{2}}\phi(x) = \lambda \int_0^1 g(x,t)(r(x)r(t))^{\frac{1}{2}}(r(t))^{\frac{1}{2}}\phi(t)\,dt + (r(x))^{\frac{1}{2}}f(x) \quad (0 \leqslant x \leqslant 1),$$

and consequently as

$$\hat{\phi}(x) = \lambda \int_0^1 k(x,t)\hat{\phi}(t)\,dt + \hat{f}(x) \quad (0 \leqslant x \leqslant 1), \tag{2.20}$$

in which the kernel

$$k(x,t) = g(x,t)(r(x)r(t))^{\frac{1}{2}} \quad (0 \leqslant x, t \leqslant 1) \tag{2.21}$$

is symmetric and the new unknown function and new free term are given respectively by

$$\hat{\phi}(x) = (r(x))^{\frac{1}{2}}\phi(x), \quad \hat{f}(x) = (r(x))^{\frac{1}{2}}f(x). \tag{2.22}$$

Obviously k can be written in the form of (2.17) as

$$k(x,t) = \begin{cases} \hat{\psi}_1(t)\hat{\psi}_2(x) & (t \leqslant x), \\ \hat{\psi}_1(x)\hat{\psi}_2(t) & (x \leqslant t), \end{cases}$$

where $\hat{\psi}_1 = r^{\frac{1}{2}}\psi_1$ and $\hat{\psi}_2 = r^{\frac{1}{2}}\psi_2$, showing that the transformation of (2.18) to symmetric form simply requires the scaling of the functions ϕ, f, ψ_1 and ψ_2 by $r^{\frac{1}{2}}$.

Case 2

We now suppose that ψ_1 and ψ_2, as defined by (2.14), are linearly dependent in $[0, 1]$, and then the simple technique given in Case 1 for deriving an integral equation from (2.11) fails.

A simple illustration of this phenomenon occurs in the boundary value problem

$$\begin{aligned} \phi''(x) + \lambda\phi(x) = 0 & \quad (0 < x < 1), \\ \phi'(0) = 0, & \quad\quad\quad \phi'(1) = 0, \end{aligned}$$

for which ψ_1 and ψ_2 are both constants. It is obvious that this situation arises simply because one solution of the boundary value problem is $\phi(x) = \text{constant}$ $(0 \leqslant x \leqslant 1)$ with $\lambda = 0$. More generally, if the homogeneous $(h = 0)$ version of the boundary value problem (2.11) has a non-trivial solution with $\lambda = 0$ then the associated functions ψ_1 and ψ_2 defined by (2.14) are linearly dependent and both satisfy both of the given boundary conditions. Indeed this is the crux of the distinction between our two cases; Case 1 applies if the homogeneous version of the given boundary value problem has only the trivial solution with $\lambda = 0$, and Case 2 applies if there is a non-trivial solution with $\lambda = 0$.

In the light of these remarks it is clear how we should revise the definitions of ψ_1 and ψ_2 to deal with Case 2. Let

$$\left. \begin{aligned} (L\psi_1)(x) = 0 & \quad (0 < x < 1), \quad a_0\psi_1(0) + b_0\psi_1'(0) = 0, \\ & \quad\quad\quad a_1\psi_1(1) + b_1\psi_1'(1) = 0, \quad \text{(a)} \\ (L\psi_2)(x) = 0 & \quad (0 < x < 1), \quad\quad\quad\quad\quad\quad\quad\quad \text{(b)} \end{aligned} \right\} \quad (2.23)$$

where ψ_2 is any solution of $L\phi = 0$ such that ψ_1 and ψ_2 are linearly independent in $[0, 1]$. It follows that

$$a_0\psi_2(0) + b_0\psi_2'(0) \neq 0, \quad a_1\psi_2(1) + b_1\psi_2'(1) \neq 0. \quad (2.24)$$

The enforced linear independence of ψ_1 and ψ_2 by this means leaves our earlier approach intact up to a point. The solution of $L\phi = F$ is again given by

$$\phi(x) = D_1 \psi_1(x) + D_2 \psi_2(x) + \psi_1(x) \int_x^1 \psi_2(t) F(t) \, dt$$

$$+ \psi_2(x) \int_0^x \psi_1(t) F(t) \, dt \quad (0 \leqslant x \leqslant 1)$$

where the newly defined ψ_1 and ψ_2 are still assumed to satisfy

$$p(\psi_1 \psi_2' - \psi_1' \psi_2) = -1. \tag{2.6}$$

The point of departure from the previous derivation arises on the application of the boundary conditions (2.11b) which, because of (2.23a) and (2.24), imply that

$$D_2 = 0, \quad D_2 = - \int_0^1 \psi_1(t) F(t) \, dt,$$

and that D_1 is arbitrary. We therefore have to take account of the necessary condition

$$\int_0^1 \psi_1(t) F(t) \, dt = 0 \tag{2.25}$$

in the subsequent development.

Thus far we have derived the solution of $L\phi = F$ and the associated boundary conditions in the form

$$\phi(x) = D_1 \psi_1(x) + \int_0^1 g(x, t) F(t) \, dt \quad (0 \leqslant x \leqslant 1), \tag{2.26}$$

where

$$g(x, t) = \begin{cases} \psi_1(t)\psi_2(x) & (t \leqslant x), \\ \psi_1(x)\psi_2(t) & (x \leqslant t). \end{cases} \tag{2.17}$$

If F is regarded as given, it is inevitable that an arbitrary amount of ψ_1 should appear in the solution for ϕ, because of its definition (2.23a). However this solution exists only if the given F satisfies the necessary condition (2.25).

Recalling that in our circumstances F is just an abbreviation for $\lambda r \phi - h$, we see that F is a given function only if $\lambda = 0$. In this case substituting h for F in (2.26) gives the (non-unique) solution of the boundary value problem provided that

$$\int_0^1 \psi_1(t) h(t) \, dt = 0.$$

If this condition is not satisfied then the inhomogeneous boundary value problem has no solution with $\lambda = 0$.

The case $\lambda \neq 0$ is more intricate, for then $F = \lambda r \phi - h$ depends on ϕ and therefore the condition (2.25) is effectively imposed on ϕ. It is convenient to consider the homogeneous problem with $h = 0$ first, in which case (2.26) may be written as

$$\phi(x) = D_1 \psi_1(x) + \lambda \int_0^1 g(x, t) r(t) \phi(t) \, dt \quad (0 \leqslant x \leqslant 1), \qquad (2.27)$$

and ϕ must satisfy

$$\int_0^1 \psi_1(x) r(x) \phi(x) \, dx = 0. \qquad (2.28)$$

Applying this condition to the function ϕ given by the right hand side of (2.27) obviously determines D_1 but at the expense of destroying the symmetry present in the kernel, because only the variable x is actively involved in the process. To maintain a balance between the two variables and preserve the symmetry we therefore use (2.28) to rewrite (2.27) as

$$\phi(x) = D_1 \psi_1(x)$$
$$+ \lambda \int_0^1 g(x, t) r(t) \left\{ \phi(t) - \left(\int_0^1 \psi_1(s) r(s) \phi(s) \, ds \right) \psi_1(t) \right\} dt$$
$$(0 \leqslant x \leqslant 1). \qquad (2.29)$$

Anticipating another consequence of applying (2.28), we see that it is now prudent to normalise $\psi_1(x)$ so that

$$\int_0^1 \psi_1^2(x) r(x) \, dx = 1. \qquad (2.30)$$

If we now apply (2.28) to the right hand side of (2.29) we find, after some rearrangement, that

$$D_1 = \lambda \int_0^1 r(t) \phi(t) \left\{ - \int_0^1 g(s, t) r(s) \psi_1(s) \, ds \right.$$
$$+ \left. \psi_1(t) \int_0^1 r(s) \psi_1(s) \, ds \int_0^1 g(\xi, s) r(\xi) \psi_1(\xi) \, d\xi \right\} dt.$$

Substituting this expression for D_1 into (2.29) gives the integral equation

$$\phi(x) = \lambda \int_0^1 g_1(x, t) r(t) \phi(t) \, dt \quad (0 \leqslant x \leqslant 1), \qquad (2.31)$$

where

$$g_1(x, t) = g(x, t) - \psi_1(t) \int_0^1 g(s, x) r(s) \psi_1(s) \, ds$$

$$- \psi_1(x) \int_0^1 g(s, t) r(s) \psi_1(s) \, ds$$

$$+ \psi_1(x) \psi_1(t) \int_0^1 r(s) \psi_1(s) \, ds \int_0^1 g(\xi, s) r(\xi) \psi_1(\xi) \, d\xi. \quad (2.32)$$

The symmetry $g(x, s) = g(s, x)$ has been used in the second term to produce this explicitly symmetric form for $g_1(x, t)$.

By construction the function $g_1(x, t)$ is such that

$$\int_0^1 g_1(x, t) \psi_1(x) r(x) \, dx = 0 \quad (2.33)$$

for each fixed $t \in [0, 1]$ and by symmetry

$$\int_0^1 g_1(x, t) \psi_1(t) r(t) \, dt = 0 \quad (2.34)$$

for each fixed $x \in [0, 1]$.

Note that the constraint (2.28) is now an inherent part of the integral equation (2.31) and is not required to be given as a subsidiary condition. In other words, and as usual, the integral equation contains all the conditions imposed on ϕ. The reciprocal process of deducing a boundary value problem from an integral equation in this case entails the identification of the implicit necessary condition. It also produces a feature which we have not met before.

We have previously referred to the correspondence between a boundary (or an initial) value problem and an integral equation, in the sense that each could be derived from the other. The crucial requirement, of course, is that the solution set (the aggregate of all possible solutions) of the differential equation problem and the solution set of the integral equation should be the same, so that no solutions are lost or gained in converting from one form to the other. If the solution sets are identical we describe the differential equation problem and the integral equation as *equivalent*.

This issue is raised here because we appear to have produced an integral equation – boundary value problem pair where the two are not equivalent. We are investigating the case in which $\phi = D_1 \psi_1$ is a solution of the boundary value problem with $\lambda = 0$ and clearly it is not a solution of the integral equation (2.31), where $\lambda = 0$ implies that $\phi = 0$. Of course we have supposed that $\lambda \neq 0$ in order to derive (2.31), and we have therefore certainly discarded the solution of the boundary value problem associated with $\lambda = 0$ in passing to the integral equation form. However, with the restriction $\lambda \neq 0$ adhered to, and use of the substitution $\mu = \lambda^{-1}$ (introduced in §1.2), (2.31) can be written as

$$\mu\phi(x) = \int_0^1 g_1(x, t) r(t) \phi(t) \, dt \quad (0 \leqslant x \leqslant 1). \tag{2.35}$$

Referring to (2.34), we see that $\phi = D_1 \psi_1$ is a solution of *this* integral equation, for any value of the constant D_1, with $\mu = 0$.

The conversion from boundary value problem to integral equation therefore appears to involve the transfer of a solution associated with $\lambda = 0$ to a solution associated with $\mu = 0$, and in this sense there is equivalence of the two forms.

To summarise the procedure to be followed in generating (2.31), the functions ψ_1 and ψ_2 are selected in accordance with (2.23) and the requirement of linear independence. The usual Wronskian normalisation condition (2.6) is applied, as is the normalisation (2.30) of ψ_1. The construction of $g(x, t)$ according to (2.17) is then immediate and its modified version $g_1(x, t)$ follows according to (2.32).

We return to the boundary value problem used earlier for illustration to provide an example.

Example 2.2

Let ϕ satisfy

$$\left. \begin{array}{ll} \phi''(x) + \lambda\phi(x) = 0 & (0 < x < 1), \\ \phi'(0) = 0, & \phi'(1) = 0. \end{array} \right\} \tag{2.36}$$

We have already noted that $\psi_1(x) = A$ satisfies $(L\phi)(x) = -\phi''(x) = 0$ and both boundary conditions. An acceptable second solution of $L\phi = 0$ is $\psi_2(x) = Bx$, and the Wronskian condition (2.6) gives $AB = -1$.

Therefore

$$g(x, t) = \left\{ \begin{array}{ll} -x & (t \leqslant x) \\ -t & (x \leqslant t) \end{array} \right\} = -\max(x, t).$$

Normalising ψ_1 according to (2.30) gives $A^2 = 1$ and we choose $A = 1$ (and therefore $B = -1$, although we do not need to use this fact). Since

$r(x)=1$ in this example

$$\int_0^1 g(s,x)\psi_1(s)r(s)\,ds = -A\int_0^1 \max(x,s)\,ds$$

$$= -\int_0^x x\,ds - \int_x^1 s\,ds = -\tfrac{1}{2}(1+x^2),$$

and so

$$\int_0^1 \psi_1(s)r(s)\,ds \int_0^1 g(\xi,s)\psi_1(\xi)r(\xi)\,d\xi = -A\int_0^1 \tfrac{1}{2}(1+s^2)\,ds = -\tfrac{2}{3}.$$

Assembling the elements forming $g_1(x,t)$, we obtain

$$g_1(x,t) = -\max(x,t) + \tfrac{1}{2}(1+x^2) + \tfrac{1}{2}(1+t^2) - \tfrac{2}{3}$$
$$= -\max(x,t) + \tfrac{1}{2}(x^2+t^2) + \tfrac{1}{3}.$$

The integral equation associated with (2.36) is therefore

$$\mu\phi(x) = \int_0^1 \{-\max(x,t) + \tfrac{1}{2}(x^2+t^2) + \tfrac{1}{3}\}\phi(t)\,dt \quad (0\leqslant x\leqslant 1), \quad (2.37)$$

where $\mu = \lambda^{-1}$.

To deduce (2.36) from (2.37) we first note that

$$\int_0^1 \{-\max(x,t) + \tfrac{1}{2}(x^2+t^2) + \tfrac{1}{3}\}\,dx = 0$$

for each $t \in [0,1]$, so that the solution of (2.37) satisfies

$$\int_0^1 \phi(x)\,dx = 0 \tag{2.38}$$

Differentiating (2.37) twice produces

$$\mu\phi'(x) = -\int_0^x \phi(t)\,dt + \int_0^1 x\phi(t)\,dt,$$

$$\mu\phi''(x) = -\phi(x) + \int_0^1 \phi(t)\,dt$$

and (2.36) is recovered on using (2.38), putting $\mu = \lambda^{-1}$ and noting that $\phi'(0) = 0 = \phi'(1)$.

We note that $\phi(x) = \text{constant}$ $(0\leqslant x\leqslant 1)$ is a solution of the integral equation (2.37) with $\mu = 0$ and of the boundary value problem (2.36) with $\lambda = 0$, in accordance with our earlier remarks. $\qquad\square$

To complete our discussion of Case 2 we must consider the inhomogeneous boundary value problem ($h \neq 0$) for $\lambda \neq 0$, which we discounted before (2.27) in the interests of simplicity. To reinstate h, we return to (2.25) and (2.26) and substitute $F = \lambda r\phi - h$ to give

$$\lambda \int_0^1 \psi_1(x) r(x) \phi(x) \, dx = \int_0^1 \psi_1(x) h(x) \, dx \tag{2.39}$$

and

$$\phi(x) = D_1 \psi_1(x) + \lambda \int_0^1 g(x, t) r(t) \phi(t) \, dt - \int_0^1 g(x, t) h(t) \, dt \quad (0 \leqslant x \leqslant 1), \tag{2.40}$$

which respectively extend (2.28) and (2.27) to the inhomogeneous case.

The procedure we used for $h = 0$ applies in this case also. The condition (2.39) is used in (2.40) to determine D_1, in such a way as to preserve the symmetry in the kernel. The resulting integral equation is

$$\left.\begin{array}{c} \phi(x) = \lambda \int_0^1 g_1(x, t) r(t) \phi(t) \, dt + f(x) \quad (0 \leqslant x \leqslant 1), \\[2mm] \text{where} \\[2mm] f(x) = -\int_0^1 g_1(x, t) h(t) \, dt + \dfrac{1}{\lambda} \psi_1(x) \int_0^1 \psi_1(t) h(t) \, dt, \end{array}\right\} \tag{2.41}$$

where $g_1(x, t)$ is given by (2.32) and ψ_1 is defined by (2.23a) and (2.30), as before.

The reduction of (2.31) and (2.41) to symmetric integral equations follows if $r(x) > 0$ in $[0, 1]$, as it did in the simpler Case 1. (See Problem 2.9.)

2.5 Singular boundary value problems

Our treatment of the Sturm–Liouville problem

$$\left.\begin{array}{l} (p(x)\phi'(x))' - q(x)\phi(x) + \lambda r(x)\phi(x) = h(x) \quad (0 < x < 1), \\[1mm] a_0 \phi(0) + b_0 \phi'(0) = 0, \quad a_1 \phi(1) + b_1 \phi'(1) = 0, \end{array}\right\} \tag{2.11}$$

has so far been confined to the case in which

$$\left.\begin{array}{l} p, \, p', \, q, \, r \text{ and } h \text{ are continuous in } [0, 1], \\[1mm] p > 0, \quad r > 0 \text{ in } [0, 1]. \end{array}\right\} \tag{2.4}$$

If these conditions are satisfied, as we have assumed, the boundary value problem is often described as *regular*. It is called *singular* if one or more of the requirements which constitute (2.4) holds only in (0, 1), (0, 1] or [0, 1). An end point at which (2.4) is violated is usually referred to as a *singular boundary point*.

Bessel's equation of order zero,

$$(x\phi'(x))' + \lambda x\phi(x) = 0 \quad (0 < x < 1), \tag{2.42}$$

which arises in connection with the free oscillations of a stretched circular elastic membrane, for example, has $x = 0$ as a singular boundary point since both $p(x) = x$ and $r(x) = x$ vanish there. Singular boundary value problems are widespread in practice and for completeness we need to include them in our integral equation conversion procedure. This is not difficult to achieve.

The main point is that, at a singular boundary point, a boundary condition of the form we have previously considered may be incompatible with the differential equation and an alternative, less restrictive condition has to be used which merely specifies the required behaviour of the unknown function at the end point, rather than assigning a particular value to it or to its derivative. Suitable boundary conditions associated with (2.42) are, for example,

$$\left.\begin{array}{l} \phi(x) \quad \text{bounded as} \quad x \to 0, \\ \phi(1) = 0. \end{array}\right\} \tag{2.43}$$

The conversion procedures described in §2.4 carry over to the new circumstances, the functions ψ_1 and ψ_2 now being determined in accordance with the weaker boundary conditions, as we can illustrate by means of the boundary value problem (2.42) with (2.43).

Example 2.3

Let ϕ satisfy

$$\left.\begin{array}{l} (x\phi'(x))' + \lambda x\phi(x) = 0 \quad (0 < x < 1), \\ \phi(x) \quad \text{bounded as} \quad x \to 0, \quad \phi(1) = 0. \end{array}\right\}$$

The functions ψ_1 and ψ_2 required in this case are given by

$$-(x\psi_1'(x))' = 0 \quad (0 < x < 1), \quad \psi_1(x) \quad \text{bounded as} \quad x \to 0,$$
$$-(x\psi_2'(x))' = 0 \quad (0 < x < 1), \quad \psi_2(1) = 0.$$

The solution of $(L\phi)(x) = -(x\phi'(x))' = 0$ is easily found to be $\phi(x) = A + B \log x$ and so we must take

$$\psi_1(x) = A, \quad \psi_2(x) = B \log x.$$

Obviously the Wronskian $W(\psi_1, \psi_2) = \psi_1\psi_2' - \psi_1'\psi_2$ is not defined at $x = 0$

but the Wronskian normalisation condition

$$p(\psi_1\psi_2' - \psi_1'\psi_2) = AB = -1$$

still applies. Therefore, according to our earlier theory, an integral equation satisfied by ϕ is

$$\phi(x) = \lambda \int_0^1 g(x, t)t\phi(t)\,dt \quad (0 \leqslant x \leqslant 1),$$

where

$$g(x, t) = \begin{cases} -\log x & (t \leqslant x) \\ -\log t & (x \leqslant t) \end{cases} = -\log\{\max(x, t)\}.$$

If we now put $\hat{\phi}(x) = x^{\frac{1}{2}}\phi(x)$ $(0 \leqslant x \leqslant 1)$, we obtain the symmetric integral equation

$$\hat{\phi}(x) = \lambda \int_0^1 k(x, t)\hat{\phi}(t)\,dt \quad (0 \leqslant x \leqslant 1),$$

where $k(x, t) = -(xt)^{\frac{1}{2}} \log\{\max(x, t)\}$.

The substitution $\hat{\phi}(x) = x^{\frac{1}{2}}\phi(x)$ made in the differential equation at the outset reduces it to its so-called normal form

$$\hat{\phi}''(x) + ((4x^2)^{-1} + \lambda)\hat{\phi}(x) = 0 \quad (0 < x < 1),$$

in which (2.4) is violated because $q(x) = (4x^2)^{-1}$ is discontinuous at $x = 0$. $\qquad\square$

We may infer from this particular example that our overall strategy for deriving integral equations from boundary value problems is not significantly affected by the presence of singular boundary points. One consequence is that the solution of $L\phi = 0$, and hence the Wronskian $W(\psi_1, \psi_2)$, may not be continuous in $[0, 1]$, but the condition $p(\psi_1\psi_2' - \psi_1'\psi_2) = -1$ can still be satisfied and the function $g(x, t)$ determined for $x, t \in (0, 1)$.

To conclude this section we consider an example which brings together many of the features we have introduced.

Example 2.4

Let ϕ satisfy

$$\left.\begin{array}{l} \{(1 - x^2)\phi'(x)\}' + \lambda\phi(x) = 0 \quad (-1 < x < 1), \\ \phi(x) \quad \text{bounded as } x \to \pm 1. \end{array}\right\} \tag{2.44}$$

The differential equation is Legendre's equation in its standard form; it requires us to adapt our procedure to the interval $[-1, 1]$ and this presents no difficulties. Both boundary points are singular and the 'weak' boundary conditions given are appropriate.

We first note that $(L\phi)(x) = -\{(1-x^2)\phi'(x)\}' = 0$ is satisfied by $\phi(x) = A + B\log\{(1+x)/(1-x)\}$ and therefore the function $\phi = $ constant satisfies (2.44) with $\lambda = 0$. Thus we are dealing with the more involved Case 2 of our conversion procedure, and we choose

$$\psi_1(x) = A, \quad \psi_2(x) = B\log\{(1+x)/(1-x)\} \quad (-1 < x < 1).$$

The function ψ_1 satisfies $L\psi_1 = 0$ in $[-1, 1]$ and both boundary conditions; ψ_2 satisfies $L\psi_2 = 0$ in $(-1, 1)$ and neither of the boundary conditions. Even though ψ_2 is not defined at the boundary points, the Wronskian normalisation condition

$$(1-x^2)(\psi_1(x)\psi_2'(x) - \psi_1'(x)\psi_2(x)) = -1$$

applies, giving $AB = -\frac{1}{2}$. Therefore

$$g(x, t) = \begin{cases} -\frac{1}{2}\log\{(1+x)/(1-x)\} & (t \leqslant x), \\ -\frac{1}{2}\log\{(1+t)/(1-t)\} & (x \leqslant t). \end{cases}$$

The normalisation of ψ_1, namely

$$\int_{-1}^{1} \psi_1^2(x)\, dx = 1,$$

implies that $A^2 = \frac{1}{2}$ and we therefore take $A = 2^{-\frac{1}{2}} = -B$. Straightforward calculations then give

$$\int_{-1}^{1} g(s, x)\psi_1(s)\, ds = \frac{1}{2^{\frac{1}{2}}}\log\{\tfrac{1}{2}(1-x)\}$$

and

$$\int_{-1}^{1} \psi_1(s)\, ds \int_{-1}^{1} g(\xi, s)\psi_1(\xi)\, d\xi = \tfrac{1}{2}\int_{-1}^{1}\log\{\tfrac{1}{2}(1-s)\}\, ds = -1.$$

The function $g_1(x, t)$ defined by (2.32) can now be constructed for the present example, giving

$$g_1(x, t) = \begin{cases} -\frac{1}{2}\log\{(1+x)(1-t)\} + \log 2 - \frac{1}{2} & (t \leqslant x) \\ -\frac{1}{2}\log\{(1+t)(1-x)\} + \log 2 - \frac{1}{2} & (x \leqslant t) \end{cases}. \qquad (2.45)$$

Those solutions of (2.44) corresponding to $\lambda \neq 0$ satisfy the integral equation

$$\mu\phi(x) = \int_{-1}^{1} g_1(x, t)\phi(t)\,dt \quad (-1 \leqslant x \leqslant 1) \tag{2.46}$$

where $\mu = \lambda^{-1}$ and the kernel is defined by (2.45). It is easily verified that

$$\int_{-1}^{1} g_1(x, t)\,dt = 0 \quad (0 \leqslant x \leqslant 1),$$

$$\int_{-1}^{1} g_1(x, t)\,dx = 0 \quad (0 \leqslant t \leqslant 1),$$

from which we deduce that the solutions of (2.46) satisfy

$$\int_{-1}^{1} \phi(x)\,dx = 0$$

and that one solution is $\phi = $ constant in $[-1, 1]$ with $\mu = 0$. □

Problems

2.1   Let $(L\phi)(x) = -(x\phi'(x))'$ $(0 < x < 1)$.

(i)   Show that the initial value problem

$$(L\phi)(x) = x\phi(x) \quad (0 < x < 1), \quad \phi(0) = 1, \quad \phi'(0) = 0$$

and the integral equation

$$\phi(x) = 1 + \int_0^x t \log\left(\frac{t}{x}\right)\phi(t)\,dt \quad (0 \leqslant x \leqslant 1)$$

are equivalent.

(ii)   Derive an integral equation equivalent to the initial value problem

$$(L\phi)(x) = x\phi(x) - x \log x \ (0 < x < 1), \quad \phi(0) = \phi'(0) = 0.$$

(Notice that the differential equation in (ii) follows on writing $\tilde{\phi}(x) = \phi(x) - \log x$ in $(L\tilde{\phi})(x) = x\tilde{\phi}(x)$ $(0 < x < 1)$. The initial value problem in (ii) therefore leads to a solution $\tilde{\phi}$ of Bessel's equation of zero order which is logarithmically singular at $x = 0$.)

2.2   Convert the boundary value problem

$$\phi''(x) + \lambda\phi(x) = 0 \quad (0 < x < 1), \quad \phi'(0) = \phi(0), \quad \phi'(1) = \phi(1),$$

into an integral equation.

2.3   Let $(L\phi)(x) = -\phi''(x) - \alpha\phi(x)$ $(0 < x < 1)$ where α is a real, non-zero constant.

Convert the boundary value problem

$$(L\phi)(x) = \lambda r(x)\phi(x) \quad (0 < x < 1), \quad \phi(0) = \phi(1) = 0,$$

into an integral equation in the following cases:

(i) $\alpha > 0$, $\alpha \neq (n\pi)^2$ $(n \in \mathbb{N})$.

(ii) $\alpha < 0$.

(iii) $\alpha = (n\pi)^2$ $(n \in \mathbb{N})$ and $r(x) = 1$ $(0 \leqslant x \leqslant 1)$, showing in this case that the kernel of the integral equation is given by

$$k(x, t) = \frac{1}{n\pi} \{(1 - x) \sin(n\pi t) \cos(n\pi x) - t\sin(n\pi x) \cos(n\pi t)$$

$$+ \frac{1}{2n\pi} \sin(n\pi t) \sin(n\pi x)\} \quad (0 \leqslant t \leqslant x \leqslant 1)$$

and $k(x, t) = k(t, x)$ $(0 \leqslant x, t \leqslant 1)$.

(The differential equation $L\phi = \lambda r\phi$ arises in a number of application areas. An equation of this form in which r is periodic is often referred to as Hill's equation. The particular case $(L\phi)(x) = \lambda\cos(4\pi x)\phi(x)$ $(0 < x < 1)$ is known as Mathieu's equation.)

2.4   The buckling of an elastic column of unit length under a compressive load is governed by the differential equation

$$(I(x)\phi''(x))'' + \lambda\phi''(x) = 0 \quad (0 < x < 1).$$

Here $\phi(x)$ denotes the transverse displacement of the centroid of the column at position x from its unstrained position and $I(x) > 0$ $(0 \leqslant x \leqslant 1)$ is the moment of inertia of the section at position x about an axis through the centroid and perpendicular to the $x\phi$ plane. The parameter λ is proportional to the compressive force.

For hinged ends, where $\phi(0) = \phi(1) = \phi''(0) = \phi''(1) = 0$, put $\psi = \phi''$ and show that ψ can be determined via an integral equation with the kernel

$$\frac{\{\min(x, t) - xt\}}{\{I(x)I(t)\}^{\frac{1}{2}}}.$$

(The smallest allowable value of λ gives the buckling load for the column.)

2.5   Let ϕ satisfy the differential equation

$$(L\phi)(x) = -(p(x)\phi'(x))' + q(x)\phi(x) = F(x) \quad (0 < x < 1)$$

and the inhomogeneous boundary conditions $a_0\phi(0) + b_0\phi'(0) = c_0$, $a_1\phi(1) + b_1\phi'(1) = c_1$ where a_i, b_i and c_i $(i = 1, 2)$ are given constants such that $a_0^2 + b_0^2 \neq 0$, $a_1^2 + b_1^2 \neq 0$ and $c_1^2 + c_2^2 \neq 0$. Show that if a function ψ is defined on $[0, 1]$ such that $a_0\psi(0) + b_0\psi'(0) = c_0$, $a_1\psi(1) + b_1\psi'(1) = c_1$, then $\phi = \tilde{\phi} + \psi$ where $\tilde{\phi}$ satisfies the differential equation $L\tilde{\phi} = F - L\psi$ in $(0, 1)$ and homogeneous boundary conditions.

Let ϕ satisfy

$$\phi''(x) + \lambda r(x)\phi(x) = 0 \quad (0 < x < 1)$$
$$\phi(0) = \alpha, \quad \phi'(1) = \beta\phi(1) + \gamma,$$

where α, β and γ are given constants with $\beta \neq 1$. By writing $\phi = \tilde{\phi} + \psi$, where $\psi(x) = A + Bx$ $(0 \leqslant x \leqslant 1)$ for suitably chosen values of A and B, show that ϕ satisfies the integral equation

$$\phi(x) = \alpha + (\alpha\beta + \gamma)(1 - \beta)^{-1}x$$
$$+ \lambda \int_0^1 \{\min(x, t) + \beta(1 - \beta)^{-1}xt\}r(t)\phi(t)\, dt \quad (0 \leqslant x \leqslant 1).$$

(The case in which r is constant and $\lambda r < 0$ is related to a particular problem in groundwater flow in the presence of an aquifer; $\phi(x)$ then represents the groundwater level at x, relative to a horizontal datum level.)

2.6　Shallow water theory and the assumption of quasi-one-dimensional motion can be used to show that the surface elevation $\eta(x, t) = \phi(x)\cos(\sigma t + \epsilon)$ of small amplitude time harmonic tidal waves in an estuary of slowly varying breadth $b(x)$ and mean depth $h(x)$ is determined by

$$g\{b(x)h(x)\phi'(x)\}' + \sigma^2 b(x)\phi(x) = 0 \quad (0 < x < 1)$$

where $b(x) > 0$ and $h(x) > 0$ for $0 < x \leqslant 1$. The boundary conditions are (i) $\phi(1) = c$, a given non-zero constant, representing an imposed tidal oscillation at the mouth of the estuary and (ii) $\phi'(0) = 0$ if $b(0)h(0) \neq 0$ or $\phi(x)$ bounded as $x \to 0$ if $b(0)h(0) = 0$. Show that ϕ can be determined by solving an integral equation with the symmetric kernel

$$\{b(x)b(t)\}^{\frac{1}{2}} \int_{\max(x, t)}^1 \frac{ds}{b(s)h(s)} \quad (0 \leqslant x, t \leqslant 1).$$

2.7　Convert the boundary value problem

$$\{(1 + 4x - 4x^2)\phi'(x)\}' + \lambda\phi(x) = 0 \quad (0 < x < 1),$$
$$\phi'(0) = \phi'(1) = 0,$$

to an integral equation.

(The boundary value problem may be interpreted in terms of the free oscillations of a rectangular lake having a parabolic depth profile. In this case $\lambda = \sigma^2/g$ and the allowable values of λ give the natural frequencies of oscillation σ.)

2.8　Let ϕ satisfy the regular Sturm–Liouville problem

$$(p(x)\phi'(x))' - q(x)\phi(x) + \lambda r(x)\phi(x) = 0 \quad (0 < x < 1),$$
$$\cos\theta_0\,\phi(0) - \sin\theta_0\,\phi'(0) = 0, \quad \cos\theta_1\,\phi(1) + \sin\theta_1\,\phi'(1) = 0,$$

where $0 \leqslant \theta_0 < \pi$, $0 \leqslant \theta_1 < \pi$. Assume that there is no non-trivial solution corresponding to $\lambda = 0$ and let K_{θ_0, θ_1} denote the associated integral operator, that is

$$(K_{\theta_0, \theta_1} \phi)(x) = \int_0^1 \psi_1(\min(x, t)) \psi_2(\max(x, t)) r(t) \phi(t)\, dt \quad (0 \leqslant x \leqslant 1),$$

for appropriately defined functions ψ_1 and ψ_2. Let $0 \leqslant \omega_0 < \pi, 0 \leqslant \omega_1 < \pi$ and show that

$$(K_{\omega_0, \theta_1} \phi)(x) = (K_{\theta_0, \theta_1} \phi)(x) + A_0 \int_0^1 \psi_2(x) \psi_2(t) r(t) \phi(t)\, dt \quad (0 \leqslant x \leqslant 1),$$

$$(K_{\theta_0, \omega_1} \phi)(x) = (K_{\theta_0, \theta_1} \phi)(x) + A_1 \int_0^1 \psi_1(x) \psi_1(t) r(t) \phi(t)\, dt \quad (0 \leqslant x \leqslant 1),$$

where

$$A_0 = -(\cos \omega_0 \psi_1(0) - \sin \omega_0 \psi_1'(0))(\cos \omega_0 \psi_2(0) - \sin \omega_0 \psi_2'(0))^{-1},$$
$$A_1 = -(\cos \omega_1 \psi_2(1) + \sin \omega_1 \psi_2'(1))(\cos \omega_1 \psi_1(1) + \sin \omega_1 \psi_1'(1))^{-1}.$$

2.9   Show that, if $r(x) > 0$ in $[0, 1]$, the integral equation (2.31) can be transformed into an integral equation with the symmetric kernel

$$k_1(x, t) = k(x, t) - \hat{\psi}_1(x) \int_0^1 k(s, t) \hat{\psi}_1(s)\, ds - \hat{\psi}_1(t) \int_0^1 k(s, x) \hat{\psi}_1(s)\, ds$$

$$+ \hat{\psi}_1(x) \hat{\psi}_1(t) \int_0^1 \hat{\psi}_1(s)\, ds \int_0^1 k(\xi, s) \hat{\psi}_1(\xi)\, d\xi \quad (0 \leqslant x, t \leqslant 1),$$

where $k(x, t) = \hat{\psi}_1(\min(x, t)) \hat{\psi}_2(\max(x, t))$ $(0 \leqslant x, t \leqslant 1)$, $\hat{\psi}_i(x) = \psi_i(x)(r(x))^{\frac{1}{2}}$ and

$$\int_0^1 \hat{\psi}_1^2(x)\, dx = 1.$$

2.10   Convert the singular boundary value problem

$$\begin{cases} x^2 \phi''(x) + x \phi'(x) + (\lambda x^2 - v^2) \phi(x) = 0 & (0 < x < 1), \\ \phi(x) \text{ bounded as } x \to 0, \quad \phi(1) = 0, \end{cases}$$

where $v > 0$, into an integral equation with a symmetric kernel.
(The differential equation is Bessel's equation of order v.)

2.11   Let

$$-(L\phi)(x) = (1 - x^2) \phi''(x) - 2x \phi'(x) - v^2(1 - x^2)^{-1} \phi(x) \quad (-1 < x < 1)$$

where $v > 0$. Convert the singular boundary value problem $(L\phi)(x) = \lambda \phi(x)$ $(-1 < x < 1)$ with $\phi(x)$ bounded as $x \to \pm 1$ into an integral equation.
(To solve $(L\phi)(x) = 0$ put $x = \tanh \xi$. The differential equation $L\phi = \lambda \phi$ is

sometimes called the associated Legendre equation and its solutions are known as spherical harmonics.)

2.12 Let $(L\phi)(x) = -(1-x^2)^{\frac{1}{4}} \{(1-x^2)^{\frac{1}{4}} \phi'(x)\}' (-1 < x < 1)$. Convert the following boundary value problems to integral equations.

(i) $(L\phi)(x) = \lambda\phi(x) (-1 < x < 1)$, $\phi(1) = 1$, $\phi(-1) = 0$.

(ii) $(L\phi)(x) = \lambda\phi(x) (-1 < x < 1)$, $\phi(1) = 1$, $\phi'(x)$ bounded as $x \to -1$.

(iii) $(L\phi)(x) = \lambda\phi(x) (-1 < x < 1)$, $\phi'(x)$ bounded as $x \to \pm 1$.

(In (i) and (ii) use the decomposition $\phi = \tilde{\phi} + \psi$ to reduce the boundary conditions to homogeneous type.

Notice that both boundary points are singular but that the singularity is 'weak' in the sense that $(L\phi)(x) = 0$ has two linearly independent solutions which are continuous in $[-1, 1]$. Only one of these has a continuous derivative in $[-1, 1]$ however. As a result, boundary conditions of the 'strong' type, in which specific boundary values are prescribed, can be given for ϕ. Only boundary conditions of the 'weak' form, in which the general behaviour at the boundary points is assigned, can be given for ϕ'.

The differential equation $L\phi = \lambda\phi$ is the particular case of Gegenbauer's equation

$$(1-x^2)\phi''(x) - (2v+1)x\phi'(x) + \lambda\phi(x) = 0 \quad (-1 < x < 1)$$

with $v = 0$. For $|v| < \frac{1}{2}$, Gegenbauer's equation, which can be written as $\{(1-x^2)^{v+\frac{1}{2}}\phi'(x)\}' + \lambda(1-x^2)^{v-\frac{1}{2}}\phi(x) = 0 (-1 < x < 1)$, has 'weakly' singular boundary points. For $v = \frac{1}{2}$ the equation reduces to Legendre's which, as we saw in Example 2.4, has 'strongly' singular boundary points.)

2.13 Let ϕ satisfy the differential equation

$$\{p(x)\phi'(x)\}' - q(x)\phi(x) + \lambda\phi(x) = h(x) \quad (0 < x < 1)$$

and the *mixed* boundary conditions

$$\left. \begin{array}{l} a_{11}\phi(0) + a_{12}\phi'(0) + b_{11}\phi(1) + b_{12}\phi'(1) = 0 \\ a_{21}\phi(0) + a_{22}\phi'(0) + b_{21}\phi(1) + b_{22}\phi'(1) = 0 \end{array} \right\},$$

which are distinct, and suppose that the conditions (2.4) apply. Show that ϕ satisfies an integral equation with a symmetric kernel if and only if

$$p(0)(b_{11}b_{22} - b_{12}b_{21}) = p(1)(a_{11}a_{22} - a_{12}a_{21}).$$

(Apply the given boundary conditions to (2.12) to determine D_1 and D_2. To achieve a symmetric kernel the terms in $\psi_1(x)\psi_2(t)$ and $\psi_1(t)\psi_2(x)$ must have equal coefficients.)

2.14 Let ϕ satisfy the mixed boundary value problem

$$\begin{cases} \phi''(x) + \lambda\phi(x) = 0 & (0 < x < 1), \\ \alpha\phi(0) = \beta\phi(1), & \beta\phi'(0) = \alpha\phi'(1), \end{cases}$$

where α and β are constants. Show that ϕ satisfies the integral equation

$$\phi(x) = \lambda \int_0^1 k(x,t)\phi(t)\,dt \quad (0 \leqslant x \leqslant 1)$$

where

$$(\alpha - \beta)^2 k(x,t) = \alpha\beta + \alpha(\alpha - \beta)\min(x,t) - \beta(\alpha - \beta)(1 - \max(x,t)) \quad (0 \leqslant x, t \leqslant 1)$$

if $\alpha \neq \beta$, and

$$k(x,t) = \tfrac{1}{2}\{(x-t)^2 - |x-t| + \tfrac{1}{6}\} \quad (0 \leqslant x, t \leqslant 1)$$

if $\alpha = \beta$, the function ϕ in the latter case satisfying

$$\int_0^1 \phi(x)\,dx = 0.$$

2.15   Let $(L\phi)(x) = (p(x)\phi''(x))'' - (q(x)\phi'(x))'$ $(0 < x < 1)$ where p, p', p'', q and q' are continuous in $[0,1]$ and $p > 0$ in $[0,1]$. Let ϕ satisfy the fourth order boundary value problem consisting of $(L\phi)(x) = F(x)$ $(0 < x < 1)$, $a_0\phi'(0) + b_0\phi''(0) = 0$, $a_1\phi'(1) + b_1\phi''(1) = 0$ and two further boundary conditions so that there are four distinct, non-degenerate boundary conditions in all. Show that $\psi = \phi'$ satisfies a second order Sturm–Liouville problem.

Assume now that there are linearly independent functions ψ_1 and ψ_2 satisfying $(p\psi_i')' = q\psi_i$ $(i = 1, 2)$ in $(0, 1)$,

$$a_0\psi_1(0) + b_0\psi_1'(0) = 0, \quad a_1\psi_2(1) + b_1\psi_2'(1) = 0 \text{ and } p(\psi_1\psi_2' - \psi_2\psi_1') = -1$$

in $[0, 1]$. Let r and h be continuous in $[0, 1]$. Show that the boundary value problem

$$(L\phi)(x) + \lambda r(x)\phi(x) = h(x) \quad (0 < x < 1),$$

$$\phi(0) = \phi(1) = a_0\phi'(0) + b_0\phi''(0) = a_1\phi'(1) + b_1\phi''(1) = 0,$$

can be converted into the integral equation

$$\phi(x) = \lambda \int_0^1 g(x,t)r(t)\phi(t)\,dt - \int_0^1 g(x,t)h(t)\,dt \quad (0 \leqslant x \leqslant 1),$$

where

$$g(x,t) = \frac{J(x,1)J(t,1)}{J(1,1)} - J(x,t) \quad (0 \leqslant x, t \leqslant 1)$$

and

$$J(x,t) = \int_0^x d\xi \int_0^t \psi_1(\min(\xi,\eta))\psi_2(\max(\xi,\eta))\,d\eta,$$

provided that $J(1,1) \neq 0$.

Illustrate by converting the boundary value problem

$$\phi''''(x) = \lambda\phi(x) \quad (0 < x < 1),$$
$$\phi(0) = \phi(1) = \phi'(0) = \phi'(1) = 0,$$

to an integral equation. (*cf.* Problem 1.11.)

3

Integral equations of the second kind

3.1 Introduction

In this chapter we shall investigate integral equations of the second kind

$$\phi(x) - \lambda \int_a^b k(x, t)\phi(t)\, dt = f(x) \quad (a \leqslant x \leqslant b) \tag{3.1}$$

where the kernel k is a 'well-behaved' function in a sense to be specified. If we define the function $K\phi$ by

$$(K\phi)(x) = \int_a^b k(x, t)\phi(t)\, dt \quad (a \leqslant x \leqslant b)$$

then $K(\phi_1 + \phi_2) = K\phi_1 + K\phi_2$ and if μ is a constant $K(\mu\phi) = \mu(K\phi)$, so that K is a linear transformation of a suitable vector space of functions into another vector space of functions. In these terms (3.1) may be rewritten as

$$\phi - \lambda K\phi = f. \tag{3.2}$$

In this formulation we subtract $\lambda K\phi$ from ϕ so they must both belong to the same vector space and we therefore demand that K maps our vector space of functions into itself. If we do this, and denote the identity linear map on the vector space by I, then (3.2) may be rewritten again as

$$(I - \lambda K)\phi = f. \tag{3.3}$$

Since $I - \lambda K$ is a linear map, (3.3) draws our attention to pure linear algebra and the information it may yield about solutions of our equation. If it happens that the linear map $I - \lambda K$ has an inverse, $(I - \lambda K)^{-1}$, then (3.3) will have a unique solution $\phi = (I - \lambda K)^{-1} f$. Even if $(I - \lambda K)$ does not have an inverse, standard linear algebra shows that (3.3) will have at most one solution if and only if the corresponding homogeneous equation $(I - \lambda K)\phi = 0$ has only the trivial solution. Whether or not the solution is unique, if $(I - \lambda K)$ is surjective (3.3) will have at least one solution; if $I - \lambda K$ is not surjective a solution to (3.3) will exist if and only if f belongs to the image of $(I - \lambda K)$, a trickier issue.

In the case where the vector space concerned has finite dimension, say n, so that (3.3) is essentially an equation involving the $n \times n$ matrices I and K and the vectors ϕ and f, then we known that $(I - \lambda K)$ has an inverse for most values of λ and that in the case where $I - \lambda K$ does not have an inverse we can relate the various conditions regarding the existence and uniqueness of solutions. Clearly we need to investigate to what extent similar statements can be made in the context of integral equations. The vector spaces likely to be of interest in the integral equation context will usually have infinite dimension; to see this notice that the functions given by x^n $(n = 0, 1, 2, \ldots)$ on $[0, 1]$ are linearly independent, so it is likely that any class of functions from which we seek our solutions will have infinite dimension since it will usually contain the set of all polynomials. These details we shall address later, but for the moment we note that we expect to have to derive information about the solutions of linear equations in infinite-dimensional vector spaces.

Many of the standard techniques and results of matrix theory and linear algebra in finite-dimensional vector spaces rely on arguments of counting dimensions, and most of these are of little use in the infinite-dimensional situation. The most fruitful remedy for this is to introduce the idea of distance into the vector space, in the form of a norm, and use functional analysis. One immediate advantage of this viewpoint is the existence of the notion of the continuity of the linear maps involved; with the aid of continuity it is possible to formulate questions concerning the approxima-tion of one function by another. In the all too common situation where an exact solution to a problem cannot be found directly, the technique of approximation becomes important and we shall have much to say about it in due course. The point to notice at this stage, however, is that if we wish to use approximation techniques to find the solution of $(I - \lambda K)\phi = f$, then it is an approximation to the exact solution ϕ which we seek, which is not necessarily the solution of the equation $(I - \lambda \tilde{K})\tilde{\phi} = \tilde{f}$ where $\tilde{K}$ and $\tilde{f}$ are approximations to K and f. This is more simply stated if we suppose that $I - \lambda K$ has an inverse and we consider the solution $\tilde{\phi}$ of the equation $(I - \lambda K)\tilde{\phi} = \tilde{f}$; if $\tilde{f}$ is 'close to' f, must $\tilde{\phi}$ be 'close to' ϕ? From the existence of the inverse $(I - \lambda K)^{-1}$ we see that the question is whether the closeness of $\tilde{f}$ to f implies that $(I - \lambda K)^{-1}\tilde{f}$ and $(I - \lambda K)^{-1}f$ are close, which is essentially the idea of the continuity of the solution process, that is, of the linear map $(I - \lambda K)^{-1}$.

One question which arises immediately is whether the rather more abstract reformulation of the integral equation

$$\phi(x) - \lambda \int_a^b k(x, t)\phi(t)\,\mathrm{d}t = f(x) \quad (a \leqslant x \leqslant b) \tag{3.1}$$

as a linear equation in some particular normed vector space,

$$(I - \lambda K)\phi = f, \tag{3.3}$$

is likely to prove fruitful. The principal advantage of (3.3) is that the independent variable x has been suppressed and we are forced to consider the *process* of changing the function ϕ into the function f; the details of the process can be obscured by the additional variable and technicalities appearing in (3.1). The aim of the more abstract formulation (3.3) is therefore to remove the inessential detail in order to obtain a clear view of general properties of the processes concerned. As we shall see, however, the more abstract formulation is not a panacea for our problems and we shall frequently find that a return to the original equation (3.1) can add to the more general results obtainable from our functional analysis.

All of this presupposes that we can find a suitable vector space of functions in which our various linear maps act, which forces our attention on an issue which we have hitherto avoided: what sort of function is acceptable as a solution to (3.1)? In some problems it may be clear that, for example, the only solutions of interest are continuous functions so we must incorporate this feature into our methods. In other cases we may not have any *a priori* restrictions on the acceptable solutions, other than the tacit restriction that ϕ must be of such a type that the integral in (3.1) exists.

3.2 Degenerate kernels

To set the scene, we shall suppose that $k(x, t)$ can be decomposed in a particularly simple form, as follows.

Definition 3.1

Suppose that $k(x, t)$ is a kernel defined on the square $[a, b] \times [a, b]$ and that there are finitely many functions $a_1, \ldots, a_n, b_1, \ldots, b_n$ such that

$$k(x, t) = \sum_{i=1}^{n} a_i(x)b_i(t) \quad (a \leqslant x, t \leqslant b).$$

In this case the kernel k is said to be *degenerate*. □

Suppose that

$$k(x, t) = \sum_{i=1}^{n} a_i(x)b_i(t) \quad (a \leqslant x, t \leqslant b)$$

where the functions $a_1, \ldots, a_n, b_1, \ldots, b_n$ are all continuous on $[a, b]$ and seek solutions of the integral equation

$$\phi(x) - \lambda \int_a^b k(x, t)\phi(t)\,dt = f(x) \quad (a \leqslant x \leqslant b). \tag{3.1}$$

Then

$$(K\phi)(x) = \int_a^b k(x, t)\phi(t)\,dt = \sum_{i=1}^n \alpha_i a_i(x) \quad (a \leqslant x \leqslant b) \tag{3.4}$$

where $\alpha_i = \int_a^b b_i(t)\phi(t)\,dt$, which shows that $K\phi$ is a linear combination of the functions $a_1, \ldots, a_n$. Because ϕ satisfies (3.1) if and only if

$$\phi(x) = f(x) + \lambda(K\phi)(x) \quad (a \leqslant x \leqslant b), \tag{3.5}$$

(3.4) shows that ϕ must be expressible in the form

$$\phi(x) = f(x) + \sum_{i=1}^n \gamma_i a_i(x) \quad (a \leqslant x \leqslant b)$$

for some coefficients $\gamma_1, \ldots, \gamma_n$. Substitution in (3.5) now shows that

$$\sum_{i=1}^n \gamma_i a_i(x) = \lambda \sum_{i=1}^n \left(\beta_i + \sum_{j=1}^n \alpha_{ij}\gamma_j\right) a_i(x) \quad (a \leqslant x \leqslant b) \tag{3.6}$$

where

$$\beta_i = \int_a^b f(t)b_i(t)\,dt \quad \text{and} \quad \alpha_{ij} = \int_a^b a_j(t)b_i(t)\,dt.$$

On the assumption that the functions $a_1, \ldots, a_n$ are linearly independent (and if this is not so, the terms in the expression for $k(x, t)$ can be rearranged as a sum of fewer terms so that it is), (3.6) yields the set of n simultaneous linear equations

$$\gamma_i = \lambda \sum_{j=1}^n \alpha_{ij}\gamma_j + \lambda\beta_i \quad (i = 1, 2, \ldots, n),$$

where α_{ij} and β_i are known and γ_i remain to be found. Writing $\gamma_1, \ldots, \gamma_n$ in a column vector $\mathbf{c}$ and $\beta_1, \ldots, \beta_n$ in the column $\mathbf{b}$, this set of equations becomes

$$(I - \lambda A)\mathbf{c} = \lambda\mathbf{b}, \tag{3.7}$$

where A is the $n \times n$ matrix whose (i, j)-th entry is α_{ij}.

The problem represented by (3.7) is one we can solve, the usual theory of simultaneous equations or of matrices giving two cases:

either   the matrix $(I - \lambda A)$ is non-singular and thus (3.7) has a unique
           solution $\mathbf{c} = (I - \lambda A)^{-1} \lambda \mathbf{b}$,
or   the matrix $(I - \lambda A)$ is singular and the homogeneous equation
           $(I - \lambda A)\mathbf{c} = \mathbf{0}$ has a non-trivial solution. In this case the solution to
           (3.7) is not unique and will exist only if $\mathbf{b}$ satisfies additional
           constraints.

The nature of the underlying issues here is made slightly more explicit if we
adopt the alternative formulation used in Chapter 1 where a parameter
$\mu = 1/\lambda$ is placed in front of I, giving, in place of (3.7), $(\mu I - A)\mathbf{c} = \mathbf{b}$. The
existence of an inverse to $\mu I - A$ is exactly the question of whether or not μ is
an eigenvalue of the matrix A and we know that if A is an $n \times n$ matrix it has
at most n distinct eigenvalues. Therefore the first of the two alternatives for
(3.7) holds for all but at most n distinct values of λ, these being the
reciprocals of the non-zero eigenvalues of the matrix A; the case $\lambda = 0$ is
trivial.

   Returning to (3.7), we see that if $(I - \lambda A)$ is non-singular (3.7) has a unique
solution $\mathbf{c}$, and the corresponding n-tuple $(\gamma_1, \ldots, \gamma_n)$ satisfies (3.6). It is now
easily checked that the function ϕ defined by $\phi(x) = f(x) + \sum_{i=1}^{n} \gamma_i a_i(x)$ is the
unique solution of the integral equation (3.1). Notice that, under our
assumption that $a_1, \ldots, a_n$ are continuous, ϕ is continuous provided that
f is.

Example 3.1

Solve

$$\phi(x) - \lambda \int_0^\pi \sin(x + t)\phi(t)\, dt = 1 \quad (0 \leqslant x \leqslant \pi)$$

in the case where $\lambda \neq \pm 2/\pi$.

   The usual expansion for $\sin(x + t)$ shows that the kernel is degenerate
and we see that $\phi(x) = 1 + \gamma_1 \cos x + \gamma_2 \sin x$ where γ_1 and γ_2 satisfy the
equations

$$\gamma_1 - \frac{\lambda \pi}{2} \gamma_2 = 2\lambda,$$

$$\gamma_2 - \frac{\lambda \pi}{2} \gamma_1 = 0.$$

The unique solution of these equations is

$$\gamma_1 = 8\lambda/(4 - \lambda^2 \pi^2), \quad \gamma_2 = 4\lambda^2 \pi/(4 - \lambda^2 \pi^2),$$

yielding the unique solution of the integral equation

$$\phi(x) = 1 + \frac{8\lambda}{4 - \lambda^2 \pi^2} \cos x + \frac{4\lambda^2 \pi}{4 - \lambda^2 \pi^2} \sin x,$$

provided $\lambda \neq \pm 2/\pi$.   □

Returning now to the case where the matrix $(I - \lambda A)$ in (3.7) is singular, which is equivalent to the existence of non-trivial solutions of the equation $(I - \lambda A)\mathbf{c} = \mathbf{0}$, we see that there are numbers $\gamma_1, \ldots, \gamma_n$, not all zero, for which $\gamma_i = \lambda \sum_{j=1}^{n} \alpha_{ij} \gamma_j$ $(i = 1, \ldots, n)$. Then, setting $\phi(x) = \sum_{i=1}^{n} \gamma_i a_i(x)$ yields a non-zero solution of the homogeneous integral equation

$$\phi(x) - \lambda \int_a^b k(x, t)\phi(t)\, dt = 0 \quad (a \leqslant x \leqslant b).$$

In the situation where the matrix $B = I - \lambda A$ is singular, we know from matrix theory that there are vectors $\mathbf{d}$ for which the equation $B\mathbf{c} = \mathbf{d}$ has no solution. In fact if B^* denotes the matrix obtained from B by transposing and taking the complex conjugate of the entries, then if $B\mathbf{c} = \mathbf{d}$ and $B^*\mathbf{x} = \mathbf{0}$, it follows that $\mathbf{x}^*\mathbf{d} = \mathbf{x}^*(B\mathbf{c}) = (B^*\mathbf{x})^*\mathbf{c} = 0$ so that $\mathbf{d}$ is orthogonal to every solution of the equation $B^*\mathbf{x} = \mathbf{0}$. The converse result is again supplied by matrix theory, so that $B\mathbf{c} = \mathbf{d}$ possesses a solution if and only if $\mathbf{d}$ is orthogonal to every solution of the equation $B^*\mathbf{x} = \mathbf{0}$. Denoting the entries of $\mathbf{d}$ by $\delta_1, \ldots, \delta_n$, this is equivalent to the condition that $\sum_{i=1}^{n} \bar{\xi}_i \delta_i = 0$ for all $(\xi_1, \ldots, \xi_n)$ satisfying the equations

$$\sum_{j=1}^{n} \bar{\beta}_{ji} \xi_j = 0 \quad (i = 1, 2, \ldots, n)$$

where the (i, j)-th entry of B is β_{ij}.

Returning to our equation $(I - \lambda A)\mathbf{c} = \lambda \mathbf{b}$ and presuming $\lambda \neq 0$ (since the equation is trivial if $\lambda = 0$), it possesses a solution if and only if $\sum_{i=1}^{n} \bar{\xi}_i \beta_i = 0$ for all $(\xi_1, \ldots, \xi_n)$ satisfying

$$\xi_i - \bar{\lambda} \sum_{j=1}^{n} \bar{\alpha}_{ji} \xi_j = 0 \quad (i = 1, 2, \ldots, n). \tag{3.8}$$

Noticing that (3.8) holds if and only if $\psi(x) = \sum_{i=1}^{n} \xi_i \overline{b_i(x)}$ satisfies the equation

$$\psi(x) - \bar{\lambda} \int_a^b \overline{k(t, x)}\psi(t)\, dt = 0,$$

and that for this definition of ψ, $\sum_{i=1}^{n} \bar{\xi}_i \beta_i = \int_a^b f(t)\overline{\psi(t)}\, dt$, we see that the condition, expressed directly in terms of the integral equation, is that the equation

$$\phi(x) - \lambda \int_a^b k(x, t)\phi(t)\, dt = f(x) \quad (a \leqslant x \leqslant b)$$

possesses a solution ϕ if and only if $\int_a^b f(t)\overline{\psi(t)}\, dt = 0$ for every solution ψ of the *adjoint* homogeneous equation

$$\psi(x) - \bar{\lambda} \int_a^b \overline{k(t, x)}\psi(t)\, dt = 0 \quad (a \leqslant x \leqslant b).$$

We have therefore arrived at a result known as the Fredholm Alternative.

Theorem 3.1

If $k(x, t)$ is a degenerate kernel on $[a, b] \times [a, b]$ which can be expressed in the form $k(x, t) = \sum_{i=1}^n a_i(x)b_i(t)$ where $a_1, \ldots, a_n, b_1, \ldots, b_n$ are continuous, then

either for every continuous function f on $[a, b]$ the integral equation

$$\phi(x) - \lambda \int_a^b k(x, t)\phi(t)\, dt = f(x) \quad (a \leqslant x \leqslant b) \tag{3.1}$$

possesses a unique continuous solution ϕ,

or   the homogeneous equation

$$\phi(x) - \lambda \int_a^b k(x, t)\phi(t)\, dt = 0 \quad (a \leqslant x \leqslant b)$$

has a non-trivial solution ϕ, in which case (3.1) will have (non-unique) solutions if and only if $\int_a^b f(t)\overline{\psi(t)}\, dt = 0$ for every continuous solution ψ of the equation

$$\psi(x) - \bar{\lambda} \int_a^b \overline{k(t, x)}\psi(t)\, dt = 0 \quad (a \leqslant x \leqslant b). \qquad \square$$

The discussion above proved Theorem 3.1, relating to the case where $k(x, t)$ degenerates into the sum of finitely many continuous functions $a_i(x)b_i(t)$. It is possible to allow a larger class of functions in this decomposition, producing, in exactly the same way, a theorem analogous to Theorem 3.1 where ϕ, ψ and f need not be continuous but belong instead to the chosen larger class of functions. As this will eventually be subsumed by later results, we shall do no more than make the remark.

Example 3.2

Let us return to the integral equation

$$\phi(x) - \lambda \int_0^\pi \sin(x+t)\phi(t)\,dt = 1 \quad (0 \leqslant x \leqslant \pi). \qquad (3.9)$$

In this case the corresponding homogeneous equation is

$$\phi(x) - \lambda \int_0^\pi \sin(x+t)\phi(t)\,dt = 0 \quad (0 \leqslant x \leqslant \pi) \qquad (3.10)$$

and proceeding along similar lines to Example 3.1 shows that if $\lambda \neq \pm 2/\pi$ this equation has only the trivial solution. If, however, $\lambda = 2/\pi$, ϕ will satisfy (3.10) if and only if ϕ is a constant multiple of $\sin x + \cos x$, so (3.10) certainly has non-trivial solutions. Setting $k(x, t) = \sin(x+t)$ and noticing that $k(x, t) = \overline{k(t, x)}$, Theorem 3.1 now shows us that (3.9) has a solution if and only if $\int_0^\pi 1(\sin t + \cos t)\,dt = 0$. Since $\int_0^\pi (\sin t + \cos t)\,dt = 2$, we see that (3.9) has no solution if $\lambda = 2/\pi$.

If, however, we seek solutions of

$$\phi(x) - \lambda \int_0^\pi \sin(x+t)\phi(t)\,dt = \sin x + \cos x \quad (0 \leqslant x \leqslant \pi) \qquad (3.11)$$

then we see, as before, that there is a unique continuous solution ϕ if $\lambda \neq \pm 2/\pi$. If $\lambda = 2/\pi$ the necessary and sufficient condition that (3.11) have a solution is that $\int_0^\pi (\sin t + \cos t)^2\,dt = 0$, which is not satisfied and again there is no solution. If, however, $\lambda = -2/\pi$ the solutions of the equation $\phi(x) - \lambda \int_0^\pi \sin(x+t)\phi(t)\,dt = 0$ $(0 \leqslant x \leqslant \pi)$ are the constant multiples of $\cos x - \sin x$, and, since $\int_0^\pi (\cos t - \sin t)(\cos t + \sin t)\,dt = 0$ (3.11) has solutions in this case. The general solution of (3.11) with $\lambda = -2/\pi$ is

$$\phi(x) = \tfrac{1}{2}(\cos x + \sin x) + A(\cos x - \sin x)$$

where A is an arbitrary constant. $\qquad\qquad\qquad\qquad\qquad\qquad \Box$

3.3 A different approach

Our technique in dealing with degenerate kernels was, essentially, to reduce the integral equation problem to one involving matrices. The key to this was the observation that if $k(x, t) = \sum_{i=1}^n a_i(x)b_i(t)$ and $\phi(x) - \lambda \int_a^b k(x, t)\phi(t)\,dt = f(x)$ $(a \leqslant x \leqslant b)$, then $\phi - f$ is a linear combination of the functions $a_1, \ldots, a_n$; the matrix formulation arises by considering the coefficients in the various linear combinations which arise. To deal with non-degenerate kernels one might attempt to pursue this approach by decomposing $k(x, t)$ into a series of the form $\sum_{i=1}^{\infty} a_i(x)b_i(t)$, but this tactic immediately raises questions of the nature of the convergence involved, while the consideration of the coefficients will lead us to consider infinitely many simultaneous equations (in infinitely many unknowns) or what is much the same thing, infinite matrices. Since the conversion of the integral equation problem into one about infinite matrices does not give us a conspicuously simpler problem to solve than the original, we shall proceed rather differently.

We noticed earlier that if we define K by

$$(K\phi)(x) = \int_a^b k(x, t)\phi(t)\, dt \quad (a \leqslant x \leqslant b),$$

where k is a suitable function on $[a, b] \times [a, b]$, then

$$K(\phi_1 + \phi_2) = K\phi_1 + K\phi_2 \text{ and } K(\lambda\phi) = \lambda(K\phi)$$

if λ is a constant; in other words, as a transformation between suitable vector spaces of functions, K is a linear map. We shall, therefore, introduce the ideas of functional analysis to tackle the solution of integral equations. (The basic definitions and results from functional analysis are given in Appendix A.)

Suppose that k is a continuous function on $[a, b] \times [a, b]$ and that we wish to consider the integral equation (3.1), that is,

$$\phi(x) - \lambda \int_a^b k(x, t)\phi(t)\, dt = f(x) \quad (a \leqslant x \leqslant b), \tag{3.1}$$

where f is a given continuous function and λ is a (given) constant. Let X_1 denote the set of all continuous, complex-valued functions defined on $[a, b]$ and define K_1 by $(K_1\phi)(x) = \int_a^b k(x, t)\phi(t)\, dt \ (a \leqslant x \leqslant b)$. It is easy to check that $K_1\phi$ is continuous, so K_1 is a linear map from X_1 to X_1. Therefore the continuous function ϕ will satisfy (3.1) if and only if ϕ is an element of X_1 satisfying the linear equation

$$(I - \lambda K_1)\phi = f. \tag{3.12}$$

If we can now show that the linear map $I - \lambda K_1$, as a map from X_1 to X_1, has an inverse, then (3.12) will have the unique solution $\phi = (I - \lambda K_1)^{-1} f$ and therefore (3.1) will have a unique continuous solution. Notice that (3.12) is an equation in the vector space X_1, and from it we may deduce information about those solutions of (3.1) which lie in X_1, but we can obtain no direct information about any solutions of (3.1) which are not in this space, so this approach is capable of yielding information only about continuous solutions of (3.1).

In order to be able to draw substantial conclusions about the existence of an inverse of $I - \lambda K_1$, and various related properties, we shall need an idea of convergence in X_1. We notice that, with respect to the norm

$$\|\phi\| = \sup\{|\phi(t)|: a \leqslant t \leqslant b\},$$

X_1 is a Banach space, and it is the completeness inherent in this which underlies a number of existence theorems. Now if $\phi \in X_1$

$$\|K_1 \phi\| = \sup_{a \leqslant x \leqslant b} |(K\phi)(x)| = \sup_{a \leqslant x \leqslant b} \left| \int_a^b k(x,t)\phi(t)\,dt \right|$$

$$\leqslant \left(\sup_{a \leqslant x \leqslant b} \int_a^b |k(x,t)|\,dt \right) \|\phi\|,$$

whence

$$\|K_1\| = \sup\{\|K_1\phi\|: \|\phi\| \leqslant 1\} \leqslant \sup_{a \leqslant x \leqslant b} \int_a^b |k(x,t)|\,dt$$

The supremum exists since k, being continuous on the compact set $[a,b] \times [a,b]$, is a bounded function. K_1 is therefore a bounded linear map from X_1 to itself. If we can deploy the functional analysis to give us conditions under which $I - \lambda K_1$ possesses an inverse, we will have gained useful information about the continuous solutions of (3.1). If the inverse $(I - \lambda K_1)^{-1}$ is also a bounded linear map then we gain information about the variation of the solution ϕ of (3.12) with f. To see this, suppose that $(I - \lambda K_1)\phi_i = f_i (i = 1, 2)$ so that $\phi_i = (I - \lambda K_1)^{-1} f_i$ and

$$\|\phi_1 - \phi_2\| = \|(I - \lambda K_1)^{-1}(f_1 - f_2)\|$$
$$\leqslant \|(I - \lambda K_1)^{-1}\| \, \|f_1 - f_2\|.$$

Elaborating further, suppose that $f_n \to f$ in X_1, in the sense that $\|f_n - f\| \to 0$ as $n \to \infty$, and suppose that $(I - \lambda K_1)\phi_n = f_n (n \in \mathbb{N})$. Then $\|\phi_n - \phi\| \leqslant \|(I - \lambda K_1)^{-1}\| \, \|f_n - f\|$ whence $\|\phi_n - \phi\| \to 0$ as $n \to \infty$. This shows us that if we choose a sequence of functions f_n which approximate more and more closely to f (in the sense that $\|f_n - f\| \to 0$) then the corresponding solutions ϕ_n approximate the solution ϕ of (3.12) correspondingly closely. This now needs to be translated back into the language in which we started, that of functions of a real variable. Since $\|\phi_n - \phi\| = \sup\{|\phi_n(x) - \phi(x)|: a \leqslant x \leqslant b\}$, $\|\phi_n - \phi\| \to 0$ as $n \to \infty$ is equivalent to the statement that $\phi_n \to \phi$ uniformly on $[a,b]$, so if $f_n \to f$ uniformly and $(I - \lambda K_1)^{-1}$ is a bounded linear map, then $\phi_n \to \phi$ uniformly, where $\phi_n = (I - \lambda K_1)^{-1} f_n$. The reformulation of (3.1) into the equation (3.12) in the Banach space X_1 therefore holds the prospect of substantial results about continuous solutions of (3.1) and of approximations to the functions concerned by uniformly convergent sequences.

We may choose to reformulate (3.1) in different ways. Suppose that k is a measurable function which satisfies the condition $\int_a^b \int_a^b |k(x,t)|^2 \, dx \, dt < \infty$; this includes the case where k is continuous, of course, and others besides. Let X_2 denote the space $L_2(a,b)$ of all (equivalence classes of) complex-valued, Lebesgue measurable functions ϕ satisfying the condition $\int_a^b |\phi|^2 < \infty$ (where ϕ and ψ are equivalent if they are equal almost everywhere). Then we know (Appendix B) that with $\|\phi\|_2 = \{\int_a^b |\phi|^2\}^{\frac{1}{2}}$, X_2 is a Hilbert space. Setting $(K_2 \phi)(x) = \int_a^b k(x,t)\phi(t) \, dt$, we see that, by Schwarz's inequality,

$$|(K_2 \phi)(x)|^2 \le \int_a^b |k(x,t)|^2 \, dt \int_a^b |\phi(t)|^2 \, dt,$$

whence

$$\int_a^b |(K_2 \phi)(x)|^2 \, dx \le \int_a^b \int_a^b |k(x,t)|^2 \, dt \, dx \|\phi\|_2^2$$

and

$$\|K_2 \phi\|_2 \le \|\phi\|_2 \left\{ \int_a^b \int_a^b |k(x,t)|^2 \, dt \, dx \right\}^{\frac{1}{2}}.$$

This shows us that $K_2 \phi \in X_2$ and that K_2 is a bounded linear map from X_2 to itself with $\|K_2\| \le \{\int_a^b \int_a^b |k(x,t)|^2 \, dt \, dx\}^{\frac{1}{2}}$. (We have glossed over some details here: for the integral of $|K_2 \phi|^2$ to make sense, we need to know that $K_2 \phi$ is a measurable function and that the integral defining $(K_2 \phi)(x)$ exists for almost all $x \in [a,b]$, matters we have relegated to Appendix B.) Then, using the same symbol ϕ for a function and the equivalence class in $L_2(a,b)$ containing it, (3.1) may be reformulated in X_2 as

$$(I - \lambda K_2)\phi = f. \tag{3.13}$$

Again, if we can prove that the bounded linear map $I - \lambda K_2$ in X_2 has an inverse, then (3.13) will have a unique solution $\phi = (I - \lambda K_2)^{-1} f$ in X_2. From this we deduce that all solutions of (3.1) which are square-integrable belong to the same equivalence class in $L_2(a,b)$ and so are equal almost everywhere. If $(I - \lambda K_2)^{-1}$ is a bounded linear map, $(I - \lambda K_2)\phi_n = f_n$ and $(I - \lambda K_2)\phi = f$, then we see that $\|f_n - f\|_2 \to 0$ implies that $\|\phi_n - \phi\|_2 \to 0$ as $n \to \infty$. In this case the information is more conspicuously different from the other formulation, since $\| f_n - f \|_2 \to 0$ is equivalent to $\int_a^b |f_n - f|^2 \to 0$, a weaker condition than the assumption that $f_n \to f$ uniformly. The conclusion, that $\int_a^b |\phi_n - \phi|^2 \to 0$ as $n \to \infty$, is also weaker. When $\int_a^b |\phi_n - \phi|^2 \to 0$ as $n \to \infty$ we shall say that $\phi_n \to \phi$ *in the mean* as $n \to \infty$.

Notice that the information yielded by (3.12) and (3.13) is, at least in its immediate content, different. If both linear maps $(I - \lambda K_1)$ and $(I - \lambda K_2)$ turn out to have an inverse, then (3.12) has a unique solution in X_1, and hence (3.1) has a unique continuous solution. On its own, this information does not preclude the existence of other, discontinuous, solutions. The invertibility of $(I - \lambda K_2)$ shows us that (3.13) has a unique solution in X_2 and hence that (3.1) has a unique square-integrable solution (unique in the sense that any two differ only on a set of measure zero). This, on its own, does not guarantee that a continuous solution exists, but it is a stronger uniqueness result.

Before embarking on the heart of these matters, the discussion of the existence or otherwise of an inverse of $I - \lambda K$, we need to consider which of the two formulations, in X_1 or X_2, we should adopt – or whether another should be sought. The formulation in X_1 has the advantage of giving information directly about continuous functions without the clutter of the 'almost everywhere' technicalities, but it requires that the kernel be better behaved than is necessary for the second viewpoint. The formulation in X_2 has the advantage that X_2 is a Hilbert space, which is frequently more convenient to use and has a richer associated structure, while the assumption that $\int_a^b \int_a^b |k(x, t)|^2 \, dx \, dt < \infty$ is sufficiently general that it includes most examples which arise in practice. The 'almost everywhere' qualification which accompanies a piece of information about a function (or, strictly speaking, an equivalence class of functions) in $L_2(a, b)$ will, in practice, give little difficulty, and is a technicality which appears more troublesome than it is.

Example 3.3

Suppose that $k: [a, b] \times [a, b] \to \mathbb{C}$ is continuous and define $K: L_2(a, b) \to L_2(a, b)$ by $(K\phi)(x) = \int_a^b k(x, t)\phi(t) \, dt$ $(a \leqslant x \leqslant b)$. Then if the equation $\phi - K\phi = 0$ in $L_2(a, b)$ has a solution, the integral equation

$$\phi(x) - \int_a^b k(x, t)\phi(t) \, dt = 0 \quad (a \leqslant x \leqslant b)$$

has a continuous solution and the two are equal almost everywhere.

Since $\int_a^b \int_a^b |k(x, t)|^2 \, dx \, dt < \infty$, K is certainly a bounded linear map from $L_2(a, b)$ to itself, by the work of the last few paragraphs. Suppose, then, that

$\phi \in L_2(a, b)$ satisfies $\phi - K\phi = 0$. Then there is a square-integrable function ϕ for which

$$\phi(x) - \int_a^b k(x, t)\phi(t)\, dt = 0 \quad \text{(for almost all } x \in [a, b]\text{)}.$$

Let $\tilde{\phi}(x) = \int_a^b k(x, t)\phi(t)\, dt$ $(a \leqslant x \leqslant b)$; then $\tilde{\phi}$ is continuous. To see this, choose $\epsilon > 0$ so there exists $\delta > 0$ such that if $|x - x'| < \delta$ and $|t - t'| < \delta$, $|k(x, t) - k(x', t')| < \epsilon$. Then

$$|x - x'| < \delta \Rightarrow |\tilde{\phi}(x) - \tilde{\phi}(x')| \leqslant \int_a^b |k(x, t) - k(x', t)||\phi(t)|\, dt$$

$$< \epsilon\{(b - a) \int_a^b |\phi(t)|^2\, dt\}^{\frac{1}{2}}.$$

Moreover $\phi(x) - \tilde{\phi}(x) = 0$ for almost all $x \in [a, b]$, whence for all $x \in [a, b]$ $\int_a^b k(x, t)\phi(t)\, dt = \int_a^b k(x, t)\tilde{\phi}(t)\, dt$ and $\tilde{\phi}(x) - \int_a^b k(x, t)\tilde{\phi}(t)\, dt = 0$ (for all $x \in [a, b]$). $\qquad\qquad\square$

The technique of Example 3.3, of recovering a continuous solution of an integral equation when the solution is initially known only to belong to L_2, is frequently possible in practice since the kernels which arise in examples are usually known to have better behaviour than the mere existence of $\int_a^b \int_a^b |k(x, t)|^2\, dx\, dt$, although we shall encounter discontinuous kernels. Example 3.3 also illustrates how the technical nuisance arising from elements of $L_2(a, b)$ only being defined almost everywhere can be avoided for integral equations of the second kind: we may choose the appropriate function from the equivalence class in $L_2(a, b)$ which satisfies $\phi(x) - \int_a^b k(x, t)\phi(t)\, dt = 0$ for all $x \in [a, b]$.

Rather than choose to formulate each integral equation that arises in a normed linear space appropriate to that problem, we shall recast all of our integral equations in Hilbert space, that is, in $L_2(a, b)$ (for appropriate a and b). This will eliminate the need to keep a wide stock of such normed spaces and their properties in mind, and it will give us the technical advantages afforded by Hilbert space, but we shall from time to time have to use techniques such as those in Example 3.3 to recover additional information about solutions. We shall use $\| \cdot \|$ to denote the norm of the Hilbert space with which we are concerned.

Let us take this new point of view and use it to see what we can deduce about equations with a degenerate kernel. For definiteness, let us consider the equation

$$\phi(x) - \lambda \int_a^b k(x, t)\phi(t)\, dt = f(x) \quad (a \leqslant x \leqslant b), \tag{3.14}$$

where f is a given square-integrable function, λ a given constant and $k(x, t) = \sum_{i=1}^n a_i(x) b_i(t)$ where the functions $a_1, \ldots, a_n, b_1, \ldots, b_n$ are given and square-integrable. Then, since $\int_a^b \int_a^b |k(x, t)|^2\, dx\, dt < \infty$, K is a bounded linear map of $H = L_2(a, b)$ into itself, where K is defined by $(K\phi)(x) = \int_a^b k(x, t)\phi(t)\, dt$.

For all $\phi \in H$, $K\phi$ is a linear combination of the vectors $a_1, \ldots, a_n$, so denote by E the subspace of H consisting of all linear combinations of $a_1, \ldots, a_n$. In the Hilbert space H, (3.14) becomes

$$\phi = f + \lambda K\phi \tag{3.15}$$

from which we deduce that

$$\phi = f + \lambda K\phi = f + \lambda Kf + \lambda^2 K(K\phi).$$

It is now easily checked that ϕ satisfies (3.15) if and only if $\phi = f + \lambda\psi$ where $\psi \in E$ and

$$\psi - \lambda K\psi = Kf. \tag{3.16}$$

Now $\psi, K\psi$ and Kf all belong to the finite-dimensional subspace E of H, and $I - \lambda K$ maps vectors of E into E, so the restriction of $I - \lambda K$ to E either has an inverse linear map, say L, or there is a non-zero vector $\psi \in E$ for which $(I - \lambda K)\psi = 0$. In the first case (3.16) has the unique solution $\psi = LKf$ and (3.15) has the unique solution $\phi = f + \lambda LKf$. In the second case the corresponding homogeneous equation $\phi - \lambda K\phi = 0$ has the non-trivial solution $\phi = \psi$. In the new context (of L_2 functions) this is the Fredholm Alternative again, but in this case we can move further: L, being a linear map on the finite-dimensional space E, is a bounded linear map (Theorem A6) so, letting $\|L\|$ denote the norm of L as a map from E to E, we have

$$\|LKf\| \leqslant \|L\|\,\|Kf\| \leqslant \|L\|\,\|K\|\,\|f\|.$$

Therefore, in the case where (3.15) has a unique solution, which is when $\psi - \lambda K\psi = 0$ has no non-trivial solution, the solution ϕ of (3.15) satisfies $\|\phi\| \leqslant (1 + |\lambda|\,\|L\|\,\|K\|)\|f\|$, showing that the map taking f into ϕ, that is, $(I - \lambda K)^{-1}$, is a *bounded* linear map.

Notice that if we revert to the original equation (3.14), the above calculations have shown that *either* there is a square-integrable function ϕ such that $\phi(x) - \lambda \int_a^b k(x, t)\phi(t)\, dt = f(x)$ for almost all $x \in [a, b]$ and this is almost unique in the sense that any two solutions are equal for almost all $x \in [a, b]$, *or* there is a square-integrable function ψ, which is not almost everywhere equal to zero, satisfying $\psi(x) - \lambda \int_a^b k(x, t)\psi(t)\, dt = 0$ (almost everywhere). By setting $\tilde{\phi}(x) = f(x) + \lambda \int_a^b k(x, t)\phi(t)\, dt$ in the first case and $\tilde{\psi}(x) = \lambda \int_a^b k(x, t)\psi(t)\, dt$ in the second, we obtain the unique square-integrable function $\tilde{\phi}$ which satisfies (3.14) (for *all* $x \in [a, b]$) or a non-trivial $\tilde{\psi}$ satisfying $\tilde{\psi}(x) = \lambda \int_a^b k(x, t)\tilde{\psi}(t)\, dt$ for *all* $x \in [a, b]$.

We have again established the main part of the Fredholm Alternative for integral equations arising from degenerate kernels. We may put this in a slightly different form, emphasising the linear map $(I - \lambda K)^{-1}$ associated with the solution process.

Theorem 3.2

Let K be a bounded linear map from $L_2(a, b)$ to itself with the property that $\{K\phi : \phi \in L_2(a, b)\}$ has finite dimension. Then

either the linear map $I - \lambda K$ has an inverse and $(I - \lambda K)^{-1}$ is a bounded linear map

or   the equation $\psi - \lambda K\psi = 0$ has a non-trivial solution $\psi \in L_2(a, b)$. $\square$

We may take this argument further. One of the key results we shall use from functional analysis is that if T is a bounded linear map on the Hilbert space H and $\|T\| < 1$, then $I - T$ has an inverse as a linear map, and $(I - T)^{-1}$ is also a bounded linear map. (This and the steps leading to it are theorems in Appendix A.) The idea is that if k is a kernel which can be expressed as a limit of degenerate kernels k_n in a suitable sense, then the Fredholm Alternative holds for the integral equation $\phi(x) - \lambda \int_a^b k(x, t)\phi(t)\, dt = f(x)\, (a \leqslant x \leqslant b)$. For this we need a little terminology.

Definition 3.2

Since we have to deal with linear maps of one normed vector space into another rather frequently we shall use the word *operator* to denote these; we shall mostly be concerned with operators from a Hilbert space to itself. We let $B(H)$ denote the set of all bounded operators from H to itself; Theorem A1 in Appendix A tells us that, with the norm $\|T\| = \sup\{\|T\phi\| : \|\phi\| \leqslant 1\}$, $B(H)$ is a Banach space. An operator $T: X \to Y$ is said to have *finite rank* if its image $\{T\phi : \phi \in X\}$ has finite dimension, the *rank* of T being the dimension of the subspace. An operator $T \in B(H)$ is said to be *invertible* if it has an inverse which is also in $B(H)$, that is, if there is an $S \in B(H)$ for which $ST = TS = I$. S is denoted by T^{-1}. Finally, $K \in B(H)$ is said to be *compact* if there is a sequence (K_n) of bounded finite-rank operators in $B(H)$ such that $\|K_n - K\| \to 0$ as $n \to \infty$. □

Notice that degenerate kernels give rise to finite-rank operators, in the sense that if $a_1, \ldots, a_n, b_1, \ldots, b_n$ are all square-integrable functions on $[a, b] \times [a, b]$, and $k(x, t) = \sum_{i=1}^{n} a_i(x) b_i(t)$, then the operator K on $L_2(a, b)$ given by $(K\phi)(x) = \int_a^b k(x, t)\phi(t) \, dt$, has finite rank, since its image $\{K\phi : \phi \in L_2(a, b)\}$, is contained in the subspace spanned by $a_1, \ldots, a_n$. That K is a bounded operator follows, because $a_1, \ldots, a_n, b_1, \ldots, b_n$ are square-integrable, by using Schwarz's inequality, for

$$(K\phi)(x) = \sum_{i=1}^{n} a_i(x) \int_a^b b_i(t)\phi(t) \, dt$$

and $|\int_a^b b_i(t)\phi(t) \, dt| \leqslant \|b_i\| \, \|\phi\|$ whence $\|K\phi\| \leqslant (\sum_{i=1}^{n} \|a_i\| \, \|b_i\|) \|\phi\|$. The converse, that every bounded finite-rank operator on $L_2(a, b)$ arises from a degenerate kernel is also true (Problem 3.21), so compact operators are those whose kernels are limits of degenerate kernels, in the sense above.

Theorem 3.3

Let H be the Hilbert space $L_2(a, b)$ and $k(x, t)$ be a kernel on $[a, b] \times [a, b]$ such that $(K\phi)(x) = \int_a^b k(x, t)\phi(t) \, dt$ defines a compact operator on $L_2(a, b)$. Then

either for every square-integrable function f on $[a, b]$ the integral equation

$$\phi(x) - \lambda \int_a^b k(x, t)\phi(t) \, dt = f(x) \quad (a \leqslant x \leqslant b)$$

has a unique square-integrable solution ϕ,

or the homogeneous integral equation

$$\psi(x) - \lambda \int_a^b k(x, t)\psi(t) \, dt = 0 \quad (a \leqslant x \leqslant b)$$

has a non-trivial square-integrable solution ψ.

Proof

Let (K_n) be the sequence of finite-rank operators in $B(H)$ such that $\|K_n - K\| \to 0$ as $n \to \infty$; such a sequence exists by definition of the compactness of K. Choose n so that $|\lambda| \, \|K - K_n\| < 1$, whence $I - \lambda(K - K_n)$ has an inverse which belongs to $B(H)$ by Corollary A3. The equation

$\phi - \lambda K \phi = f$ in H may be rewritten

$$(I - \lambda K)\phi = \{(I - \lambda(K - K_n)) - \lambda K_n\}\phi = f$$

or, equivalently,

$$(I - \lambda(K - K_n))(I - (I - \lambda(K - K_n))^{-1}\lambda K_n)\phi = f. \tag{3.17}$$

Let $F_n = (I - \lambda(K - K_n))^{-1}\lambda K_n$. Since K_n has finite rank so has F_n, and Theorem 3.2 shows that either $I - F_n$ has an inverse and $(I - F_n)^{-1} \in B(H)$ or $\psi - F_n\psi = 0$ has a non-trivial solution in H. In the first case, writing (3.17) in the form

$$(I - \lambda(K - K_n))(I - F_n)\phi = f,$$

we see that it has the unique solution in H

$$\phi = (I - F_n)^{-1}(I - \lambda(K - K_n))^{-1}f,$$

while in the second case

$$(I - \lambda K)\psi = (I - \lambda(K - K_n))(I - F_n)\psi = 0.$$

Apart from the usual adjustment in that the corresponding functions ϕ and ψ need only satisfy the integral equations concerned for almost all x, we have established the result. $\qquad\square$

Remark

Notice that we have proved in Theorem 3.3 that if $(I - \lambda K)\phi = f$ has a unique solution for all $f \in H$, then $(I - \lambda K)^{-1} = (I - F_n)^{-1}(I - \lambda(K - K_n))^{-1}$ is a bounded linear map.

The emphasis given here to the Fredholm Alternative needs some explanation. We wish to prove the existence and uniqueness of the solution to

$$\phi(x) - \lambda \int_a^b k(x, t)\phi(t)\, \mathrm{d}t = f(x) \quad (a \leqslant x \leqslant b).$$

To show that there is no more than one solution we need to prove that the corresponding homogeneous equation possesses only the trivial solution $\phi = 0$. The existence problem is, of course, solved if we can find an exact solution explicitly but in most important cases this is impractical and an indirect existence proof is needed. This is usually a tricky problem and the virtue of the Fredholm Alternative is that we need only establish the uniqueness of solutions to the corresponding homogeneous equation to solve also the existence issue. Even in the case where the homogeneous equation does possess non-trivial solutions, we established a necessary and sufficient condition for the existence of a solution to equation (3.1) in the case where the kernel is degenerate; this we shall in due course develop to include a wider class of kernel.

3.4 Operators

In the preceding section we saw how the techniques of functional analysis could be brought to bear on integral equations with degenerate kernels and we found that these results could easily be extended to those equations which arise from a compact operator. In this section we shall establish some simple criteria relating the kernels of integral operators to the properties of the linear maps which arise from them.

Definition 3.3

We say that a complex-valued measurable function k defined on $[a, b] \times [a, b]$ is an L_2-*kernel* if it has the property that $\int_a^b \int_a^b |k(x, t)|^2 \, dx \, dt < \infty$. □

We have already shown in §3.3 that if k is an L_2-kernel, the integral operator it generates, that is, the linear map $K: L_2(a, b) \to L_2(a, b)$ defined by

$$(K\phi)(x) = \int_a^b k(x, t)\phi(t) \, dt \quad (a \leqslant x \leqslant b)$$

is a bounded operator and

$$\|K\| \leqslant \left\{ \int_a^b \int_a^b |k(x, t)|^2 \, dx \, dt \right\}^{\frac{1}{2}}.$$

Theorem 3.4 is the most immediately significant general result about L_2-kernels.

Theorem 3.4

If k is an L_2-kernel on $[a, b] \times [a, b]$ and K is the bounded operator from $L_2(a, b)$ to itself generated by k, then K is compact.

Proof

We start from the standard results on Fourier series in one variable. Suppose that (ϕ_n) is a complete orthonormal sequence in $L_2(a, b)$, so that if $f \in L_2(a, b)$ and we set $c_n = \int_a^b f(t)\overline{\phi_n(t)} \, dt$ then $\Sigma c_n\phi_n(t)$ converges to f in the mean, that is, $\int_a^b |f(t) - \Sigma_{n=1}^N c_n\phi_n(t)|^2 \, dt \to 0$ as $N \to \infty$.

Choose $x \in [a, b]$ satisfying $\int_a^b |k(x, t)|^2 \, dt < \infty$; this holds for almost all $x \in [a, b]$. Set $f_x(t) = k(x, t) \, (a \leqslant t \leqslant b)$, so that $f_x \in L_2$, and define $c_n(x) =$

$\int_a^b f_x(t)\overline{\phi_n(t)}\,dt$. Then, by Parseval's Theorem, A8, $\sum_{n=1}^{\infty}|c_n(x)|^2 = \int_a^b|f_x(t)|^2\,dt$ and

$$\int_a^b\left|f_x(t)-\sum_{n=1}^{N}c_n(x)\phi_n(t)\right|^2 dt = \sum_{n=N+1}^{\infty}|c_n(x)|^2. \tag{3.18}$$

If we set $g_N(x)=\sum_{n=N+1}^{\infty}|c_n(x)|^2$ then we see that for our chosen x, $g_N(x)\to 0$ as $N\to\infty$ since $\sum|c_n(x)|^2$ converges.

Let $h(x)=\int_a^b|k(x,t)|^2\,dt$. Then, for almost all $x\in[a,b]$, $h(x)<\infty$ and hence $g_N(x)\to 0$ as $N\to\infty$. Moreover, for all $n\in\mathbb{N}$ and all $x\in[a,b]$, $|g_n(x)|\leqslant h(x)$ and h is an integrable function (since k is an L_2-kernel). By the Dominated Convergence Theorem, B4, then, $\int_a^b g_N(x)\,dx\to 0$ as $N\to\infty$, that is, by (3.18)

$$\int_a^b\int_a^b\left|k(x,t)-\sum_{n=1}^{N}c_n(x)\phi_n(t)\right|^2 dt\,dx\to 0 \text{ as } N\to\infty.$$

If we set $k_N(x,t)=\sum_{n=1}^{N}c_n(x)\phi_n(t)$, then the corresponding operator K_N has finite rank, since k_N is degenerate, and

$$\|K-K_N\|\leqslant\left\{\int_a^b\int_a^b|k(x,t)-k_N(x,t)|^2\,dx\,dt\right\}^{\frac{1}{2}},$$

so $\|K-K_N\|\to 0$ as $N\to\infty$. K is therefore compact. $\square$

The class of L_2-kernels is usefully large, since it includes not only continuous functions k, but a number of less well-behaved examples. If we set $k(x,t)=|x-t|^{-\alpha}$ where $0<\alpha<\frac{1}{2}$, then k is clearly discontinuous on the line $x=t$, but it is nevertheless an L_2-kernel on $[a,b]\times[a,b]$ for any two real numbers a and b. (Strictly, if k is to be a function on $[a,b]\times[a,b]$ we should define $k(x,x)$, but since, for any fixed x, this leaves only one value of t for which $k(x,t)$ is undefined, the value chosen will not affect the integral $\int_a^b k(x,t)\phi(t)\,dt$.) However, there are operators which we shall wish to consider which do not arise from L_2-kernels.

Example 3.4

Suppose that $k:[a,b]\times[a,b]\to\mathbb{C}$ is a measurable function and that there are constants A and B such that

(i) $\qquad\qquad$ for all $x\in[a,b]$ $\displaystyle\int_a^b|k(x,t)|\,dt\leqslant A$

and

(ii) $\qquad\qquad$ for all $t\in[a,b]$ $\displaystyle\int_a^b|k(x,t)|\,dx\leqslant B$.

Then the integral operator arising from k, given by $(K\phi)(x)=\int_a^b k(x, t)\phi(t)\,dt$, is a bounded operator from $L_2(a, b)$ to itself and $\|K\|\leqslant(AB)^{\frac{1}{2}}$.

This is just a matter of an ingenious use of Schwarz's inequality. We notice that $|k(x, t)\phi(t)|=(|k(x, t)|^{\frac{1}{2}})(|k(x, t)|^{\frac{1}{2}}|\phi(t)|)$ and apply Schwarz's inequality to the product on the right. Then if $x\in[a, b]$, and $\phi\in L_2(a, b)$,

$$|(K\phi)(x)|^2 = \left|\int_a^b k(x, t)\phi(t)\,dt\right|^2 \leqslant \left(\int_a^b |k(x, t)\phi(t)|\,dt\right)^2$$

$$\leqslant \left(\int_a^b |k(x, t)|\,dt\right)\left(\int_a^b |k(x, t)||\phi(t)|^2\,dt\right)$$

$$\leqslant A\int_a^b |k(x, t)||\phi(t)|^2\,dt,$$

whence

$$\|K\phi\|^2 = \int_a^b |(K\phi)(x)|^2\,dx$$

$$\leqslant A\int_a^b \left(\int_a^b |k(x, t)||\phi(t)|^2\,dt\right)dx$$

$$\leqslant A\int_a^b \left(\int_a^b |k(x, t)||\phi(t)|^2\,dx\right)dt \qquad (3.19)$$

$$\leqslant A\int_a^b B|\phi(t)|^2\,dt$$

$$= AB\|\phi\|^2.$$

Thus, for all $\phi\in L_2(a, b)$, $\|K\phi\|\leqslant(AB)^{\frac{1}{2}}\|\phi\|$.

The reversal of the order of integration in deducing (3.19) from the previous line is justified by Fubini's Theorem, B8.   □

Definition 3.4

A kernel of the form arising in Example 3.4 we shall call a *Schur kernel*.   □

The most important example of a Schur kernel is that of a weakly singular kernel, that is, one of the form $k(x, t)=m(x, t)/|x-t|^\alpha$ where m is a bounded function and $0<\alpha<1$. If, for all $x, t\in[a, b]$, $|m(x, t)|\leqslant M$, then we see that for $a \leqslant x \leqslant b$,

$$\int_a^b |k(x,t)| \, \mathrm{d}t \leqslant M\left(\int_a^x (x-t)^{-\alpha} \, \mathrm{d}t + \int_x^b (t-x)^{-\alpha} \, \mathrm{d}t \right)$$
$$= M((x-a)^{1-\alpha} + (b-x)^{1-\alpha})/(1-\alpha)$$
$$\leqslant M2^\alpha (b-a)^{1-\alpha}/(1-\alpha).$$

Since $|x-t|^{-\alpha}$ is symmetric in x and t, condition (ii) is now automatically true. Notice that if $\frac{1}{2} \leqslant \alpha \leqslant 1$ these kernels need not be L_2-kernels; consider the case where $m = 1$.

Many of the Schur kernels turn out to generate compact operators. Before showing this we shall prepare the ground with some additional properties of compact operators.

Theorem 3.5

Let H be a Hilbert space and $K, L \in B(H)$. Then:

(i)   if K and L are compact so are $K + L$ and λK (where λ is a scalar);
(ii) if K is compact, KL and LK are compact;
(iii) if (K_n) is a sequence of compact operators and $\|K_n - K\| \to 0$ as $n \to \infty$ then K is compact.

Proof

(i)   Let (K_n) and (L_n) be sequences of operators of finite rank with $\|K_n - K\| \to 0$ and $\|L_n - L\| \to 0$ as $n \to \infty$. Then $\|(K_n + L_n) - (K+L)\| \to 0$ and $\|\lambda K_n - \lambda K\| \to 0$, so, since $K_n + L_n$ and λK_n have finite rank, $K + L$ and λK are compact.

(ii) Let K_n have finite rank and $\|K_n - K\| \to 0$. Then $K_n L$ and LK_n have finite rank (for if the image of K_n is spanned by $a_1, \ldots, a_m$, then the images of $K_n L$ and LK_n are spanned by $a_1, \ldots, a_m$ and $La_1, \ldots, La_m$ respectively). Also $\|K_n L - KL\| \leqslant \|K_n - K\| \|L\|$ and $\|LK_n - LK\| \leqslant \|L\| \|K_n - K\|$ so KL and LK are compact.

(iii) For each n, since K_n is compact there is a finite-rank operator, L_n say, with $\|K_n - L_n\| < 1/n$. Then if $\|K_n - K\| \to 0$, $\|L_n - K\| \to 0$ as $n \to \infty$ and K is compact.   □

One application of part (iii) is as follows: if $k(x,t)$ is a kernel whose associated operator we wish to show compact, we show that for all $\epsilon > 0$, $k(x,t)$ can be expressed as $k(x,t) = h_1(x,t) + h_2(x,t)$ where h_1 is known to give rise to a compact operator and h_2 gives rise to an operator H_2 with $\|H_2\| < \epsilon$.

Example 3.5

If $k(x, t) = m(x, t)|x - t|^{-\alpha}$ $(a \leqslant x, t \leqslant b)$, where m is bounded and measurable and $0 < \alpha < 1$, then the integral operator from $L_2(a, b)$ to itself with this kernel is compact.

Since a bounded kernel, that is, a kernel which is a bounded function, is an L_2-kernel and therefore compact we choose an approximation to k by a function which is equal to k if $|x - t|^{-\alpha}$ does not exceed a given constant and is zero elsewhere. Let

$$h_1(x, t) = \begin{cases} k(x, t) & \text{if } |x - t| \geqslant \epsilon, \\ 0 & \text{otherwise,} \end{cases}$$

and

$$h_2(x, t) = \begin{cases} 0 & \text{if } |x - t| \geqslant \epsilon, \\ k(x, t) & \text{otherwise.} \end{cases}$$

Then if $M = \sup\{|m(x, t)| : a \leqslant x, t \leqslant b\}$ we see that if $a \leqslant x, t \leqslant b, |h_1(x, t)| \leqslant M\epsilon^{-\alpha}$ so h_1 is an L_2-kernel and the operator it generates, H_1, is compact.

Now for $x \in [a, b]$,

$$\int_0^1 |h_2(x, t)| \, dt \leqslant \int_{x-\epsilon}^{x+\epsilon} M|x - t|^{-\alpha} \, dt = 2M\epsilon^{1-\alpha}/(1 - \alpha)$$

and, by symmetry, if $t \in [a, b]$

$$\int_0^1 |h_2(x, t)| \, dx \leqslant 2M\epsilon^{1-\alpha}/(1 - \alpha),$$

so h_2 is a Schur kernel and if H_2 is the operator on $L_2(a, b)$ which it generates, $\|H_2\| \leqslant 2M\epsilon^{1-\alpha}/(1 - \alpha)$. Since $\|H_2\|$ can be made as small as desired by a suitable choice of ϵ, and H_1 is compact, we can construct a sequence K_n of compact operators from $L_2(a, b)$ to itself with $\|K_n - K\| \to 0$ as $n \to \infty$ where $K = H_1 + H_2$ is the operator generated by k. $\quad\square$

It is worth noticing here that L_2-kernels and the weakly singular kernels dealt with above need not be bounded functions, yet the operators they generate are compact and they therefore satisfy the hypotheses of Theorem 3.3. In this case, and in many which follow, our methods will yield results about integral equations which arise from such kernels, as promised in Chapter 1.

Example 3.6

To dispel the notion that all 'reasonable' kernels generate compact operators, we give an example where the operator is not compact. For $\phi \in L_2(0, 1)$ and $x \in (0, 1]$ define

$$(K\phi)(x) = \frac{1}{x} \int_0^x \phi(t)\, dt.$$

Let $\chi(t) = 1$ if $t \leqslant 1$ and $\chi(t) = 0$ if $t > 1$, so that for $x \in (0, 1]$,

$$|(K\phi)(x)|^2 = \frac{1}{x^2} \left| \int_0^x \phi(t)\, dt \right|^2$$

$$\leqslant \frac{1}{x^2} \int_0^1 t^{-\frac{1}{2}} \chi\left(\frac{t}{x}\right) dt \int_0^1 t^{\frac{1}{2}} \chi\left(\frac{t}{x}\right) |\phi(t)|^2\, dt.$$

Therefore

$$\int_0^1 |(K\phi)(x)|^2\, dx \leqslant \int_0^1 \left\{ \frac{1}{x^2} \int_0^1 t^{-\frac{1}{2}} \chi\left(\frac{t}{x}\right) dt \int_0^1 t^{\frac{1}{2}} \chi\left(\frac{t}{x}\right) |\phi(t)|^2\, dt \right\} dx$$

$$= \int_0^1 \left\{ \frac{1}{x^3} \int_0^1 \left(\frac{t}{x}\right)^{-\frac{1}{2}} \chi\left(\frac{t}{x}\right) dt \int_0^1 t \left(\frac{t}{x}\right)^{-\frac{1}{2}} \chi\left(\frac{t}{x}\right) |\phi(t)|^2\, dt \right\} dx$$

$$= \int_0^1 \left\{ \frac{1}{x^2} \int_0^{1/x} u^{-\frac{1}{2}} \chi(u)\, du \int_0^1 t \left(\frac{t}{x}\right)^{-\frac{1}{2}} \chi\left(\frac{t}{x}\right) |\phi(t)|^2\, dt \right\} dx$$

(substituting $u = t/x$)

$$\leqslant \left(\int_0^\infty u^{-\frac{1}{2}} \chi(u)\, du \right) \int_0^1 \left\{ \int_0^1 \left(\frac{t}{x}\right)^{-\frac{1}{2}} \chi\left(\frac{t}{x}\right) \frac{t}{x^2}\, dx \right\} |\phi(t)|^2\, dt$$

$$\leqslant C \int_0^1 \left\{ \int_t^\infty v^{-\frac{1}{2}} \chi(v)\, dv \right\} |\phi(t)|^2\, dt$$

(substituting $v = t/x$)

$$\leqslant C^2 \|\phi\|^2$$

where $C = \int_0^\infty u^{-\frac{1}{2}} \chi(u)\, du = 2$. Therefore, for all $\phi \in L_2(0, 1)$, $\|K\phi\| \leqslant 2\|\phi\|$ and K is a bounded operator.

To see that K is not compact, notice that the equation

$$\phi(x) - \frac{1}{2x} \int_0^x \phi(t)\, dt = 0 \quad (0 < x < 1) \tag{3.20}$$

is satisfied only by $\phi(t) = At^{-\frac{1}{2}}$ (for some constant A) and since this function is not in $L_2(0, 1)$ (3.20) has only the trivial solution $\phi = 0$ (a.e.) in the Hilbert space $L_2(0, 1)$. Thus

$$(I - \tfrac{1}{2}K)\phi = 0$$

has only the trivial solution. However, the linear map $I - \frac{1}{2}K$ does not have a bounded inverse, for if we set $\phi_\alpha(t) = \sqrt{(1 - 2\alpha)}t^{-\alpha}$ then $\|\phi_\alpha\| = 1$ (for $\alpha < \frac{1}{2}$) but

$$\|(I - \tfrac{1}{2}K)\phi_\alpha\| = \left\|\left(1 - \frac{1}{2(1 - \alpha)}\right)\phi_\alpha\right\| \to 0 \text{ as } \alpha \to \tfrac{1}{2} - .$$

Now if $I - \frac{1}{2}K$ had a bounded inverse, say L, then for all $\phi \in L_2(0, 1)$,

$$\|\phi\| = \|L(I - \tfrac{1}{2}K)\phi\| \leqslant \|L\| \, \|(I - \tfrac{1}{2}K)\phi\|$$

whence $\|(I - \frac{1}{2}K)\phi_\alpha\| \geqslant 1/\|L\|$. But this violates the result proved in the course of Theorem 3.3 (and remarked on after the proof) that if K is compact then either $I - \lambda K$ has a bounded inverse or the equation $\psi - \lambda K\psi = 0$ has a non-trivial solution. $\qquad\square$

We may sum up our position so far by noticing that in the case where the kernel k generates a compact operator on $L_2(a, b)$, which includes a wide class of kernels, the integral equation

$$\phi(x) - \lambda \int_a^b k(x, t)\phi(t) \, \mathrm{d}t = f(x) \quad (a \leqslant x \leqslant b)$$

will have a unique solution, and the solution operator $(I - \lambda K)^{-1}$ will be bounded, provided the corresponding homogeneous equation

$$\phi(x) - \lambda \int_a^b k(x, t)\phi(t) \, \mathrm{d}t = 0 \quad (a \leqslant x \leqslant b)$$

has only the trivial solution $L_2(a, b)$. Notice that we have given, in Example 3.6, a non-compact operator for which the Fredholm Alternative does not hold.

3.5 The Neumann series

Our results so far have centred on the Fredholm Alternative, that certain operators are invertible if an associated homogeneous equation has only the trivial solution. In this section we shall establish criteria which guarantee the existence of certain inverses without the need to investigate associated equations.

The idea which motivates the Neumann series is the result that if z is a complex number and $|z| < 1$ then $(1-z)^{-1} = \sum_{n=0}^{\infty} z^n$. Analogy (and, perhaps, some optimism) suggests that we try to interpret this in the context of operators. The results of Appendix A, which we have already used, show us that if H is a Hilbert space, $T \in B(H)$ and $\Sigma \|T^n\|$ converges, then $(I - T)^{-1} = \sum_{n=0}^{\infty} T^n$, the series on the right converging to an element of $B(H)$.

Consider the integral equation

$$\phi(x) - \lambda \int_a^b k(x, t)\phi(t)\, dt = f(x) \quad (a \leqslant x \leqslant b) \tag{3.1}$$

where $f \in L_2(a, b)$ is given, k is of such a form that it gives rise to a bounded operator on $L_2(a, b)$ and solutions ϕ in $L_2(a, b)$ are sought. In operator form this equation may be rewritten

$$(I - \lambda K)\phi = f$$

and its solution, when $\|\lambda K\| < 1$, is

$$\phi = (I - \lambda K)^{-1} f = \sum_{n=0}^{\infty} \lambda^n K^n f, \tag{3.21}$$

using the results outlined above. The only new point in this is to notice the series expansion for ϕ. Let $S_N = \sum_{n=0}^{N} \lambda^n K^n$ so that $\|S_N - (I - \lambda K)^{-1}\| \to 0$ as $N \to \infty$ whence $\|S_N f - (I - \lambda K)^{-1} f\| \leqslant \|S_N - (I - \lambda K)^{-1}\| \|f\| \to 0$ as $N \to \infty$. It is worth noting that if we start with the equation $\phi = f + \lambda K\phi$ and substitute $f + \lambda K\phi$ for ϕ on the right hand side, we obtain $\phi = f + \lambda Kf + \lambda^2 K^2 \phi$, and repeated application yields $\phi = \sum_{n=0}^{N-1} \lambda^n K^n f + \lambda^N K^N \phi$. The expression on the right hand side will tend to $(I - \lambda K)^{-1} f$ provided the series $\Sigma \lambda^n K^n f$ converges and $\|\lambda^N K^N \phi\| \to 0$, both being true under our assumption that $\|\lambda K\| < 1$. Notice here that the sense in which the series $\Sigma \lambda^n K^n f$ converges is that of the norm of the space under consideration, so that in this case

$$\int_a^b |\phi(t) - s_n(t)|^2\, dt \to 0 \quad \text{as } n \to \infty$$

where $s_n(t) = \sum_{j=0}^{n} \lambda^j (K^j f)(t)$. This shows us that equation (3.1) has a unique solution provided $|\lambda|$ is sufficiently small, provided $|\lambda| < 1/\|K\|$ in fact, interpreting the trivial case $\|K\| = 0$ appropriately as for all λ. If k is a kernel for which we know only that K is a bounded linear map, this information is new. If K is compact we could observe that the conclusion would follow on noticing that $\phi - \lambda K\phi = 0$ implies $\|\phi\| \leq \|\lambda K\| \|\phi\|$ which, in the case $\|\lambda K\| < 1$, implies in turn that $\phi = 0$ and we could have deduced the existence of the solution ϕ from the Fredholm Alternative, although the series expansion is new. The result just proved has the merit of simplicity.

Let us be specific. Let

$$k(x, t) = \begin{cases} x(1-t) & (0 \leq x \leq t \leq 1), \\ t(1-x) & (0 \leq t \leq x \leq 1), \end{cases}$$

so that k is an L_2-kernel (and therefore K is a compact operator). $\|K\| \leq (\int_0^1 \int_0^1 |k(x, t)|^2 \, dx \, dt)^{\frac{1}{2}} = 1/(90)^{\frac{1}{2}}$, whence $|\lambda| < (90)^{\frac{1}{2}} \Rightarrow \|\lambda K\| < 1$. Therefore

$$\phi(x) - \lambda \int_0^1 k(x, t)\phi(t) \, dt = f(x) \quad (0 \leq x \leq 1)$$

has a unique solution $\phi \in L_2(0, 1)$ if $|\lambda| < (90)^{\frac{1}{2}}$. This is not the best possible constant obtainable here since we have not calculated $\|K\|$ but obtained an estimate for it; the best possible result in this direction is that the equation has a unique solution for all λ satisfying $|\lambda| < \pi^2$.

The series expansion $\Sigma \lambda^n K^n$ for $I - \lambda K$ is called the *Neumann series*. It not only gives us the results we might wish concerning the existence of $(I - \lambda K)^{-1}$ but can be used to find approximations to it, and, in particular, to $(I - \lambda K)^{-1}f$. We have, still subject to our assumption $\|\lambda K\| < 1$,

$$\phi = \sum_{n=0}^{\infty} \lambda^n K^n f = \sum_{n=0}^{N-1} \lambda^n K^n f + \sum_{n=N}^{\infty} \lambda^n K^n f.$$

Now

$$\left\| \sum_{n=N}^{\infty} \lambda^n K^n f \right\| \leq \sum_{n=N}^{\infty} |\lambda^n| \|K^n f\| \leq \sum_{n=0}^{\infty} |\lambda|^n \|K^n\| |\lambda|^N \|K^N f\|$$

$$\leq \frac{|\lambda|^N \|K^N f\|}{1 - |\lambda| \|K\|} \tag{3.22}$$

since $\|K^{m+n}\| \leq \|K^m\| \|K^n\|$ for all m and n, and hence $\Sigma |\lambda|^n \|K^n\|$ is dominated by the geometric series $\Sigma (|\lambda| \|K\|)^n$. Therefore, (3.22) shows us that the solution ϕ of (3.1) can be approximated by $\Sigma_{n=0}^{N-1} \lambda^n K^n f$, with the error estimate

$$\left\| \phi - \sum_{n=0}^{N-1} \lambda^n K^n f \right\| \leq \frac{|\lambda|^N \|K^N f\|}{1 - |\lambda| \|K\|}. \tag{3.23}$$

Example 3.7

In Example 1.1 we saw that the initial value problem

$$\phi''(x) - x\phi(x) = 0 \ (0 \leqslant x \leqslant 1), \quad \phi(0) = 1, \quad \phi'(0) = 0$$

is equivalent to the integral equation

$$\phi(x) - \int_0^x t(x-t)\phi(t) \, dt = 1 \quad (0 \leqslant x \leqslant 1).$$

Denoting by $(K\phi)(x)$ the expression $\int_0^x t(x-t)\phi(t) \, dt$, we see that, as a map from $L_2(0, 1)$ to itself, $\|K\| \leqslant 1/(6\sqrt{5})$. (Use the fact that K is an L_2-kernel.) Also, with $f = 1$, we have $(Kf)(x) = \frac{1}{6}x^3$, $(K^2 f)(x) = \frac{1}{180}x^6$ so by (3.22)

$$\phi(x) = 1 + \tfrac{1}{6}x^3 + \tfrac{1}{180}x^6 + \epsilon(x)$$

where $\epsilon(x) = \Sigma_{n=3}^{\infty}(K^n f)(x)$ and $\|\epsilon\| \leqslant \dfrac{\|K^3 f\|}{1 - \|K\|}$. In this case $K^3 f = \frac{1}{180} \times \frac{1}{72}x^9$ and $\|K^3 f\| = 1/(180 \times 72 \times (19)^{\frac{1}{2}})$, so, using our knowledge that $\|K\| \leqslant 1/6\sqrt{5}$ we see that $\|\epsilon\| < 0.00002$. Thus, in terms of our chosen norm in $L_2(0, 1)$, we see that the difference between ϕ and the function $1 + x^3/6 + x^6/180$ is a function ϵ where $(\int_0^1 |\epsilon(x)|^2 \, dx)^{\frac{1}{2}} < 1/50\,000$. □

In the light of the example just done it is worth pointing out two features. The first is that our knowledge of the error ϵ is in terms of $\int|\epsilon|^2$, and we do not know directly that the value of $\epsilon(x)$ is small for every x, only that it is, in the mean, small. The analysis can be modified to obtain results in terms of other types of convergence, either by formulating the problems in a different Banach space or by various other means of recovering this information which we shall leave aside until later.

The other point to notice is that, in our calculation above, the successive terms f, Kf, $K^2 f, \ldots$ appear to be decreasing in magnitude rather faster than we have assumed. If we use the inequality $\|K^{n+1}\| \leqslant \|K\| \, \|K^n\|$, then we see that the norms of the successive terms are forced to decrease by a factor of $1/(6\sqrt{5})$, whereas they actually decrease much faster. For some operators K such behaviour is peculiar to the choice of the function f, but in other cases it is a property of the operator.

Suppose that $k: [a, b] \times [a, b] \to \mathbb{C}$ is a bounded measurable function and let $(K\phi)(x) = \int_a^x k(x, t)\phi(t) \, dt$, so that K is a compact operator on $L_2(a, b)$

since k is an L_2-kernel. Then

$$(K^2\phi)(x) = K(K\phi)(x) = \int_a^x k(x, s)(K\phi)(s)\, ds$$

$$= \int_a^x \left\{ \int_a^s k(x, s)k(s, t)\phi(t)\, dt \right\} ds$$

$$= \int_a^x k_2(x, t)\phi(t)\, dt$$

where $k_2(x, t) = \int_t^x k(x, s)k(s, t)\, ds$. The inversion of the order of integration in this calculation is justified since $\int_a^x \int_a^s |k(x, s)k(s, t)\phi(t)|\, dt\, ds < \infty$, which may be seen by applying Schwarz's inequality and the boundedness of k. Now, writing $M = \sup\{|k(x, t)| : a \leqslant t \leqslant x \leqslant b\}$ we see that

$$|k_2(x, t)| \leqslant \int_t^x M^2 = M^2(x - t) \quad (a \leqslant t \leqslant x \leqslant b).$$

Proceeding in this way gives

$$(K^n\phi)(x) = \int_a^x k_n(x, t)\phi(t)\, dt \quad (a \leqslant x \leqslant b)$$

where, for each natural number n,

$$k_{n+1}(x, t) = \int_t^x k_n(x, s)k(s, t)\, ds \quad (a \leqslant t \leqslant x \leqslant b).$$

From this we see by induction that, for all n,

$$|k_n(x, t)| \leqslant \frac{M^n(x - t)^{n-1}}{(n - 1)!} \quad (a \leqslant t \leqslant x \leqslant b).$$

Therefore, setting $k_n(x, t) = 0$ for $t > x$, so that $(K^n\phi)(x) = \int_a^b k_n(x, t)\phi(t)\, dt$ we have

$$\|K^n\| \leqslant \left(\int_a^b \int_a^b |k_n(x, t)|^2\, dx\, dt \right)^{\frac{1}{2}} = \frac{(b - a)^n}{(2n(2n - 1))^{\frac{1}{2}}} \frac{M^n}{(n - 1)!}.$$

In particular, $\Sigma \|K^n\|$ converges, whatever the value of M. In this case the term on the right hand side in (3.23) tends to 0 very rapidly as $N \to \infty$ and a very accurate approximation to ϕ may be obtained by summing relatively few terms.

Considering the equation

$$\phi(x) - \lambda \int_a^x k(x, t)\phi(t)\, dt = f(x) \quad (a \leqslant x \leqslant b) \qquad (3.24)$$

where k is as above, bounded and measurable, we see that for all $\lambda \Sigma \|\lambda^n K^n\|$ converges, whence the series $\Sigma(\lambda K)^n$ converges in $B(H)$, where $H = L_2(a, b)$, and $(I - \lambda K)^{-1} = \Sigma_{n=0}^\infty \lambda^n K^n$. Therefore, in this case, the equation (3.24) has the unique solution

$$\phi = (I - \lambda K)^{-1} f = \sum_{n=0}^\infty \lambda^n K^n f$$

for all $f \in L_2(a, b)$ and all scalars λ; no exceptional cases occur and there are no non-trivial solutions to the homogeneous equation

$$\phi(x) - \lambda \int_a^x k(x, t)\phi(t) \, dt = 0 \quad (a \leqslant x \leqslant b)$$

for any scalar λ.

Integral operators K of the form $(K\phi)(x) = \int_a^x k(x, t)\phi(t) \, dt$ are called *Volterra operators*. In many cases such operators have the property deduced above, that $I - \lambda K$ is invertible for all scalars λ, but this depends on the nature of the kernel k. Example 3.6, however, where $k(x, t) = 1/x$, shows that it is possible to choose k so that K does not possess this property, for in this case if $\phi = 1$, $K\phi = \phi$ and $\|K^n\phi\| \not\to 0$. We have shown that $I - \lambda K$ is invertible for all λ in the case where k is a bounded function. The same conclusion holds when k is an L_2-kernel, as we shall now prove.

Suppose that k is an L_2-kernel on $\{(x, t): a \leqslant t \leqslant x \leqslant b\}$, i.e.

$$\int_a^b \int_a^x |k(x, t)|^2 \, dt \, dx < \infty,$$

and define $(K\phi)(x) = \int_a^x k(x, t)\phi(t) \, dt$. (We do not need to consider $k(x, t)$ for $t > x$ since K is a Volterra operator.) Let $A(x) = (\int_a^x |k(x, t)|^2 \, dt)^{\frac{1}{2}}$ and $B(t) = (\int_t^b |k(x, t)|^2 \, dx)^{\frac{1}{2}}$ so that A and B belong to $L_2(a, b)$. Denote by $k_n(x, t)$ the kernel for which $(K^n\phi)(x) = \int_a^x k_n(x, t)\phi(t) \, dt$. Then, for $a \leqslant t \leqslant x \leqslant b$,

$$|k_2(x, t)| = \left| \int_t^x k(x, s)k(s, t) \, ds \right| \leqslant \left(\int_t^x |k(x, s)|^2 \, ds \right)^{\frac{1}{2}} \left(\int_t^x |k(s, t)|^2 \, ds \right)^{\frac{1}{2}}$$

$$\leqslant A(x)B(t).$$

Also, for $a \leqslant t \leqslant x \leqslant b$

$$|k_4(x, t)| = \left| \int_t^x k_2(x, s)k_2(s, t) \, dt \right| \leqslant A(x)B(t) \int_t^x A(s)B(s) \, ds$$

$$= A(x)B(t)(\Phi(x) - \Phi(t)).$$

where $\Phi(x) = \int_a^x A(s)B(s)\,ds$. Φ is continuous from its definition, and bounded since $0 \leqslant \Phi(x) \leqslant \int_a^b A(s)B(s)\,ds \leqslant (\int_a^b A^2(s)\,ds)^{\frac{1}{2}}(\int_a^b B^2(s)\,ds)^{\frac{1}{2}}$. Continuing in this vein yields

$$|k_6(x,t)| = \left| \int_t^x k_4(x,s)k_2(s,t)\,ds \right| \leqslant A(x)B(t) \int_t^x A(s)B(s)(\Phi(x) - \Phi(s))\,ds$$

$$= A(x)B(t)(\Phi(x) - \Phi(t))^2/2,$$

and, generally,

$$|k_{2n}(x,t)| \leqslant A(x)B(t)\frac{(\Phi(x) - \Phi(t))^{n-1}}{(n-1)!} \quad (a \leqslant t \leqslant x \leqslant b).$$

Recalling that Φ is non-negative and bounded above, we see that $|k_{2n}(x,t)| \leqslant A(x)B(t)\Phi(b)^{n-1}/(n-1)!$ and

$$\|K^{2n}\| \leqslant \left(\int_a^b \int_a^x |k_{2n}(x,t)|^2\,dt\,dx \right)^{\frac{1}{2}} \leqslant \frac{\Phi(b)^{n-1}}{(n-1)!}\left(\int_a^b A^2(x)\,dx \right)^{\frac{1}{2}}\left(\int_a^b B^2(t)\,dt \right)^{\frac{1}{2}}.$$

From this we see that $\Sigma\,\|\lambda^{2n}K^{2n}\|$ converges for all λ and since $\|K^{2n+1}\| \leqslant \|K\|\,\|K^{2n}\|$, $\Sigma\,\|\lambda^n K^n\|$ also converges, giving the desired result that $(I - \lambda K)^{-1}$ exists for all scalars λ.

Example 3.8

The initial value problem

$$\left. \begin{array}{l} \dfrac{d^n y}{dx^n} + a_{n-1}(x)\dfrac{d^{n-1}y}{dx^{n-1}} + \cdots + a_0(x)y(x) = h(x) \quad (a \leqslant x \leqslant b), \\[2mm] y(a) = \alpha_0, \quad y'(a) = \alpha_1, \ldots, y^{(n-1)}(a) = \alpha_{n-1}, \end{array} \right\} \tag{3.25}$$

where $\alpha_0, \ldots, \alpha_{n-1}$ are given real numbers and $a_0, \ldots, a_{n-1}, h$ are given continuous functions, possesses a unique solution y.

First, notice that if y satisfies the equations (3.25), it must be n times differentiable and so $y, y', \ldots, y^{(n-1)}$ are all continuous. From this and the continuity of $a_0, \ldots, a_{n-1}$ and h it follows that $y^{(n)}$ is continuous. Let $\phi = y^{(n)}$. Then, if y satisfies (3.25),

$$y^{(n-1)}(x) = \alpha_{n-1} + \int_a^x \phi(t)\,dt,$$

$$y^{(n-2)}(x) = \alpha_{n-2} + \alpha_{n-1}(x-a) + \int_a^x (x-t)\phi(t)\,dt,$$

and so on, using the formula

$$\Phi_k(x) = \int_a^x \frac{(x-t)^{k-1}}{(k-1)!} \phi(t)\,dt$$

to obtain the function Φ_k whose kth derivative is ϕ and which satisfies $\Phi_k(a) = \Phi_k'(a) = \cdots = \Phi_k^{(k-1)}(a) = 0$ (Problem 1.9). Then ϕ is a continuous function satisfying

$$\phi(x) + \int_a^x k(x,t)\phi(t)\,dt = f(x) \quad (a \leqslant x \leqslant b) \tag{3.26}$$

where

$$k(x,t) = \sum_{k=1}^n a_{n-k}(x) \frac{(x-t)^{k-1}}{(k-1)!}$$

and

$$f(x) = h(x) - \sum_{k=1}^n \sum_{j=1}^k \alpha_{n-j} a_{n-k}(x)(x-a)^{k-j}/(k-j)!$$

By using the Fundamental Theorem of Calculus, it is easy to check that if ϕ is continuous and satisfies (3.26) then the function y defined by

$$y(x) = \int_a^x \frac{(x-t)^{n-1}}{(n-1)!} \phi(t)\,dt + \sum_{j=0}^{n-1} \alpha_j \frac{(x-a)^j}{j!}$$

is n times differentiable and satisfies (3.25). The existence and uniqueness of the solution of (3.25) is therefore equivalent to the existence and uniqueness of continuous solutions to (3.26).

The calculations above show that k and f in (3.26) are continuous, so, in particular, k is bounded and $f \in L_2(a,b)$. Equation (3.26) may then be rewritten in the form $(I - \lambda K)\phi = f$ with $\lambda = -1$ and $(K\phi)(x) = \int_a^x k(x,t)\phi(t)\,dt$. Therefore (3.26) has a unique square-integrable solution ϕ, and since k and f are continuous and

$$\phi(x) = f(x) - \int_a^x k(x,t)\phi(t)\,dt \quad (a \leqslant x \leqslant b)$$

ϕ is also continuous. Therefore (3.26) has a unique continuous solution ϕ and (3.25) has a unique solution y. $\qquad\square$

The property which gave rise to the existence of $(I - \lambda K)^{-1}$ for all scalars λ in the case where K is a Volterra operator was the convergence of $\sum \|\lambda^n K^n\|$ for all λ. This property is a particular case of a more widespread phenomenon. By the root test, the series $\Sigma \| \lambda^n K^n \|$ converges if $\lim_{n \to \infty}$ $\| \lambda^n K^n \|^{1/n} = \lim_{n \to \infty} |\lambda| \| K^n \|^{1/n} < 1$, which suggests that we consider the behaviour of $\| K^n \|^{1/n}$. Noticing that, for every Hilbert space H and all $T \in B(H)$, $\| T^{m+n} \| \leqslant \| T^m \| \| T^n \|$, we establish the following technical lemma.

Lemma 3.6

Suppose that (α_n) is a sequence of non-negative numbers with the property that for all m, $n \in \mathbb{N}$ $\alpha_{m+n} \leqslant \alpha_m \alpha_n$. Then

$$\lim_{n \to \infty} \alpha_n^{1/n} = \inf_{n \in \mathbb{N}} \alpha_n^{1/n}.$$

Proof

Since $\alpha_n \geqslant 0$ for all n, the existence of the infimum is not in doubt. Let $\alpha = \inf \alpha_n^{1/n}$. Let $\epsilon > 0$ and choose N so that $\alpha_N^{1/N} < \alpha + \epsilon$.

For all $n \in \mathbb{N}$, $\alpha_{2n}^{1/2n} \leqslant (\alpha_n \alpha_n)^{1/2n} = \alpha_n^{1/n}$ and, by induction, if $k \in \mathbb{N}$ $\alpha_{kn}^{1/kn} \leqslant \alpha_n^{1/n}$ (since $\alpha_{kn} \leqslant \alpha_n^k$). It follows, then, that $\alpha_n^{1/n} \leqslant \alpha_N^{1/N} < \alpha + \epsilon$ for all n which are multiples of N.

If $n > N$, then we can write $n = kN + r$ where $k \in \mathbb{N}$ and $r \in \{1, 2, \dots, N\}$. Then

$$\alpha_n = \alpha_{kN+r} \leqslant \alpha_{kN} \alpha_r \leqslant \alpha_N^k A$$

where $A = \max(\alpha_1, \dots, \alpha_N)$. Now $A > 0$ unless $\alpha_1 = 0$, in which case $\alpha_n = 0$ for all $n \in \mathbb{N}$ and the result is trivial, so we may presume $A > 0$. Then for $n > N$

$$\alpha_n^{1/n} \leqslant \alpha_N^{k/n} A^{1/n} = (\alpha_N^{1/N})^{kN/n} A^{1/n} < (\alpha + \epsilon)^{kN/n} A^{1/n}$$

and since $kN/n \to 1$ as $n \to \infty$, $(\alpha + \epsilon)^{kN/n} A^{1/n} \to \alpha + \epsilon$ as $n \to \infty$. We can, therefore, choose N_1 such that for all $n \geqslant N_1$, $(\alpha + \epsilon)^{kN/n} A^{1/n} < \alpha + 2\epsilon$ and then for all $n \geqslant N_1$, $\alpha \leqslant \alpha_n^{1/n} < \alpha + 2\epsilon$. Since $\epsilon > 0$ was arbitrary, $\alpha_n^{1/n} \to \alpha$ as $n \to \infty$. $\square$

Definition 3.5

Let H be a Hilbert space and $T \in B(H)$. The *spectral radius* of T, $\rho(T)$, is defined to be $\rho(T) = \lim_{n \to \infty} \| T^n \|^{1/n}$. $\square$

The significance of the name spectral radius will become clear later. Notice that $0 \leqslant \rho(T) \leqslant \| T \|$, and that equality is possible in either inequality. For if $(T\phi)(x) = \int_a^x \phi(t) \, dt$, acting as a bounded linear map in $L_2(0, 1)$, our work above shows that $\| T^n \| \leqslant 1/((2n(2n-1))^{\frac{1}{2}} (n-1)!)$ so $\rho(T) = 0$ (let $k(x, t) = 1$ $0 \leqslant t \leqslant x \leqslant 1$). If $T = I$ we have $\rho(T) = \| T \| = 1$.

We see that if $|\lambda|\rho(T)<1$, $\Sigma \lambda^n T^n$ converges, since $\Sigma \|\lambda^n T^n\|$ converges by the root test and therefore $(I-\lambda T)^{-1}$ exists as an element of $B(H)$. That is, for $|\lambda|<1/\rho(T)$ (if $\rho(T)\neq0$), $(I-\lambda T)^{-1}$ exists.

We conclude this chapter with a small result which will, nevertheless, be put to good use in later chapters.

Lemma 3.7

Suppose that H is a Hilbert space and that $S, T \in B(H)$. Then $(I-ST)$ is invertible if and only if $(I-TS)$ is invertible, whence, for non-zero μ, $ST-\mu I$ is invertible if and only if $TS-\mu I$ is invertible.

Proof

Assume that $(I-ST)$ is invertible. Then

$$(I-TS)(I+T(I-ST)^{-1}S)=I-TS+T(I-ST)^{-1}S-TST(I-ST)^{-1}S$$
$$=I-TS+T(I-ST)(I-ST)^{-1}S$$
$$=I,$$

and a similar calculation shows that $(I+T(I-ST)^{-1}S)(I-TS)=I$. Therefore $I-TS$ is invertible and $(I-TS)^{-1}=I+T(I-ST)^{-1}S$. By symmetry, then, $I-TS$ is invertible if and only if $(I-ST)$ is invertible.

For the last part, suppose $\mu\neq0$. Then

$$ST-\mu I \text{ is invertible} \Leftrightarrow I-\mu^{-1}ST \text{ is invertible}$$
$$\Leftrightarrow I-\mu^{-1}TS \text{ is invertible}$$
$$\Leftrightarrow \mu I-TS \text{ is invertible}. \qquad \square$$

Remark

The proof of Lemma 3.7 may seem mysterious, with the formula for $(I-TS)^{-1}$ apparently plucked from nowhere. If, however, we suppose $\|S\| \|T\| <1$ and use the Neumann series for $(I-TS)^{-1}$, that is, $\Sigma_{n=0}^{\infty} (TS)^n$, it is not difficult to relate this to $\Sigma_{n=0}^{\infty}(ST)^n$. This yields the formula for $(I-TS)^{-1}$ in terms of $(I-ST)^{-1}$ and, as used in the lemma, the restriction on $\|S\| \|T\|$ is unnecessary.

Example 3.9

Suppose that $k(x,t)$ is a kernel on $[0,1] \times [0,1]$ which gives rise to a bounded linear map $K: L_2(0,1) \to L_2(0,1)$ defined by $(K\phi)(x)=\int_0^1 k(x,t)\phi(t)\,dt$

$(0 \leqslant x \leqslant 1)$, and let $r: [0, 1] \to \mathbb{C}$ be continuous. Then the equation

$$\phi(x) - \lambda \int_0^1 k(x, t) r(t) \phi(t) \, dt = f(x) \quad (0 \leqslant x \leqslant 1)$$

possesses a unique solution $\phi \in L_2(0, 1)$ for all $f \in L_2(0, 1)$ and the solution operator which maps f into ϕ is bounded, if and only if the same is true of the equation

$$\phi(x) - \lambda r(x) \int_0^1 k(x, t) \phi(t) \, dt = f(x) \quad (0 \leqslant x \leqslant 1).$$

To see this, we notice that

$$(R\phi)(x) = r(x) \phi(x) \quad (0 \leqslant x \leqslant 1)$$

defines a bounded linear map, and so $I - \lambda KR$ and $I - \lambda RK$ are either both invertible or both non-invertible.

In the case where $r(x) \geqslant 0$ $(x \in [0, 1])$ we may consider the additional equation

$$\phi(x) - \lambda \int_0^1 k(x, t) (r(x) r(t))^{\frac{1}{2}} \phi(t) \, dt = f(x) \quad (0 \leqslant x \leqslant 1)$$

corresponding to the operator $I - \lambda SKS$ (where $(S\phi)(x) = (r(x))^{\frac{1}{2}} \phi(x)$); $I - \lambda SKS$, $I - \lambda KS^2$ and $I - \lambda S^2 K$ are either all invertible or all non-invertible.

The situation is slightly simpler in the case where K is compact and the invertibility of $I - \lambda KR$ can be determined by using the Fredholm Alternative, for it is easily checked that if ϕ satisfies $\phi - \lambda KR\phi = 0$ then $R\phi$ satisfies $\psi - \lambda RK\psi = 0$, and that if $\psi - \lambda RK\psi = 0$ then $K\psi$ satisfies $\phi - \lambda KR\phi = 0$.

These remarks give some theoretical structure to the sort of device used after Example 2.1, and subsequently in Chapter 2, to symmetrise the kernels arising from ordinary differential equations. $\qquad\qquad\square$

Problems

3.1   (i)   Solve the integral equation

$$\phi(x) - \int_0^1 (x + t) \phi(t) \, dt = 1 \quad (0 \leqslant x \leqslant 1).$$

(ii)   Solve

$$\phi(x) - \int_0^1 (x + t) \phi(t) \, dt = x^2 \quad (0 \leqslant x \leqslant 1).$$

3.2 Show that if f is a given continuous function and $\lambda \neq 2\sqrt{3}/(\sqrt{3}\pm 2)$ the integral equation

$$\phi(x) - \lambda \int_0^1 (x+t)\phi(t)\,dt = f(x) \quad (0 \leqslant x \leqslant 1)$$

has a unique solution, and find that solution.

3.3 Solve the integral equation

$$\phi(x) = \lambda \int_0^1 \cos\{\alpha(x-t)\}\phi(t)\,dt + f(x) \quad (0 \leqslant x \leqslant 1)$$

in the cases
(i) $f(x) = 0$ $(0 \leqslant x \leqslant 1)$, $\alpha \in \mathbb{R}$;
(ii) $f(x) = 1$ $(0 \leqslant x \leqslant 1)$, $\alpha = n\pi$, $n \in \mathbb{N}$.

3.4 Suppose that f is a real-valued continuous function on $[0, 1]$. Show that the integral equation

$$\phi(x) - \int_0^1 (12xt + 6(t-x) - 2)\phi(t)\,dt = f(x) \quad (0 \leqslant x \leqslant 1)$$

has a solution if and only if $\int_0^1 (1-2t)f(t)\,dt = 0$. Show also that if f satisfies the stated condition the solution of the integral equation is not unique. Find the solution in the case $f = 1$.

3.5 Define the integral operator K on $L_2(0, 1)$ by

$$(K\phi)(x) = \int_0^1 (1 - \max(x, t))\phi(t)\,dt.$$

By observing that the kernel is an L_2-kernel, show that $\|K\| \leqslant 1/6^{\frac{1}{2}}$ and therefore that the homogeneous equation

$$\phi(x) - \lambda \int_0^1 (1 - \max(x, t))\phi(t)\,dt = 0 \quad (0 \leqslant x \leqslant 1)$$

has no non-trivial solution in $L_2(0, 1)$ if $|\lambda| < 6^{\frac{1}{2}}$. Deduce that if $|\lambda| < 6^{\frac{1}{2}}$ and f is a square-integrable function, the equation

$$\phi(x) - \lambda \int_0^1 (1 - \max(x, t))\phi(t)\,dt = f(x) \quad (0 \leqslant x \leqslant 1)$$

has a unique square-integrable solution ϕ.

3.6 Suppose that $k: [-1, 1] \to \mathbb{R}$ is a measurable function and that $\int_{-1}^1 |k(s)|\,ds < \infty$. Define K by

$$(K\phi)(x) = \int_0^1 k(x-t)\phi(t)\,dt \quad (0 \leqslant x \leqslant 1)$$

and show that K is a bounded linear map from $L_2(0, 1)$ to itself with

$\|K\| \leqslant \int_{-1}^{1} |k(s)|\, ds$. (Hint: the kernel is a Schur kernel.) Now define k_n by $k_n(s) = k(s)$ if $|k(s)| \leqslant n$ and $k_n(s) = 0$ if $|k(s)| > n$, so that k_n is a bounded function and the operator K_n defined by $(K_n\phi)(x) = \int_0^1 k_n(x-t)\phi(t)\, dt$ $(0 \leqslant x \leqslant 1)$ is compact. Deduce that $\|K_n - K\| \to 0$ and thus that K is compact.

3.7 Suppose that $k : [0, \infty) \to \mathbb{C}$ is a measurable function and that $\int_0^\infty |k(u)| u^{-\frac{1}{2}}\, du$ exists. Use the technique of Example 3.6 to show that the operator K defined by

$$(K\phi)(x) = \frac{1}{x} \int_0^1 k\left(\frac{t}{x}\right) \phi(t)\, dt \quad (0 < x \leqslant 1)$$

is a bounded linear map from $L_2(0, 1)$ to itself and that $\|K\| \leqslant \int_0^\infty |k(u)| u^{-\frac{1}{2}}\, du$.

By considering the function ϕ where $\phi(t) = t^{-\frac{1}{2}}$ $(a \leqslant t \leqslant 1)$ and $\phi(t) = 0$ for $t < a$, or otherwise, show that if for all $u > 0$ $k(u) \geqslant 0$ then $\|K\| = \int_0^\infty k(u) u^{-\frac{1}{2}}\, du$.

3.8 Example 3.6 shows that not all operators of the form considered in Problem 3.7 are compact. Show that if $\int_0^\infty |k(u)|^2\, du < \infty$ and $0 < \alpha < 1$, then the operator defined by

$$(K\phi)(x) = \frac{1}{x^\alpha} \int_0^1 k\left(\frac{t}{x}\right) \phi(t)\, dt \quad (0 < x \leqslant 1)$$

is compact. (It has an L_2-kernel.) Deduce that for $0 < \alpha < 1$ the operator K_α given by

$$(K_\alpha \phi)(x) = \frac{1}{x^\alpha} \int_0^x \phi(t)\, dt \quad (0 < x \leqslant 1)$$

is compact.

3.9 Let $(T\phi)(x) = \int_0^1 \phi(t)/(x+t)\, dt$ $(0 < x \leqslant 1)$. Prove that T is a bounded operator from $L_2(0, 1)$ to itself and that $\|T\| \leqslant \pi$.

3.10 Suppose that U is a bounded linear map from $L_2(0, 1)$ to itself whose inverse is also bounded, and let $K : L_2(0, 1) \to L_2(0, 1)$ be compact and linear. By considering $I + U^{-1}K$, or otherwise, show that for $f \in L_2(0, 1)$ either the equation $U\phi + K\phi = f$ has a unique solution $\phi \in L_2(0, 1)$ or the homogeneous equation $U\phi + K\phi = 0$ has a non-trivial solution.

Use this to show that the integral equations

$$(1+x)\phi(x) - \int_0^x \phi(t)\, dt = f(x) \quad (0 \leqslant x \leqslant 1)$$

and

$$\phi(1-x) - \int_0^x \phi(t)\, dt = f(x) \quad (0 \leqslant x \leqslant 1)$$

both possess a unique solution ϕ in $L_2(0, 1)$ provided $f \in L_2(0, 1)$.

3.11 Let p and f be continuous functions and α be constant. Show that the initial value problem

$$\left.\begin{aligned} \phi'(x) - p(x)\phi(x) &= f(x) \\ \phi(0) &= \alpha \end{aligned}\right\} \quad (0 < x < 1)$$

is equivalent to the integral equation

$$\phi(x) - \int_0^x p(t)\phi(t)\, dt = \alpha + F(x) \quad (0 \leqslant x \leqslant 1), \qquad (3.27)$$

where $F(x) = \int_0^x f(t)\, dt$. Let K be the operator given by $(K\phi)(x) = \int_0^x p(t)\phi(t)\, dt$ and show that

$$(K^n\phi)(x) = \int_0^x \frac{(P(x) - P(t))^{n-1}}{(n-1)!} p(t)\phi(t)\, dt \quad (0 \leqslant x \leqslant 1, n \in \mathbb{N}),$$

where $P(x) = \int_0^x p(t)\, dt$. Obtain the Neumann series for the solution ϕ of (3.27) and, by observing that the uniform convergence of $\Sigma(P(x) - P(t))^{n-1}p(t)f(t)/(n-1)!$ allows us to interchange the integration and summation, show that

$$\phi(x) = \alpha + F(x) + \int_0^x (\alpha + F(t))e^{P(x) - P(t)}p(t)\, dt$$

$$= \alpha e^{P(x)} + \int_0^x f(t)e^{P(x) - P(t)}\, dt \quad (0 \leqslant x \leqslant 1).$$

3.12 Show that the boundary value problem

$$\left.\begin{aligned} \phi''(x) - \lambda x\phi(x) &= 0 \quad (0 < x < 1) \\ \phi(0) &= \phi'(1) = 0 \end{aligned}\right\} \qquad (3.28)$$

has a non-trivial solution ϕ if and only if the integral equation

$$\phi(x) + \lambda \int_0^1 t \min(x, t)\phi(t)\, dt = 0 \quad (0 \leqslant x \leqslant 1)$$

possesses a non-trivial solution. By showing that the operator K defined by $(K\phi)(x) = \int_0^1 t \min(x, t)\phi(t)\, dt$ has $\|K\| \leqslant 2/(3\sqrt{5})$ deduce that if (3.28) has a non-trivial solution then $|\lambda| > 3\sqrt{5}/2$.

3.13 Let $(K\phi)(x) = \int_0^x (xt)^{\frac{1}{2}}\phi(t)\, dt$ $(0 \leqslant x \leqslant 1)$. Find the kernel k_n for which

$$(K^n\phi)(x) = \int_0^x k_n(x, t)\phi(t)\, dt \quad (0 \leqslant x \leqslant 1)$$

and deduce that the (unique) solution of the equation

$$\phi - K\phi = f$$

satisfies

$$\phi(x) = f(x) + \int_0^x (xt)^{\frac{1}{2}} \, e^{(x^2-t^2)/2} f(t) \, dt \quad (0 \leqslant x \leqslant 1).$$

3.14 Let $(K\phi)(x) = \int_0^x x\phi(t) \, dt \ (0 \leqslant x \leqslant 1)$ and deduce that the integral equation

$$\phi(x) - \int_0^x x\phi(t) \, dt = 1 \quad (0 \leqslant x \leqslant 1)$$

has a unique solution ϕ. By letting $f = 1$ and finding $K^n f$ obtain an expression for ϕ in the form of a convergent series.

3.15 Show that the initial value problem $\phi''(x) - (\sin x)\phi(x) = 0$, $\phi(0) = 0$, $\phi'(0) = 1$ is equivalent to

$$\phi(x) - \int_0^x (x-t)(\sin t)\phi(t) \, dt = x \quad (0 \leqslant x \leqslant 1).$$

Show that $\phi(x) = x(1 - \sin x) + 2(1 - \cos x) + \epsilon(x) \quad (0 \leqslant x \leqslant 1)$ where $(\int_0^1 |\epsilon(x)|^2 \, dx)^{\frac{1}{2}} < 0.0028$.

3.16 (i) Let ϕ satisfy $\phi = f + \lambda K\phi$ in $[0, 1]$ where $f(x) = 1 \ (0 \leqslant x \leqslant 1)$ and $(K\phi)(x) = \int_0^x t \log(t/x)\phi(t) \, dt \ (0 \leqslant x \leqslant 1)$. Show by induction that $(K^n f)(x) = (-\frac{1}{4}x^2)^n/(n!)^2 \ (n \in \mathbb{N})$ and hence prove that

$$\phi(x) = 1 + \sum_{n=1}^{\infty} (-\frac{1}{4}\lambda x^2)^n/(n!)^2 \quad (0 \leqslant x \leqslant 1)$$

satisfies the given integral equation. Show also that, with $\lambda = 1$, $\phi(x) = 1 - \frac{1}{4}x^2 + \frac{1}{64}x^4 + \epsilon(x) \ (0 \leqslant x \leqslant 1)$ where $\int_0^1 |\epsilon(x)|^2 \, dx < 1.4 \times 10^{-4}$.

(ii) Let $\phi = f + \lambda K\phi$ in $(0, 1]$ where $f(x) = \log x \ (0 < x \leqslant 1)$ and K is as defined in (i). Show that

$$(K^n f)(x) = (-\frac{1}{4}x^2)^n \left\{ \log x - 1 - \frac{1}{2} - \cdots - \frac{1}{n} \right\} / (n!)^2 \quad (n \in \mathbb{N})$$

and deduce the solution of the integral equation. Show that, with $\lambda = 1$,

$$\phi(x) = \left(1 - \frac{x^2}{4} + \frac{x^4}{64}\right) \log x + \frac{x^2}{4} - \frac{3x^4}{128} + \epsilon(x) \quad (0 < x \leqslant 1),$$

where $\int_0^1 |\epsilon(x)|^2 \, dx < 2.7 \times 10^{-4}$.

(With $\lambda = 1$ the two integral equations are related to Bessel's differential equation of zero order, as Problem 2.1 shows. The solution of the equation in (i) is in fact the Bessel function of the first kind of order zero, J_0. The solution of the equation in (ii) is equal to $\frac{1}{2}\pi Y_0 - (\gamma - \log 2)J_0$, where Y_0 is the second-kind Bessel function of order zero and $\gamma = 0.5772 \ldots$ is Euler's constant. An integral equation whose solution is Y_0 follows by linearity. The truncated solutions provide a means of calculating the values of the Bessel functions in the given intervals. In the case $\lambda = -1$ the associated differential equation is the modified Bessel equation of zero order, $x\phi'' + \phi' - x\phi = 0$ $(x > 0)$. The solution of the equation in (i) is the modified Bessel function I_0 and the solution of the equation in (ii) is equal to $-K_0 - (\gamma - \log 2)I_0$, where the modified Bessel function K_0 is a (singular) solution of the modified Bessel equation. Approximations to I_0 and K_0 can of course be derived on any interval $(0, x_0)$.)

3.17 Show that the integral equation

$$\phi(x, y) + \int_x^y dt \int_x^t \phi(s, t)\, ds = \tfrac{1}{2}(f(x) + f(y)) \quad (x \in \mathbb{R},\ x \leqslant y) \qquad (3.29)$$

is equivalent to the initial value problem

$$\phi_{xy}(x, y) - \phi(x, y) = 0 \quad (x \in \mathbb{R},\ x \leqslant y)$$

$$\phi(x, x) = f(x), \quad \phi_x(x, x) - \phi_y(x, x) = 0 \quad (x \in \mathbb{R}).$$

Let $(K\phi)(x, y) = -\displaystyle\int_x^y dt \int_x^t \phi(s, t)\, ds$ $(x \in \mathbb{R},\ x \leqslant y \leqslant x + a)$ and show that

$$(K^n\phi)(x, y) = (-1)^n \int_x^y dt \int_x^t \frac{(y - t)^{n-1}(s - x)^{n-1}}{((n-1)!)^2} \phi(s, t)\, ds.$$

By showing that, as a map from $L_2(\mathbb{R} \times [0, a])$ to itself, $\|K^n\| \leqslant a^n/n!$ deduce that (3.29) has a unique solution ϕ. Deduce that the solution of (3.29) is

$$\phi(x, y) = \tfrac{1}{2}(f(x) + f(y)) - \tfrac{1}{2}\int_x^y dt \int_x^t J_0(2(y - t)^{\frac{1}{2}}(s - x)^{\frac{1}{2}})(f(s) + f(t))\, ds$$

where $J_0(z) = \Sigma_{n=0}^\infty (-1)^n \dfrac{z^{2n}}{4^n(n!)^2}$ (the Bessel function of the first kind of order zero).

3.18 The following simple transformations give rise to bounded linear maps from $L_2(0, 1)$ to itself; check the details.

(i) If $a: [0, 1] \to \mathbb{R}$ is a bounded measurable function and $(A\phi)(x) = a(x)\phi(x)$ $(x \in [0, 1])$, then $\|A\| = \text{ess sup}\{|a(x)|: 0 \leqslant x \leqslant 1\}$. Here ess sup denotes the essential supremum, the essential supremum of a set S being $\sup\{\lambda: \mu\{x \in S: x \geqslant \lambda\} \neq 0\}$ where μ denotes Lebesgue measure. (In fact, in these terms, a need only be an essentially bounded function.)

(ii) Let $h: [0, 1] \to [0, 1]$ be a bijective function such that h and h^{-1} are differentiable and have continuous derivatives. Then

$$(U\phi)(x) = \phi(h(x))(|h'(x)|)^{\frac{1}{2}} \quad (0 \leqslant x \leqslant 1)$$

defines a linear map from $L_2(0, 1)$ to itself with the property that for all $\phi \in L_2(0, 1)$ $\|U\phi\| = \|\phi\|$. U is invertible.

3.19 Let H be a Hilbert space and S, $T \in B(H)$. Show that if $ST = TS$ then
$\rho(ST) \leqslant \rho(S)\rho(T)$.

3.20 Let $V \in B(L_2(0,1))$ be defined by $(V\phi)(x) = \int_0^x \phi$. From our work on Volterra
operators we know $\rho(V) = 0$. Let U be defined by $(U\phi)(x) = \phi(1-x)$ (*cf.*
Problem 3.18). By observing that if $\phi(x) = \cos(\frac{1}{2}\pi x)$ $(0 \leqslant x \leqslant 1)$ then $UV\phi = \frac{2}{\pi}\phi$,
show that $\rho(UV) \geqslant \frac{2}{\pi}$, and therefore that the requirement that $ST = TS$ in
Problem 3.19 cannot be omitted. Show also that $\rho(UVU) = 0$.

3.21 Suppose that $K \in B(L_2(a,b))$ has rank N, and let $\{\phi_1, \ldots, \phi_N\}$ be an
orthonormal basis for $\mathcal{R}(K)$. By observing that for all $\phi \in L_2(a,b)$ $K\phi = \Sigma_{n=1}^{N}(K\phi, \phi_n)\phi_n$, or otherwise, show that there are functions $\psi_1, \ldots, \psi_N$ in
$L_2(a,b)$ for which, for all $\phi \in L_2(a,b)$,

$$(K\phi)(x) = \int_a^b \left(\sum_{n=1}^{N} \phi_n(x)\psi_n(t) \right) \phi(t) \, dt$$

almost everywhere in (a,b). (Thus every operator of finite rank in $L_2(a,b)$
arises from a degenerate kernel.)

4

Compact operators

4.1 Introduction

This chapter is the core of the book. We have seen that if the function k is sufficiently well-behaved then the operator K defined by $(K\phi)(x) = \int_a^b k(x, t)$ $\phi(t) \, dt$ is a compact operator from $L_2(a, b)$ to itself, so results about compact operators will produce corresponding information about integral equations. Just as for general matrices, the amount that can be said about general compact operators is limited, and the most substantial conclusions follow if the operator has some symmetry, symmetry in this case being self-adjointness. It turns out that there is a powerful classification theorem for compact self-adjoint operators which describes the action of the operators in terms of their eigenvalues and eigenvectors.

The benefit of the theory of this chapter is that it expresses compact self-adjoint operators in a standard form, which is known to exist for each operator. This reduces many problems to one of evaluation of information about parameters already known to exist, and the techniques can be used to provide specific information about integral operators. We shall, in fact, return frequently to the structure theorem for compact self-adjoint operators to find the basis for a variety of techniques in subsequent chapters.

Many of the results given here are true in greater generality than we shall state them. In particular, the theory can be extended to compact operators on Banach spaces, although the additional complication is considerable and the results on self-adjointness do not extend satisfactorily.

4.2 General properties

Definition 4.1

Let H be a Hilbert space over the field of scalars $\mathscr{F}$ (where $\mathscr{F} = \mathbb{R}$ or $\mathbb{C}$) and let $T \in B(H)$. We say that $\mu \in \mathscr{F}$ is an *eigenvalue* of T if and only if there is a non-zero element $\phi \in H$ for which $T\phi = \mu\phi$; ϕ is said to be an *eigenvector* corresponding to μ. The *spectrum* of T, denoted by $\sigma(T)$, is the set

$$\sigma(T) = \{\mu \in \mathscr{F} : T - \mu I \text{ is not invertible}\}. \qquad \square$$

That an eigenvalue must belong to the field of scalars appropriate to the problem is implicit in the equation $T\phi = \mu\phi$. This contrasts with the approach often taken in matrix theory where the field of scalars may not be made explicit, or where various devices are used to consider complex eigenvalues when we would wish the scalars to be real. The issue is well illustrated by the linear map T of $\mathbb{R}^2$ to itself whose matrix with respect to the standard basis is $\left(\begin{smallmatrix} 0 & 1 \\ -1 & 0 \end{smallmatrix}\right)$. This linear map has no eigenvalues, since there are no real numbers μ for which $T\phi = \mu\phi$ has a non-trivial solution ϕ. The corresponding linear map from $\mathbb{C}^2$ to $\mathbb{C}^2$ with the same matrix has eigenvalues $\pm i$. Finite dimensional linear algebra uses eigenvalues as the main tool with which to describe the structure of linear maps and for this to be successful it is necessary to ensure that there are no eigenvalues concealed by an inappropriate choice of scalars. Similar issues occur with the more general structure of linear maps in Hilbert space and we shall occasionally need to ensure that the set of scalars we choose is $\mathbb{C}$.

At this point we need to face up to the unfortunate duality of notation alluded to in Chapter 1. When we consider integral equations with a scalar parameter λ, this normally occurs with the λ multiplying the integral operator, for example,

$$\phi(x) - \lambda \int_a^b k(x, t)\phi(t)\,\mathrm{d}t = f(x) \quad (a \leqslant x \leqslant b).$$

This arises from the formulation of the problem and is accepted practice. In linear algebra and its analytical relation, functional analysis, it is normal for the parameter to multiply the identity operator; we say μ is an eigenvalue of K if $K\phi = \mu\phi$ has a non-trivial solution. Thus in the one case we consider the operator $I - \lambda K$ and in the other $K - \mu I$. There is clearly no problem in translating from one to the other, in that information about $K - \mu I$ may be obtained from that about $I - \lambda K$ by setting $\lambda = 1/\mu$ and vice versa (provided μ and λ are non-zero), but there is an irritating need to make this conversion. In this book we shall artificially emphasise the position of the parameter by the convention that λ will be reserved for a scalar multiplying the linear map in question and μ will be reserved for a scalar multiplying the identity.

Some of the results already encountered give us information about the spectrum of a bounded linear map. Suppose H is a Hilbert space and

$T \in B(H)$. Then if $|\mu| > \|T\|$ so that $\|\frac{1}{\mu}T\| < 1$ then $I - \frac{1}{\mu}T$ is invertible (Corollary A3) and so is $T - \mu I = (-\mu)(I - \frac{1}{\mu}T)$. From the remarks following Definition 3.5 we see that if $|\mu| > \rho(T)$, $T - \mu I$ is invertible. Therefore $\sigma(T) \subset \{\mu \in \mathscr{F} : |\mu| \leqslant \rho(T)\}$, where $\mathscr{F}$ is the field of scalars associated with H. It is clear that if μ is an eigenvalue then $\mu \in \sigma(T)$; the converse result is false, in general. Let $(K\phi)(x) = \int_0^x \phi(t)\,dt$ define $K \in B(H)$ where $H = L_2(0, 1)$; we know K is compact, and since $K\phi$ is continuous, for each $\phi \in H$, K is not surjective, hence not invertible, and $0 \in \sigma(K)$. But it is easy to verify that 0 is not an eigenvalue of K. In a finite-dimensional Hilbert space H, $\sigma(T)$ coincides with the set of eigenvalues of T.

There are a few simple results about invertibility and the spectrum which are best stated formally.

Lemma 4.1

Let H be a Hilbert space, $S \in B(H)$ and suppose that S is invertible. Then if $T \in B(H)$ and $\|S - T\| < 1/\|S^{-1}\|$, T is invertible. The set, G, of invertible elements of $B(H)$ is therefore open and $\theta : G \to G$ defined by $\theta(T) = T^{-1}$ is continuous.

Proof

$T = S - (S - T) = S(I - S^{-1}(S - T))$. If $\|S - T\| < 1/\|S^{-1}\|$, it follows that $\|S^{-1}(S - T)\| < 1$ and so $I - S^{-1}(S - T)$ is invertible. Then T is invertible and $T^{-1} = (I - S^{-1}(S - T))^{-1}S^{-1}$. Also notice that if $\|S^{-1}(S - T)\| < 1$, then

$$\|(I - S^{-1}(S - T))^{-1}\| \leqslant \sum_{n=0}^{\infty} \|S^{-1}(S - T)\|^n \leqslant \frac{1}{1 - \|S^{-1}\|\,\|S - T\|}.$$

Then if $T \in B(H)$ satisfies $\|T - S\| < 1/\|S^{-1}\|$, so $T \in G$,

$$\|\theta(T) - \theta(S)\| = \|T^{-1} - S^{-1}\| = \|S^{-1}(S - T)T^{-1}\|$$

$$\leqslant \frac{\|S^{-1}\|^2 \|S - T\|}{1 - \|S^{-1}\|\,\|S - T\|}$$

(the last inequality since $T^{-1} = (I - S^{-1}(S - T))^{-1}S^{-1}$). From this it is readily seen that for all $\epsilon > 0$ there is a $\delta > 0$ such that $\|S - T\| < \delta \Rightarrow \|\theta(T) - \theta(S)\| < \epsilon$. $\qquad\square$

Corollary 4.2

Let H be a Hilbert space over $\mathscr{F}$ and $T \in B(H)$. $\sigma(T)$ is a closed, bounded (i.e. compact) subset of $\mathscr{F}$, and the function $\mu \mapsto (T - \mu I)^{-1}$ is continuous on $\mathscr{F} \setminus \sigma(T)$.

Proof

The function $f: \mathscr{F} \to B(H)$ given by $f(\mu) = T - \mu I$ is continuous, and $\mathscr{F} \setminus \sigma(T) = f^{-1}(G)$ in the notation of Lemma 4.1, so, since G is open, $\mathscr{F} \setminus \sigma(T)$ is open and $\sigma(T)$ is closed. By composition of two continuous functions $\mu \mapsto (T - \mu I)^{-1}$ is continuous on $\mathscr{F} \setminus \sigma(T)$. We already know that $\sigma(T)$ is bounded. $\qquad\square$

These results are essentially technicalities which we shall use occasionally, though the last is of some direct use. Suppose that $\lambda_0 \neq 0$ and that $I - \lambda_0 K$ is invertible, that is, $1/\lambda_0 \notin \sigma(K)$ where K is the integral operator on $H = L_2(a, b)$ arising from the L_2-kernel $k(x, t)$. Then the integral equation

$$\phi(x) - \lambda \int_a^b k(x, t)\phi(t)\, \mathrm{d}t = f(x) \quad (a \leqslant x \leqslant b)$$

has a unique solution $\phi_\lambda = (I - \lambda K)^{-1} f$ in $L_2(a, b)$, provided λ is sufficiently close to λ_0. The function ϕ_λ depends continuously on λ (in the sense that $\|\phi_\lambda - \phi_{\lambda_0}\| < \epsilon$ if $|\lambda - \lambda_0|$ is small enough). This sort of consideration is significant when we wish to calculate an approximation to ϕ_{λ_0} where the procedures used may introduce small changes in the parameter λ_0, which would be expected in a numerical algorithm, for example.

Turning to the study of the spectral properties of compact operators, we recall from Chapter 3 that if $K \in B(H)$ is compact and we choose F to be of finite rank with $\|K - F\| < 1$,

$$\begin{aligned}(I - K) &= (I - (K - F))(I - (I - (K - F))^{-1}F) \\ &= (I - (K - F))(I - F_1)\end{aligned} \tag{4.1}$$

where F_1 has finite rank and $I - (K - F)$ is invertible. From this we deduced the Fredholm Alternative, Theorem 3.3, that either $I - K$ is invertible or $(I - K)\phi = 0$ has a non-trivial solution ϕ. In other words, if K is compact, either 1 is an eigenvalue of K or $K - I$ is invertible (so $1 \notin \sigma(K)$). By observing that if $\mu \neq 0$, $K - \mu I = \mu(\frac{1}{\mu}K - I)$ and that $\frac{1}{\mu}K$ is compact, we see that either $K - \mu I$ is invertible or μ is an eigenvalue of K. This gives our first spectral properties of compact operators.

Theorem 4.3

Let H be a Hilbert space and $K \in B(H)$ be compact. Every non-zero element of $\sigma(K)$ is an eigenvalue. $\qquad\square$

Theorem 4.4

Let H be a Hilbert space and $K \in B(H)$ be compact. Then if μ is a non-zero eigenvalue of K the set of eigenvectors of K corresponding to μ has finite dimension.

Proof

From (4.1) we see that $K\phi = \phi \Leftrightarrow F_1\phi = \phi$. Since F_1 has finite rank, the set of eigenvectors of K corresponding to the eigenvalue 1 is a subspace of the image of F_1, which has finite dimension. Since ϕ is an eigenvector of K corresponding to the eigenvalue μ if and only if it is an eigenvector of $\frac{1}{\mu}K$ corresponding to the eigenvalue 1, the result follows for all non-zero μ. $\qquad\square$

Theorem 4.4 shows us that if H has infinite dimension, then the identity operator I on H is not compact, since the set of eigenvectors corresponding to the eigenvalue 1 is too large. Theorem 3.5(ii) now shows that, in an infinite dimensional space, a compact operator K cannot be invertible, so that $0 \in \sigma(K)$. 0 need not be an eigenvalue (as we saw above) and, if it is, it need not have a finite dimensional set of eigenvectors. (Consider $K \in B(L_2(0, 1))$ given by $(K\phi)(x) = \int_0^1 \phi(t)\,dt$.)

We have deduced some of the properties of an individual non-zero point of $\sigma(K)$, where K is compact. To see how many such points there can be we first need a purely technical lemma.

Lemma 4.5

Let H be a Hilbert space and $K \in B(H)$ be compact. If (ϕ_n) is an orthonormal sequence and (ψ_n) is bounded, then $(K\psi_n, \phi_n) \to 0$ as $n \to \infty$. More generally, if (ϕ_n) is a bounded sequence which has the property that for all $\phi \in H$ $(\phi_n, \phi) \to 0$ as $n \to \infty$ and (ψ_n) is bounded then $(K\psi_n, \phi_n) \to 0$ as $n \to \infty$.

Remark

Here, as subsequently, we shall denote the inner product in a Hilbert space by $(.\,,.)$.

Proof

Notice first that if (ϕ_n) is orthonormal then, by Bessel's inequality, if $\phi \in H$ then $\Sigma|(\phi_n, \phi)|^2$ converges; therefore $(\phi_n, \phi) \to 0$ as $n \to \infty$, and we see that the result about orthonormal sequences follows from the second result.

Let $F \in B(H)$ have finite rank and suppose that $\{\theta_1, \theta_2, \ldots, \theta_r\}$ is an orthonormal basis for $\{F\phi: \phi \in H\}$. Let (ϕ_n) have the property that for all $\phi \in H$ $(\phi_n, \phi) \to 0$ as $n \to \infty$. For each $n \in \mathbb{N}$ there are scalars $\alpha_1^{(n)}, \ldots, \alpha_r^{(n)}$ for which $F\psi_n = \alpha_1^{(n)}\theta_1 + \cdots + \alpha_r^{(n)}\theta_r$ and for each $i = 1, 2, \ldots, r$, $|\alpha_i^{(n)}| \leqslant \|F\psi_n\| \leqslant \|F\|M$ where $M = \sup\{\|\psi_n\|: n \in \mathbb{N}\}$. Therefore

$$|(F\psi_n, \phi_n)| = \left| \sum_{i=1}^{r} \alpha_i^{(n)}(\theta_i, \phi_n) \right| \leqslant \|F\|M \sum_{i=1}^{r} |(\theta_i, \phi_n)|$$

and since $(\theta_i, \phi_n) \to 0$ as $n \to \infty$ for each i, $(F\psi_n, \phi_n) \to 0$ as $n \to \infty$.

To deal with the general case, let K be compact. Let $\epsilon > 0$ and choose a finite rank operator F with $\|K - F\| < \epsilon$. Because F has finite rank the last paragraph shows that there exists N such that for all $n \geqslant N$ $|(F\psi_n, \phi_n)| < \epsilon$, and then

$$\text{for all } n \geqslant N, |(K\psi_n, \phi_n)| \leqslant |((K-F)\psi_n, \phi_n)| + |(F\psi_n, \phi_n)|$$
$$< \|K - F\|MM' + \epsilon \leqslant (MM' + 1)\epsilon$$

where $M' = \sup\{\|\phi_n\|: n \in \mathbb{N}\}$. Since $\epsilon > 0$ was arbitrary, $(K\psi_n, \phi_n) \to 0$ as $n \to \infty$. $\qquad\square$

Before moving on we should explain the rather obscure condition which has appeared in Lemma 4.5. The version of the lemma in which (ϕ_n) is orthonormal will be used immediately in what follows. The property that for all $\phi \in H, (\phi_n, \phi) \to 0$ as $n \to \infty$ is a well-known form of convergence from which it can be proved that (ϕ_n) is a bounded sequence, making part of our assumptions redundant. This proof that (ϕ_n) is bounded would, however, lead us well aside from our main theme and the only situation where we shall use the result is in Lemma 6.11 in a context where the boundedness of the sequence concerned is already known.

Theorem 4.6

Let H be a Hilbert space and $K \in B(H)$ be compact. If (μ_n) is an (infinite) sequence of distinct eigenvalues of K then $\mu_n \to 0$ as $n \to \infty$.

Proof

By omitting one term, if necessary, we may presume that, for all n, $\mu_n \neq 0$. Choose an eigenvector ϕ_n of norm 1 corresponding to each eigenvalue μ_n. The sequence (ϕ_n) need not be orthonormal, but we can obtain an orthonormal sequence (ψ_n) from it by the Gram–Schmidt process. Then for each n, ψ_n is a linear combination of $\phi_1, \ldots, \phi_n$ and ϕ_n is a linear combination of $\psi_1, \ldots, \psi_n$. (This uses the result from linear algebra that

$\phi_1, \ldots, \phi_n$, being eigenvectors corresponding to distinct eigenvalues, form a linearly independent set.) For each n, there is a (non-zero) scalar α_n for which $\psi_n = \alpha_n \phi_n +$ linear combination of $\phi_1, \ldots, \phi_{n-1}$. Then, since $K\phi_i = \mu_i \phi_i$ for all i,

$$
\begin{aligned}
K\psi_n &= \alpha_n \mu_n \phi_n + \text{linear combination of } \phi_1, \ldots, \phi_{n-1} \\
&= \mu_n \psi_n \quad + \text{linear combination of } \phi_1, \ldots, \phi_{n-1} \\
&= \mu_n \psi_n \quad + \text{linear combination of } \psi_1, \ldots, \psi_{n-1}
\end{aligned}
$$

so that $(K\psi_n, \psi_n) = \mu_n$. Since (ψ_n) is orthonormal $(K\psi_n, \psi_n) \to 0$ and so $\mu_n \to 0$, by Lemma 4.5. $\qquad\square$

Theorem 4.7

Let H be a Hilbert space of infinite dimension and $K \in B(H)$ be compact. Then the spectrum of K consists of 0 together with a finite or countably infinite set of eigenvalues. Each non-zero point of $\sigma(K)$ is an eigenvalue with a finite-dimensional set of corresponding eigenvectors and if there are infinitely many eigenvalues, $\{\mu_n : n \in \mathbb{N}\}$, then $\mu_n \to 0$ as $n \to \infty$.

Proof

The only part remaining to be proved is the countability of the set of eigenvalues. Every eigenvalue has modulus at most $\|K\|$ and, for each $m \in \mathbb{N}$, the set $\{z \in \mathbb{C} : 1/m \leq |z| \leq \|K\|\}$ can contain only finitely many eigenvalues of K, by Theorem 4.6. The set of all non-zero eigenvalues is $\bigcup_{m=1}^{\infty} \{\mu \in \sigma(K) : 1/m \leq |\mu| \leq \|K\|\}$, the union of countably many finite sets. $\qquad\square$

We may sum up our general results on the spectrum of a compact operator by saying that if $\mu \neq 0$, then either μ is an eigenvalue of K or μ does not belong to the spectrum of K, which is a form of the Fredholm Alternative. The set of points μ for which $K - \mu I$ is not invertible, the spectrum of K, is a bounded set which is at most countable. In this sense the values of μ belonging to $\sigma(K)$ may be regarded as 'exceptional' since they form a 'small' set.

4.3 Adjoint operators

When we first tackled integral equations arising from degenerate kernels we reduced the problem to one involving matrices. In order to find a criterion for the existence of a solution of the equation

$$\phi(x) - \lambda \int_a^b k(x, t)\phi(t)\, \mathrm{d}t = f(x) \quad (a \leqslant x \leqslant b)$$

in the case where the corresponding homogeneous equation possesses non-trivial solutions we used the conjugate transpose of the matrix. The adjoint of an operator is the Hilbert space generalisation of this idea.

Definition 4.2

Let H be a Hilbert space and $T \in B(H)$. The *adjoint* of T is the operator $T^* \in B(H)$ with the property that, for all $\phi, \psi \in H$, $(T\phi, \psi) = (\phi, T^*\psi)$. $\square$

The existence and uniqueness of the adjoint, as well as its elementary properties, are assumed; for details see Appendix A. We shall, however, need the following less obvious results.

Lemma 4.8

Let H be a Hilbert space. If $F \in B(H)$ has finite rank then so has F^*, and if $K \in B(H)$ is compact so is K^*.

Proof

Suppose that F has rank N, and choose an orthonormal basis $\{\phi_1, \ldots, \phi_N\}$ for the image of F. Extend $\{\phi_1, \ldots, \phi_N\}$ to a complete orthonormal set $\{\phi_i : i \in I\}$ in H. Then if $i \in I \setminus \{1, 2, \ldots, N\}$

$$\text{for all } \psi \in H \quad 0 = (F\psi, \phi_i) = (\psi, F^*\phi_i).$$

Thus we see that, for $i \in I \setminus \{1, 2, \ldots, N\}$, $F^*\phi_i = 0$. Since (ϕ_i) is a complete orthonormal set, if $\phi \in H$, $\phi = \Sigma_{i \in I}\, \xi_i \phi_i$ whence $F^*\phi = \Sigma_{i=1}^N \xi_i F^*\phi_i$, for some scalars ξ_i. Thus F^* has finite rank.

If K is compact, there is a sequence (K_n) of finite rank operators with $\|K_n - K\| \to 0$. Then $\|K_n^* - K^*\| \to 0$ showing that K^* is compact. $\square$

Remark

Using the identity $(T^*)^* = T$, it is easy to deduce from the above that F has finite rank if and only if F^* does, and that they both have the same rank. Equally easily, it follows that K is compact if and only if K^* is compact.

The principal attribute of the adjoint which will concern us is its role in interchanging properties relating to injectivity with those relating to surjectivity. For this it is convenient to introduce some notation.

Definition 4.3

Let H be a Hilbert space and $T \in B(H)$. By the *nullspace* of T, $\mathcal{N}(T)$, we mean the space of vectors annihilated by T, so $\mathcal{N}(T) = \{\phi \in H: T\phi = 0\}$. The *image* (or range) of T is $\mathcal{R}(T) = \{T\phi: \phi \in H\}$. $\qquad\square$

The key result relating these ideas (Theorem A13) is that if $T \in B(H)$ then

$$\mathcal{N}(T) = \mathcal{R}(T^*)^\perp \text{ and } \overline{\mathcal{R}(T)} = \mathcal{N}(T^*)^\perp,$$

where the bar denotes closure. The first of these equations shows that if T^* is surjective, then T is injective, while the second shows that if T^* is injective, then $\overline{\mathcal{R}(T)} = H$. The second case does not guarantee that T is surjective, only that every point of H is the limit of a sequence of points of $\mathcal{R}(T)$. For this to be greatly useful we need to know that $\mathcal{R}(T)$ is a closed subspace.

Lemma 4.9

Let H be a Hilbert space, $K \in B(H)$ be compact and $\mu \neq 0$. Then $\mathcal{R}(K - \mu I)$ is a closed subspace of H.

Proof

By dividing throughout by μ, we may, without loss, presume that $\mu = 1$. Since K is compact, we can choose a finite-rank operator $F \in B(H)$ for which $\|F - K\| < 1$ and then

$$K - I = (F_1 - I)(I - (K - F))$$

where $F_1 = F(I - (K - F))^{-1}$, a finite-rank operator. Since $I - (K - F)$ is invertible, it is surjective, so $\mathcal{R}(K - I) = \mathcal{R}(F_1 - I)$.

Let $\psi \in \overline{\mathcal{R}(F_1 - I)}$ so there is a sequence (ϕ_n) in H with $(F_1 - I)\phi_n \to \psi$. From this, $F_1(F_1 - I)\phi_n \to F_1 \psi$ so $F_1 \psi$ is a limit of a sequence of points in the finite-dimensional space $\mathcal{R}(F_1(F_1 - I))$. Since all finite-dimensional subspaces are closed, there is a vector $\phi \in H$ with $F_1 \psi = F_1(F_1 - I)\phi$, and thus $(F_1 - I)(F_1 \phi - \psi) = \psi$, showing that $\psi \in \mathcal{R}(F_1 - I)$. Since ψ was a typical member of $\overline{\mathcal{R}(F_1 - I)}$, $\mathcal{R}(F_1 - I)$ is closed. $\qquad\square$

Example 4.1

Suppose that H is a Hilbert space and $K \in B(H)$ is compact. The equation

$$(I - \lambda K)\phi = f \qquad (4.2)$$

has a unique solution ϕ in the case where the corresponding homogeneous equation

$$(I - \lambda K)\phi = 0 \tag{4.3}$$

has only the trivial solution. If (4.3) possesses non-trivial solutions then (4.2) has solutions (which will not be unique) if and only if $(f, \psi) = 0$ for all ψ satisfying

$$(I - \bar{\lambda} K^*)\psi = 0. \tag{4.4}$$

We shall call (4.4) the *adjoint homogeneous equation* to (4.2) and (4.3).

The case $\lambda = 0$ is easy, so we shall presume $\lambda \neq 0$, when we may rewrite (4.2) as

$$(K - \lambda^{-1} I)\phi = -\lambda^{-1} f.$$

Now we know from Theorem 4.3 that either $1/\lambda$ is an eigenvalue of K or it does not belong to $\sigma(K)$. This is just the Fredholm Alternative that either (4.3) has a non-trivial solution or $(I - \lambda K)$ has an inverse (and hence (4.2) has a unique solution $\phi = (I - \lambda K)^{-1} f$).

If (4.3) has a non-trivial solution, so that $\mathcal{N}(I - \lambda K) \neq \{0\}$ we have to work a little harder. Then (4.2) will have a solution if and only if $f \in \mathcal{R}(I - \lambda K) = \mathcal{R}(K - \lambda^{-1} I)$. Now, since K is compact and $\lambda^{-1} \neq 0$, $\mathcal{R}(K - \lambda^{-1} I) = \overline{\mathcal{R}(K - \lambda^{-1} I)} = \mathcal{N}((K - \lambda^{-1} I)^*)^{\perp} = \mathcal{N}(K^* - \bar{\lambda}^{-1} I)^{\perp}$. Therefore $f \in \mathcal{R}(K - \lambda^{-1} I)$ if and only if $(f, \psi) = 0$ for all ψ in $\mathcal{N}(K^* - \bar{\lambda}^{-1} I)$, that is, for all ψ satisfying (4.4). $\qquad \square$

To make use of Example 4.1 in integral equations, we need to be able to identify the adjoint of an integral operator. Suppose that k is an L_2-kernel and that $(K\phi)(x) = \int_a^b k(x, t)\phi(t) \, dt$ is the corresponding operator on $L_2(a, b)$. Then, for all ϕ and $\psi \in L_2(a, b)$

$$\begin{aligned}
(K\phi, \psi) &= \int_a^b \left\{ \int_a^b k(x, t)\phi(t) \, dt \right\} \overline{\psi(x)} \, dx \\
&= \int_a^b \left\{ \int_a^b k(x, t)\phi(t)\overline{\psi(x)} \, dx \right\} dt \tag{4.5} \\
&= \int_a^b \phi(t) \overline{\left\{ \int_a^b \overline{k(x, t)}\psi(x) \, dx \right\}} \, dt \\
&= \int_a^b \phi(t) \overline{(L\psi)(t)} \, dt = (\phi, L\psi),
\end{aligned}$$

where

$$(L\psi)(x) = \int_a^b \overline{k(t, x)}\psi(t)\,dt.$$

The crux of this manipulation is the reversal of the order of integration at (4.5) which is justified by Fubini's Theorem since

$$\int_a^b \left\{ \int_a^b |k(x, t)\phi(t)\overline{\psi(x)}|\,dt \right\} dx \leq \int_a^b \left\{ \int_a^b |k(x, t)|^2\,dt \int_a^b |\phi(t)|^2\,dt \right\}^{\frac{1}{2}} |\psi(x)|\,dx$$

$$\leq \|\phi\| \left(\int_a^b \left\{ \int_a^b |k(x, t)|^2\,dt \right\} dx \int_a^b |\psi(x)|^2\,dx \right)^{\frac{1}{2}}$$

(using Schwarz's inequality twice), and the last quantity is finite. From this we see that K^* is the operator whose kernel is that function whose value at (x, t) is $\overline{k(t, x)}$. Notice that K^* also arises from an L_2-kernel.

The same argument as above (essentially Fubini's Theorem) shows that if $k(x, t)$ is a Schur kernel, then the kernel of the adjoint of the operator it generates is again the function whose value at (x, t) is $\overline{k(t, x)}$ and this result is true of all kernels which have the property that $|k(x, t)|$ generates a bounded operator on $L_2(a, b)$. For kernels which do not have this last property, the adjoint of the operator generated by k need not be the operator generated by $\overline{k(t, x)}$.

We are now in a position to see the full force of the Fredholm Alternative.

Theorem 4.10

Let $k(x, t)$ be either an L_2-kernel or a Schur kernel on $[a, b] \times [a, b]$ which generates a compact operator, and λ be a complex number. Then

either for all $f \in L_2(a, b)$ the integral equation

$$\phi(x) - \lambda \int_a^b k(x, t)\phi(t)\,dt = f(x) \quad (a \leq x \leq b) \qquad (4.6)$$

has a unique solution, and there is a constant M such that for all f, $\|\phi\| \leq M\|f\|$ (where ϕ is the solution corresponding to f and $\|\cdot\|$ denotes the L_2 norm),

or the homogeneous equation

$$\phi(x) - \lambda \int_a^b k(x, t)\phi(t)\,dt = 0 \quad (a \leq x \leq b) \qquad (4.7)$$

has a non-trivial solution $\phi \in L_2(a, b)$. In this case (4.6) has a (non-unique) solution if and only if $\int_a^b f(t)\overline{\psi(t)}\,dt = 0$ for all $\psi \in L_2(a,b)$ which satisfy

$$\psi(x) - \bar{\lambda} \int_a^b \overline{k(t,x)}\psi(t)\,dt = 0 \quad (a \leqslant x \leqslant b). \tag{4.8}$$

□

At this point we should notice that since the operator K generated by k in Theorem 4.10 is compact, and since ϕ satisfies (4.7) if and only if ϕ is an eigenvector of K corresponding to the eigenvalue $1/\lambda$, the set of solutions of (4.7) has finite dimension. Suppose this dimension is n and that $\{\phi_1, \ldots, \phi_n\}$ is a basis for the set of solutions of (4.7). Then if ϕ satisfies (4.6) so does $\phi + \Sigma_{i=1}^n \alpha_i \phi_i$, for any choice of scalars $\alpha_1, \ldots, \alpha_n$.

Since K above is compact, so is K^*, and hence the set of solutions of (4.8) will also have finite dimension. Call the dimension m and let $\{\psi_1, \ldots, \psi_m\}$ be a basis of the set of solutions of (4.8). Then, since every solution ψ of (4.8) is of the form $\Sigma_{i=1}^m \alpha_i \psi_i$, $\int_a^b f(t)\overline{\psi(t)}\,dt = 0$ for all solutions ψ of (4.8) if and only if $\int_a^b f(t)\overline{\psi_i(t)}\,dt = 0$ for $i = 1, 2, \ldots, m$, giving us a finite number of conditions to check. It turns out that $m = n$, that is, that the sets of solutions of (4.7) and (4.8) have the same dimension.

Theorem 4.11

Let H be a Hilbert space and $K \in B(H)$ be compact. If $\mu \neq 0$, then $\mathcal{N}(K - \mu I)$ and $\mathcal{N}(K^* - \bar{\mu}I)$ have the same dimension.

Proof

Let $\mu \neq 0$. Then

$$K - \mu I \text{ injective} \Rightarrow K - \mu I \text{ invertible (Theorem 4.3)}$$
$$\Rightarrow K - \mu I \text{ surjective}$$
$$\Rightarrow (K - \mu I)^* \text{ injective (Theorem A13)}$$
$$\Rightarrow (K - \mu I)^* \text{ surjective (Theorem 4.3)}$$
$$\Rightarrow K - \mu I \text{ injective (Theorem A13)}.$$

This clearly shows that if one of the subspaces $\mathcal{N}(K - \mu I)$ and $\mathcal{N}(K^* - \bar{\mu}I)$ has dimension zero, so has the other.

Suppose that $\mathcal{N}(K - \mu I)$ and $\mathcal{N}(K^* - \bar{\mu}I)$ are both non-zero and choose non-zero vectors $\phi_1 \in \mathcal{N}(K - \mu I)$ and $\psi_1 \in \mathcal{N}(K^* - \bar{\mu}I)$. We define a new operator K_1 which we wish to coincide with K on the set of all vectors orthogonal to ϕ_1 and to be such that $(K_1 - \mu I)\phi_1 = \psi_1$. (Notice that $\psi_1 \in \mathcal{N}(K^* - \bar{\mu}I) = \mathcal{R}(K - \mu I)^\perp$ so ψ_1 is not in the image of $K - \mu I$.) Define

K_1 by

$$K_1\phi = K\phi \quad \text{if } (\phi, \phi_1) = 0,$$
$$K_1\phi_1 = \mu\phi_1 + \psi_1,$$

and extend K_1 by linearity to all of H. Then $K_1\phi = K\phi + (\phi, \phi_1)\psi_1/\|\phi_1\|^2$. Thus K_1 is bounded, and since it is the sum of the compact operator K and an operator of rank 1, it is also compact. Moreover

$$(K_1 - \mu I)\phi = 0 \Leftrightarrow (K - \mu I)\phi + (\phi, \phi_1)\psi_1/\|\phi_1\|^2 = 0$$
$$\Leftrightarrow (K - \mu I)\phi = 0 \text{ and } (\phi, \phi_1) = 0$$
$$(\text{since } \psi_1 \in \mathscr{R}(K - \mu I)^\perp)$$
$$\Leftrightarrow \phi \in \mathscr{N}(K - \mu I) \cap \{\phi_1\}^\perp,$$

so the dimension of $\mathscr{N}(K_1 - \mu I)$ is one less than that of $\mathscr{N}(K - \mu I)$. Similarly, noticing that $(K_1 - \mu I)\phi_1 = \psi_1$, we see that $\mathscr{R}(K_1 - \mu I)$ is the direct sum of $\mathscr{R}(K - \mu I)$ and the set of scalar multiples of ψ_1. Then

$$\mathscr{N}(K_1^* - \bar{\mu}I) = \mathscr{R}(K_1 - \mu I)^\perp = \mathscr{R}(K - \mu I)^\perp \cap \{\psi_1\}^\perp$$
$$= \mathscr{N}(K^* - \bar{\mu}I) \cap \{\psi_1\}^\perp,$$

and, since $\psi_1 \in \mathscr{N}(K^* - \bar{\mu}I)$, we see that the dimension of $\mathscr{N}(K_1^* - \bar{\mu}I)$ is one less than that of $\mathscr{N}(K^* - \bar{\mu}I)$.

If one of $\mathscr{N}(K_1 - \mu I)$ and $\mathscr{N}(K_1^* - \bar{\mu}I)$ is $\{0\}$, so is the other (since K_1 is compact and $\mu \neq 0$, by the argument at the start of the proof); in this case $\mathscr{N}(K - \mu I)$ and $\mathscr{N}(K^* - \bar{\mu}I)$ both have dimension 1. If not, repeat the process to define another compact operator K_2 such that $\mathscr{N}(K_2 - \mu I)$ and $\mathscr{N}(K_2^* - \bar{\mu}I)$ have dimension two less than $\mathscr{N}(K - \mu I)$ and $\mathscr{N}(K^* - \bar{\mu}I)$ respectively. After finitely many steps one, and hence both, of the nullspaces will be zero, showing that $\mathscr{N}(K - \mu I)$ and $\mathscr{N}(K^* - \bar{\mu}I)$ have the same dimension. $\qquad \square$

4.4 The Spectral Theorem

As in finite-dimensional linear algebra, the adjoint operation gives the sort of symmetry which admits a simpler structure theory. The type of operator we shall consider first is that which is equal to its own adjoint; these we shall call *self-adjoint*. Notice that if k is an L_2-kernel or a Schur kernel, the corresponding integral operator on $L_2(a, b)$ is self-adjoint if and only if $k(x, t) = \overline{k(t, x)}$ for almost all x and t; such a kernel we shall call *hermitian*. In the common situation where k is real-valued, this corresponds to $k(x, t)$ being symmetric in x and t, and k is then said to be a *symmetric* kernel.

Lemma 4.12 is just linear algebra.

Lemma 4.12

Let H be a Hilbert space and $T \in B(H)$ be self-adjoint. Then the eigenvalues of T are real and eigenvectors corresponding to distinct eigenvalues are orthogonal.

Proof

Let $T = T^*$ and $T\phi = \mu\phi$ where $\phi \neq 0$. Then

$$\mu(\phi, \phi) = (\mu\phi, \phi) = (T\phi, \phi) = (\phi, T^*\phi) = (\phi, T\phi) = \bar{\mu}(\phi, \phi)$$

so, since $(\phi, \phi) \neq 0$, $\mu = \bar{\mu}$.

If we now suppose that $T\phi_i = \mu_i \phi_i$ with $\phi_i \neq 0$ ($i = 1, 2$), then

$$\mu_1(\phi_1, \phi_2) = (T\phi_1, \phi_2) = (\phi_1, T\phi_2) = \bar{\mu}_2(\phi_1, \phi_2) = \mu_2(\phi_1, \phi_2)$$

so, if $\mu_1 \neq \mu_2$, then $(\phi_1, \phi_2) = 0$. $\square$

This proves that the spectrum of a compact self-adjoint operator is real. Although we shall be concerned in this chapter almost entirely with compact operators we shall later need to consider non-compact examples so we insert the more general case here.

Lemma 4.13

Let H be a Hilbert space and $T \in B(H)$ be self-adjoint. Then $\sigma(T) \subset \mathbb{R}$.

Proof

With our definition of spectrum, if H is a real Hilbert space $\sigma(T) \subset \mathbb{R}$ by definition, so we shall presume H is a complex space. The result to be proved is that if $\mu \in \mathbb{C} \backslash \mathbb{R}$ then $T - \mu I$ is invertible.

Let $\mu = \mu_1 + i\mu_2$ where μ_1 and μ_2 are real and $\mu_2 \neq 0$. By Lemma 4.2 μ is not an eigenvalue of T so $T - \mu I$ is injective. By the same reasoning $T - \bar{\mu}I$ is injective so $\mathcal{R}(T - \mu I)$ is dense by Theorem A13. We need to show that $\mathcal{R}(T - \mu I)$ is closed.

Because $T = T^*$, if $\phi \in H$ $(T\phi, \phi) = (\phi, T\phi) = \overline{(T\phi, \phi)}$ so $(T\phi, \phi) \in \mathbb{R}$. Therefore, splitting $((T - \mu I)\phi, \phi)$ into real and imaginary parts,

$$|((T - \mu I)\phi, \phi)| = |((T - \mu_1 I)\phi, \phi) - i\mu_2(\phi, \phi)|$$
$$\geq |\mu_2| \|\phi\|^2$$

whence $|\mu_2| \|\phi\|^2 \leqslant \|(T-\mu I)\phi\| \|\phi\|$ by Schwarz's inequality. Therefore,

$$|\mu_2| \|\phi\| \leqslant \|(T-\mu I)\phi\| \quad (\phi \in H). \tag{4.9}$$

Now suppose $\psi \in \overline{\mathscr{R}(T-\mu I)}$ so that there is a sequence (ϕ_n) in H for which $(T-\mu I)\phi_n \to \psi$ as $n \to \infty$. Then $((T-\mu I)\phi_n)$, and hence (ϕ_n), is a Cauchy sequence. Therefore, for some $\phi \in H$, $\phi_n \to \phi$ as $n \to \infty$ and we see that $(T-\mu I)\phi = \psi$, that is, $\psi \in \mathscr{R}(T-\mu I)$. This shows us that $T-\mu I$ is surjective and therefore bijective, so that it has an inverse as a linear map. Equation (4.9) now shows that if we denote the linear inverse by $(T-\mu I)^{-1}$ then

$$\|(T-\mu I)^{-1}\phi\| \leqslant |\mu_2|^{-1} \|\phi\| \quad (\phi \in H),$$

that is, $(T-\mu I)^{-1}$ is bounded and $\mu \notin \sigma(T)$. $\qquad \square$

The next lemma relates the norm of a self-adjoint operator to its spectrum. This is a result relating the analysis (the norm) to the algebraic properties of the operator (the spectrum) and will allow us to trade off the two sorts of property against each other. Notice that the following lemma guarantees that the spectrum of a self-adjoint operator is non-empty.

Lemma 4.14

Suppose that H is a non-zero Hilbert space and $T \in B(H)$ is self-adjoint. Then at least one of $\pm \|T\|$ belongs to $\sigma(T)$ and $\|T\| = \sup\{|\mu|: \mu \in \sigma(T)\}$.

Proof

Since we know $\mu \in \sigma(T) \Rightarrow |\mu| \leqslant \|T\|$ the second part will follow if we can show that one of $\pm \|T\|$ is in $\sigma(T)$. Let $\alpha = \|T\|$. Then there is a sequence (ϕ_n) of vectors in H of norm 1 for which $\|T\phi_n\| \to \alpha$ as $n \to \infty$. Then, noting that $(T^2\phi_n, \phi_n)$ and α are real,

$$\begin{aligned}
\|(T^2 - \alpha^2 I)\phi_n\|^2 &= (T^2\phi_n, T^2\phi_n) - 2\alpha^2(T^2\phi_n, \phi_n) + \alpha^4(\phi_n, \phi_n) \\
&= \|T^2\phi_n\|^2 - 2\alpha^2 \|T\phi_n\|^2 + \alpha^4 \quad \text{(since } T = T^*) \\
&\leqslant \|T\|^2 \|T\phi_n\|^2 - 2\alpha^2 \|T\phi_n\|^2 + \alpha^4 \to 0 \text{ as } n \to \infty.
\end{aligned}$$

Therefore, since, for all n, $\|\phi_n\| = 1$, $T^2 - \alpha^2 I$ cannot have an inverse in $B(H)$. Since $T^2 - \alpha^2 I = (T-\alpha I)(T+\alpha I)$ at least one of $T \mp \alpha I$ is not invertible, and thus at least one of $\pm \alpha$ is in $\sigma(T)$. $\qquad \square$

Theorem 4.15 (The Spectral Theorem)

Let H be a Hilbert space and $K \in B(H)$ be compact and self-adjoint. Then there is a, possibly finite, sequence (μ_n) of non-zero eigenvalues of K and a corresponding orthonormal sequence (ϕ_n) of eigenvectors such that for each $\phi \in H$, $K\phi = \sum_{n=1}^{\infty} \mu_n(\phi, \phi_n)\phi_n$, where the sum is a finite sum if there are only finitely many eigenvalues. Moreover if we define $K_N \in B(H)$ by $K_N\phi = \sum_{n=1}^{N} \mu_n(\phi, \phi_n)\phi_n$ then in the case where there are infinitely many eigenvalues $\|K - K_N\| \to 0$ as $N \to \infty$.

Proof

Since K is compact the non-zero points of $\sigma(K)$ are eigenvalues of which there are at most countably many, and the sequence of distinct eigenvalues tends to zero if there are infinitely many. Suppose that (v_n) is the sequence of distinct non-zero eigenvalues; by rearranging if necessary, we may presume that, for all n, $|v_{n+1}| \leqslant |v_n|$. Since K is self-adjoint, v_n is real for each n.

For each n, v_n is an eigenvalue of K and the set of corresponding eigenvectors has finite dimension, d_n say. Choose $\{\phi_1, \ldots, \phi_{d_1}\}$ to be an orthonormal basis for $\{\phi : K\phi = v_1\phi\}$, and, letting $m_n = d_1 + d_2 + \cdots + d_n$, choose $\{\phi_{m_n+1}, \ldots, \phi_{m_{n+1}}\}$ to be an orthonormal basis for $\{\phi : K\phi = v_{n+1}\phi\}$. The set $\{\phi_1, \phi_2, \ldots\}$ (terminating if there are only finitely many eigenvalues) is orthonormal; for each vector has norm 1 (it belongs to the orthonormal basis of the appropriate set of eigenvectors) while if $i \neq j$ $(\phi_i, \phi_j) = 0$, either because ϕ_i and ϕ_j are eigenvectors corresponding to distinct eigenvalues or because they belong to an orthonormal basis of the set of eigenvectors corresponding to the same eigenvalue.

Let $\mu_n = v_k$ if $m_{k-1} < n \leqslant m_k$ (setting $m_0 = 0$), so that, for all n, $K\phi_n = \mu_n\phi_n$. In what follows we shall assume that there are infinitely many eigenvalues; if there are only finitely many $K_N = K$ for some $N \geqslant 0$ (where $K_0 = 0$ and corresponds to there being no non-zero eigenvalues) and there is no convergence issue.

K_N is self-adjoint since, for all $\phi, \psi \in H$,

$$(K_N\phi, \psi) = \sum_{n=1}^{N} \mu_n(\phi, \phi_n)(\phi_n, \psi) = \sum_{n=1}^{N} (\phi, \mu_n(\psi, \phi_n)\phi_n) = (\phi, K_N\psi).$$

K_N is compact (since of finite rank), and $\|K - K_N\| = \sup\{|v| : v \in \sigma(K - K_N)\}$. Let v be an eigenvalue of $K - K_N$ and $\phi \neq 0$ be a corresponding eigenvector. Then $(K - K_N)\phi \in \mathcal{R}(K - K_N) = \mathcal{N}(K^* - K_N^*)^{\perp} = \mathcal{N}(K - K_N)^{\perp}$. Since $K\phi_i - K_N\phi_i = 0$ $(i = 1, 2, \ldots, N)$ this shows that

$$(v\phi, \phi_i) = ((K - K_N)\phi, \phi_i) = 0 \quad (i = 1, 2, \ldots, N)$$

so, since $v \neq 0$, $(\phi, \phi_i) = 0$ $(i = 1, 2, \ldots, N)$ and $K_N\phi = 0$. Thus $K\phi = v\phi$ and v is one of the eigenvalues of K. Let $v = v_k$. Then ϕ is a non-zero linear combination of $\phi_{m_{k-1}+1}, \ldots, \phi_{m_k}$ whence there is some i with $m_{k-1} + 1 \leqslant i \leqslant m_k$

for which $(\phi, \phi_i) \neq 0$. From this it follows that $m_k > N$ and thus $v = v_k = \mu_{m_k}$ for some $m_k > N$, and so $v \in \{\mu_{N+1}, \mu_{N+2}, \ldots\}$. (If there are only k distinct non-zero eigenvalues of K, and $N = m_k$, then this last conclusion would be replaced by the conclusion that no such non-zero v existed, hence $\sigma(K - K_N)$ contained no non-zero numbers and $\|K - K_N\| = 0$.) Therefore $\|K - K_N\| \leqslant \sup\{|\mu_n| : n > N\} = |\mu_{N+1}|$ so that $\|K - K_N\| \to 0$ as $N \to \infty$. From this it follows that if $\phi \in H$ $K_N\phi \to K\phi$ as $N \to \infty$, that is,

$$K\phi = \sum_{n=1}^{\infty} \mu_n(\phi, \phi_n)\phi_n. \qquad \square$$

The Spectral Theorem is capable of being developed and applied in many ways, and much of the rest of the book is devoted to this. Before we make this development, it is worth seeing what sort of information the theorem will yield immediately.

Example 4.2

Consider the boundary value problem

$$\left.\begin{array}{l} \phi''(x) + \lambda\phi(x) = 0 \quad (0 < x < 1), \\ \phi(0) = \phi(1) = 0. \end{array}\right\} \qquad (4.10)$$

In Chapter 2 we saw that this is equivalent to the integral equation

$$\phi(x) - \lambda \int_0^1 k(x, t)\phi(t)\,\mathrm{d}t = 0 \quad (0 \leqslant x \leqslant 1)$$

where

$$k(x, t) = \begin{cases} x(1 - t) & (0 \leqslant x \leqslant t \leqslant 1) \\ t(1 - x) & (0 \leqslant t \leqslant x \leqslant 1). \end{cases}$$

Since, for all $x, t \in [0, 1]$, $k(x, t) = \overline{k(t, x)}$, the integral operator on $L_2(0, 1)$ arising from k is self-adjoint, and compact because k is an L_2-kernel. Letting K denote this operator, it is obvious that K is not the zero operator, so it possesses at least one non-zero eigenvalue. Since k is not a degenerate kernel (for a few moments' thought shows that there is no way of expressing it in degenerate form), we see that there must be infinitely many such eigenvalues (μ_n), for otherwise there would be an N for which $K = K_N$ in the notation of the proof of the Spectral Theorem. Finally, the corresponding sequence of eigenvectors, (ϕ_n), can be chosen to be orthonormal.

Translating the information just stated into properties of the solutions of (4.10), we see that there are infinitely many values of λ (those of the form

$\lambda = \mu_n^{-1}$) for which (4.10) has a non-trivial solution and that the corresponding solutions can be arranged in an orthonormal sequence.

Of course, in this simple example we can obtain this information directly, without recourse to the subtleties of functional analysis, and we obtain the familiar results that (4.10) has a non-trivial solution if and only if $\lambda = n^2 \pi^2$ ($n \in \mathbb{N}$), the corresponding solution $\phi_n(x)$ being a scalar multiple of $\sin(n\pi x)$. The significance of our methods is that they will yield the general conclusions about existence and orthogonality of solutions even where explicit solutions cannot be found. As this sort of boundary value problem is at the heart of the method of 'separation of variables' for partial differential equations, and this method relies on expressing the functions involved in terms of series of the solutions ϕ_n of the boundary value problem, there is practical significance here. (Notice that the solutions of (4.10), the functions $\sin(n\pi x)$, can be used to expand a given function in the classical Fourier series.) $\qquad\Box$

Returning to the theoretical structure given by the Spectral Theorem, notice that, for K compact and self-adjoint, the expansion $K\phi = \Sigma_{n=1}^{\infty} \mu_n(\phi, \phi_n)\phi_n$ shows that the effect of K may be found by decomposing ϕ into its various components $(\phi, \phi_n)\phi_n$ in the 'direction' ϕ_n, the component of $K\phi$ in the direction ϕ_n being obtained purely from the corresponding component of ϕ_n. In other words, the action of K may be considered by considering the components separately. This is stated more formally below.

Corollary 4.16

Let K and (ϕ_n) be as in the statement of the Spectral Theorem. Then if $\phi \in H$, there is a vector $\phi_0 \in H$ such that $\phi = \phi_0 + \Sigma_{n=1}^{\infty}(\phi, \phi_n)\phi_n$ and $K\phi_0 = 0$. (If K has only finitely many eigenvalues the sum is a finite sum.)

Proof

Since (ϕ_n) is orthonormal, $\Sigma|(\phi, \phi_n)|^2$ converges and thus the series $\Sigma(\phi, \phi_n)\phi_n$ converges in H. Let $\phi_0 = \phi - \Sigma_{n=1}^{\infty}(\phi, \phi_n)\phi_n$. Then for all m, $(\phi_0, \phi_m) = (\phi, \phi_m) - (\phi, \phi_m) = 0$ so $K\phi_0 = 0$. $\qquad\Box$

Corollary 4.17

If $K \in B(H)$ is compact and self-adjoint, then there is a complete orthonormal set in H consisting of eigenvectors of K.

Proof

Let (ϕ_n) be the orthonormal sequence in H obtained from the Spectral Theorem, and extend (ϕ_n) to a complete orthonormal set; call the set S. Then if $\phi \in S$ either $\phi = \phi_n$ for some n or $(\phi, \phi_n) = 0$ for all n. In the latter case $K\phi = 0$ so in both cases ϕ is an eigenvector of K. $\qquad\square$

In finite dimensions, a linear map can be represented by a diagonal matrix if and only if there is a basis consisting of eigenvectors. Corollary 4.17 is the infinite-dimensional version of this, though we shall not have call to refer to the 'matrix' of an operator with respect to a basis.

Example 4.3

The Spectral Theorem can be used to obtain an explicit formula for the solution of an integral equation. Suppose we wish to solve

$$\phi(x) - \lambda \int_a^b k(x, t)\phi(t)\, dt = f(x) \quad (a \leqslant x \leqslant b) \tag{4.11}$$

where k gives rise to a compact self-adjoint operator on $L_2(a, b)$, $f \in L_2(a, b)$ is given and we seek a solution $\phi \in L_2(a, b)$. Then, with K the corresponding integral operator, we have $K\phi = \sum_{n=1}^{\infty} \mu_n(\phi, \phi_n)\phi_n$ where (μ_n) is a sequence of non-zero real numbers and (ϕ_n) is orthonormal. As an equation in $L_2(a, b)$, (4.11) is

$$\phi - \lambda K\phi = f. \tag{4.12}$$

Taking the inner product of both sides of (4.12) with ϕ_m, and using the spectral expansion of $K\phi$ yields

$$(1 - \lambda \mu_m)(\phi, \phi_m) = (f, \phi_m). \tag{4.13}$$

Thus provided $1 - \lambda \mu_m \neq 0$, $(\phi, \phi_m) = (f, \phi_m)/(1 - \lambda \mu_m)$. If, for all m, $1 - \lambda \mu_m \neq 0$ we see that

$$\phi = f + \lambda K\phi = f + \sum_{n=1}^{\infty} \frac{\lambda \mu_n(f, \phi_n)}{1 - \lambda \mu_n} \phi_n. \tag{4.14}$$

This shows that, if there is a solution, (4.14) defines it. It is easily checked that the series on the right of (4.14) does indeed converge in $L_2(a, b)$ because $(\lambda \mu_n/(1 - \lambda \mu_n))$ is a bounded sequence (it tends to 0 as $n \to \infty$) and thus $\sum |\lambda \mu_n(f, \phi_n)/(1 - \lambda \mu_n)|^2$ converges since $\sum |(f, \phi_n)|^2$ does, and the convergence follows from Theorem A9. The condition that, for all n, $1 - \lambda \mu_n \neq 0$ is just an alternative statement that $1/\lambda$ is not an eigenvalue of K, that is, that $I - \lambda K$ is invertible. (The case $\lambda = 0$ is trivial.) If we express (4.14) in terms of functions this yields

$$\phi(x) = f(x) + \sum_{n=1}^{\infty} \frac{\lambda \mu_n}{1 - \lambda \mu_n} \left(\int_a^b f(t)\overline{\phi_n(t)}\, dt \right) \phi_n(x) \quad (a \leqslant x \leqslant b), \tag{4.15}$$

where the series converges in the mean and equality holds almost everywhere. If the convergence of the series is sufficiently good that we may interchange the summation and integration, the right hand side becomes $f(x) + \int_a^b r_\lambda(x,t) f(t) \, dt$, where $r_\lambda(x,t) = \sum_{n=1}^{\infty} \lambda \mu_n (1 - \lambda \mu_n)^{-1} \phi_n(x) \overline{\phi_n(t)}$ is the resolvent kernel corresponding to the kernel k.

If, in the summation of (4.14), there are values of m for which $1 - \lambda \mu_m = 0$, let $A = \{m : 1 - \lambda \mu_m \neq 0\}$ and $B = \{m : 1 - \lambda \mu_m = 0\}$. Then B is a finite set and if $m \in B$, (4.13) shows that if (4.11) has a solution then $(f, \phi_m) = 0$. Since $\{\phi_m : m \in B\}$ is a basis for the set of ϕ satisfying

$$\phi - \lambda K \phi = \phi - \bar{\lambda} K^* \phi = 0$$

(since λ is real and $K = K^*$) we have the usual condition for the existence of a solution. If f satisfies this condition the solution is

$$\phi = f + \sum_{n \in A} \frac{\lambda \mu_n (f, \phi_n)}{1 - \lambda \mu_n} \phi_n + \sum_{n \in B} \alpha_n \phi_n \tag{4.16}$$

where $\{\alpha_n : n \in B\}$ are arbitrarily chosen constants. This gives an explicit illustration of the Fredholm Alternative. $\qquad\square$

Example 4.4

Consider the equation

$$\phi(x) - \lambda \int_0^1 k(x,t) \phi(t) \, dt = f(x) \quad (0 \leqslant x \leqslant 1), \tag{4.17}$$

where

$$k(x,t) = \begin{cases} t(1-x) & (0 \leqslant t \leqslant x \leqslant 1), \\ x(1-t) & (0 \leqslant x \leqslant t \leqslant 1). \end{cases}$$

As before, $f \in L_2(0,1)$ is given and we seek a solution $\phi \in L_2(0,1)$.

Letting K be the integral operator generated by k, we see that K is compact and self-adjoint (the latter since $k(x,t) = k(t,x)$) so we seek the eigenvalues and eigenvectors of K. Suppose that $\mu \neq 0$ and that ϕ is an eigenvector corresponding to μ. Then

$$\mu \phi(x) = \int_0^x (1-x) t \phi(t) \, dt + \int_x^1 x(1-t) \phi(t) \, dt \quad (0 \leqslant x \leqslant 1). \tag{4.18}$$

Since $\phi \in L_2(0,1)$, the right hand side is continuous in x, hence ϕ is continuous (since $\mu \neq 0$). (This presumes that we have chosen ϕ to satisfy (4.18) for all x. If this is not initially true we may redefine ϕ on a set of measure zero, which will not affect the value of the right hand side of (4.18) so it becomes true for the redefined function.) Since ϕ is continuous, the right hand side of (4.18) is differentiable, and thus so is ϕ, and

$$\mu\phi'(x) = -\int_0^x t\phi(t)\,dt + \int_x^1 (1-t)\phi(t)\,dt \quad (0 \leqslant x \leqslant 1).$$

By similar reasoning we see that $\mu\phi'' = -\phi$. From (4.18) we see that $\phi(0) = \phi(1) = 0$. We can solve the differential equation, obtaining $\phi(x) = A\cosh(x/|\mu|^{\frac{1}{2}}) + B\sinh(x/|\mu|^{\frac{1}{2}})$ or $A\cos(x/\mu^{\frac{1}{2}}) + B\sin(x/\mu^{\frac{1}{2}})$ according as $\mu < 0$ or $\mu > 0$. Imposing the boundary conditions shows that $A = B = 0$ in the first case, while in the second $A = 0$ and $B\sin(1/\mu^{\frac{1}{2}}) = 0$ so there is a non-trivial solution if and only if $1/\mu^{\frac{1}{2}} = n\pi$ or $\mu = 1/(n^2\pi^2)$ $(n = 1, 2, 3, \ldots)$. Therefore set $\mu_n = 1/(n^2\pi^2)$ and $\phi_n(x) = 2^{\frac{1}{2}}\sin(n\pi x)$. These are the sequences of eigenvalues and eigenvectors of K predicted by the Spectral Theorem.

Provided $\lambda \neq 1/(n^2\pi^2)$ for all $n \in \mathbb{N}$, (4.15) now shows that

$$\phi(x) = f(x) + \sum_{n=1}^{\infty} \frac{2\lambda}{n^2\pi^2 - \lambda} \int_0^1 f(t)\sin(n\pi t)\,dt\,\sin(n\pi x) \quad (0 \leqslant x \leqslant 1).$$

In this case we notice that, since $|\sin(n\pi t)\sin(n\pi x)| \leqslant 1$ for all x and t, the series $\Sigma 2\lambda \sin(n\pi t)\sin(n\pi x)/(n^2\pi^2 - \lambda)$ is uniformly convergent and therefore we may interchange summation and integration to obtain

$$\phi(x) = f(x) + \int_0^1 r_\lambda(x, t)f(t)\,dt \quad (0 \leqslant x \leqslant 1)$$

where

$$r_\lambda(x, t) = \sum_{n=1}^{\infty} \frac{2\lambda \sin(n\pi t)\sin(n\pi x)}{n^2\pi^2 - \lambda} \quad (0 \leqslant x, t \leqslant 1),$$

a continuous function.

If $\lambda = 1/(m^2\pi^2)$ for some natural number m, then (4.17) has a solution if and only if $\int_0^1 \sin(m\pi t)f(t)\,dt = 0$ and if this is the case the solution is (by (4.16))

$$\phi(x) = f(x) + \alpha\sin(m\pi x) + \int_0^1 s_\lambda(x, t)f(t)\,dt \quad (0 \leqslant x \leqslant 1),$$

where

$$s_\lambda(x, t) = \sum_{\substack{n=1 \\ n \neq m}}^{\infty} \frac{2\lambda \sin(n\pi x)\sin(n\pi t)}{n^2\pi^2 - \lambda},$$

and α is an arbitrary constant. $\qquad\square$

Example 4.5

The Spectral Theorem even provides useful information about solutions of integral equations of the first kind. Let k be the kernel of Example 4.4, with $k(x, t) = \min(x, t) - xt$ for $0 \leqslant x, t \leqslant 1$, and consider the equation

$$\int_0^1 k(x, t)\phi(t)\, dt = f(x) \quad (0 \leqslant x \leqslant 1). \tag{4.19}$$

Re-writing this as $K\phi = f$, where K is the corresponding integral operator in $L_2(0, 1)$, and expanding $K\phi$ as $\sum_{n=1}^{\infty} \mu_n(\phi, \phi_n)\phi_n$ via the Spectral Theorem, we see that if ϕ satisfies (4.19), then $\mu_n(\phi, \phi_n) = (f, \phi_n)$. Letting $c_n = \int_0^1 f(t) \sin(n\pi t)\, dt = (f, \phi_n)/2^{\frac{1}{2}}$ we deduce that

$$c_n = \frac{1}{n^2\pi^2} \int_0^1 \phi(t) \sin(n\pi t)\, dt.$$

This places a restriction on the set of functions f for which (4.19) can have a solution $\phi \in L_2(0, 1)$. For if $\phi \in L_2(0, 1)$, $\sum |\int_0^1 \phi(t) \sin(n\pi t)\, dt|^2$ converges (by Bessel's inequality) and therefore $\sum |n^2 c_n|^2$ converges. The Riesz-Fischer Theorem, A9, classifies $L_2(0, 1)$ as the set of functions of the form $\sum a_n \sin(n\pi x)$ where $\sum |a_n|^2$ converges, so the condition we have imposed on f restricts it to a proper subset of $L_2(0, 1)$. It is easy to check that if $\sum |n^2 c_n|^2$ converges, then the element $\phi \in L_2(0, 1)$ which is the sum of $\sum 2n^2\pi^2 c_n \sin(n\pi x)$ satisfies $K\phi = f$.

A little caution is needed here. The 'solution', ϕ, is a solution of the equation $K\phi = f$ in $L_2(0, 1)$, and therefore if we recall that elements of $L_2(0, 1)$ are only defined almost everywhere there is a function ϕ, square-integrable on $[0, 1]$, such that

$$\int_0^1 k(x, t)\phi(t)\, dt = f(x)$$

for almost all $x \in [0, 1]$. Choosing another element of the equivalence class of functions which constitutes the member of $L_2(0, 1)$ will produce a function $\tilde{\phi}$ equal to ϕ except on a set of measure zero, and $\int_0^1 k(x, t)\tilde{\phi}(t)\, dt$ and $\int_0^1 k(x, t)\phi(t)\, dt$ will be equal. Our solution ϕ satisfies (4.19) for almost all $x \in [0, 1]$ but not necessarily all x; the device used to find solutions holding for all x, which we used with equations of the second kind, is not available here.   □

Example 4.6

Let us consider the general case of an equation of the first kind

$$\int_a^b k(x, t)\phi(t)\,dt = f(x) \quad (a \leqslant x \leqslant b) \tag{4.20}$$

where f is square-integrable, k is a hermitian kernel which generates a compact self-adjoint operator on $L_2(a, b)$ and seek a solution ϕ which is square-integrable and such that (4.20) holds for almost all $x \in [a, b]$. Rewriting (4.20) in operator notation gives $K\phi = f$ whence, letting (μ_n) and (ϕ_n) be the sequences of eigenvalues and eigenvectors, respectively, in the spectral expansion for $K\phi$, we wish to solve

$$\sum_{n=1}^{\infty} \mu_n(\phi, \phi_n)\phi_n = f. \tag{4.21}$$

Suppose (4.21) has a solution $\phi \in L_2(a, b)$. Then if $(\phi_n, \psi) = 0$ for $n = 1, 2, \ldots$, it follows that $(f, \psi) = 0$. This is a familiar condition in a new guise: since $(\phi_n, \psi) = 0$ for all n is equivalent to the condition $K\psi = 0$ and since $K = K^*$, the condition is that $f \in \mathcal{N}(K^*)^\perp$, whence $f \in \overline{\mathcal{R}(K)}$. (This condition does not appear in Example 4.5 since it there turns out that the sequence $(2^{\frac{1}{2}}\sin(n\pi x))$ is a complete orthonormal sequence.) We also note from (4.21) that $(f, \phi_n) = \mu_n(\phi, \phi_n)$ so that since $\phi \in L_2(a, b)$ and hence $\Sigma|(\phi, \phi_n)|^2$ converges, f must satisfy $\Sigma|(f, \phi_n)|^2/|\mu_n|^2 < \infty$, imposing a growth condition on the Fourier coefficients of f.

It is simple to check that if f satisfies the two conditions of the last paragraph, then (4.21) does indeed have a solution $\phi \in L_2(a, b)$. ☐

The Spectral Theorem may be used to deduce information about non-self-adjoint compact operators. Let $K \in B(H)$ be compact. The operator K^*K is compact and self-adjoint, so we may apply the Spectral Theorem to it. Notice first, however, that the eigenvalues of K^*K are non-negative since $K^*K\phi = \mu\phi \Rightarrow \mu(\phi, \phi) = (K^*K\phi, \phi) = (K\phi, K\phi) \geqslant 0$. Therefore in the spectral expansion $K^*K\phi = \Sigma\mu_n(\phi, \phi_n)\phi_n$, we may write $\mu_n = v_n^2$ where $v_n > 0$. (Recall that each μ_n is non-zero.) Then $\|K\phi_n\|^2 = (K^*K\phi_n, \phi_n) = v_n^2$ so $\|K\phi_n\| = v_n$, and if we let $\psi_n = v_n^{-1} K\phi_n$, (ψ_n) is orthonormal, since $\|\psi_n\| = 1$ and if $m \neq n$

$$(\psi_m, \psi_n) = (K\phi_m, K\phi_n)/(v_m v_n) = (K^*K\phi_m, \phi_n)/(v_m v_n)$$
$$= (v_m/v_n)(\phi_m, \phi_n) = 0.$$

By Corollary 4.16 we can express a typical vector $\phi \in H$ in the form $\phi = \phi_0 + \Sigma_{n=1}^{\infty}(\phi, \phi_n)\phi_n$ (the sum being, as always, finite in the case where

K^*K has only finitely many non-zero eigenvalues), where $K^*K\phi_0 = 0$. Now $K^*K\phi_0 = 0 \Rightarrow \|K\phi_0\|^2 = (K^*K\phi_0, \phi_0) = 0$ so $K\phi_0 = 0$ and

$$K\phi = \sum_{n=1}^{\infty} (\phi, \phi_n)K\phi_n = \sum_{n=1}^{\infty} v_n(\phi, \phi_n)\psi_n.$$

Also, if $K_N\phi = \sum_{n=1}^{N} v_n(\phi, \phi_n)\psi_n$ then, if K^*K has infinitely many eigenvalues,

$$\|(K - K_N)\phi\|^2 = \left\| \sum_{n=N+1}^{\infty} v_n(\phi, \phi_n)\psi_n \right\|^2 = \sum_{n=N+1}^{\infty} v_n^2 |(\phi, \phi_n)|^2$$
$$\leqslant v_{N+1}^2 \|\phi\|^2,$$

so that $\|K - K_N\| \to 0$ as $N \to \infty$.

We have proved the following.

Theorem 4.18

Let H be a Hilbert space and $K \in B(H)$ be compact. Then there are two orthonormal sequences (ϕ_n) and (ψ_n) and a sequence of positive numbers (v_n) such that for all $\phi \in H$, $K\phi = \sum_{n=1}^{\infty} v_n(\phi, \phi_n)\psi_n$. Moreover, if (v_n) is an infinite sequence and K_N is defined by $K_N\phi = \sum_{n=1}^{N} v_n(\phi, \phi_n)\psi_n$ then $\|K - K_N\| \to 0$ as $N \to \infty$. The numbers v_n^2 are the non-zero eigenvalues of K^*K, each eigenvalue being repeated the same number of times as the dimension of the corresponding set of eigenvectors (as in the Spectral Theorem). $\qquad\square$

The expansion given by Theorem 4.18 is just as convenient as the analogous one for self-adjoint operators when solving equations of the first kind, but its usefulness for equations of the second kind is very limited.

Example 4.7

Define $(K\phi)(x) = \int_0^x \phi(t)\,dt$ $(0 \leqslant x \leqslant 1)$; K is a compact operator on $L_2(0, 1)$. In this case K is not self-adjoint and, indeed, we know from our work on Volterra operators in §3.5 that $\sigma(K) = \{0\}$. Since K arises from the L_2-kernel $k(x, t)$ where

$$k(x, t) = \begin{cases} 1 & \text{if } 0 \leqslant t \leqslant x \leqslant 1, \\ 0 & \text{if } 0 \leqslant x < t \leqslant 1, \end{cases}$$

K^* is defined by $(K^*\phi)(x) = \int_0^1 \overline{k(t, x)}\phi(t)\,dt = \int_x^1 \phi(t)\,dt$.

Then

$$((K^*K)\phi)(x) = \int_x^1 ds \int_0^s \phi(t)\, dt = \int_0^1 dt \int_{\max(x,\,t)}^1 \phi(t)\, ds$$

$$= \int_0^1 (1 - \max(x, t))\phi(t)\, dt \quad (0 \leqslant x \leqslant 1).$$

Moreover, if $\phi \in L_2$ satisfies

$$\mu\phi(x) = (K^*K\phi)(x) = \int_x^1 ds \int_0^s \phi(t)\, dt \quad (0 \leqslant x \leqslant 1), \tag{4.22}$$

we see that the right hand side is continuous with respect to x (since $\phi \in L_2$), whence ϕ is continuous if $\mu \neq 0$. This in turn shows the right hand side to be differentiable and $\mu\phi'(x) = -\int_0^x \phi(t)\, dt$, whence $\mu\phi''(x) = -\phi(x)$ $(0 \leqslant x \leqslant 1)$. From (4.22) we see that $\phi(1) = 0$, while the equation for $\mu\phi'(x)$ shows $\phi'(0) = 0$. From this, solving the differential equation, we see that $\mu = \mu_n = 1/((n - \frac{1}{2})^2\pi^2)$ for some $n \in \mathbb{N}$ and we choose $\phi_n(x) = 2^{\frac{1}{2}} \cos((n - \frac{1}{2})\pi x)$, so that (μ_n) and (ϕ_n) are the sequences of eigenvalues and eigenvectors for K^*K. Now let $v_n = \mu_n^{\frac{1}{2}} = 1/((n - \frac{1}{2})\pi)$ and $\psi_n = v_n^{-1}K\phi_n$, so that

$$\psi_n(x) = 2^{\frac{1}{2}} \sin((n - \frac{1}{2})\pi x) \quad (0 \leqslant x \leqslant 1).$$

Theorem 4.18 shows us that $K\phi = \sum_{n=1}^\infty v_n(\phi, \phi_n)\psi_n$, in this notation.

Consider now the equation $K\phi = f$, that is, the equation

$$\int_0^x \phi(t)\, dt = f(x) \quad \text{(for almost all } x \text{ in } [0, 1]).$$

This will have a solution if and only if (i) $(f, \psi) = 0$ for all ψ satisfying $(\psi, \psi_n) = 0$ for all n and (ii) $\sum |v_n^{-1}(f, \psi_n)|^2 < \infty$. The first condition is trivially satisfied by all $f \in L_2(0, 1)$, since

$$(\psi, \psi_n) = 0 \text{ for all } n \Rightarrow K^*\psi = 0 \Rightarrow \psi = 0 \text{ (almost everywhere)}.$$

Thus a square-integrable function f on $[0, 1]$ is equal almost everywhere to the indefinite integral of a square-integrable function if and only if $\sum |(n - \frac{1}{2})c_n|^2 < \infty$ where

$$c_n = \int_0^1 f(t) \sin((n - \frac{1}{2})\pi t)\, dt \quad (n \in \mathbb{N}).$$

Since

$$f(x) = \sum_{n=1}^\infty 2c_n \sin((n - \frac{1}{2})\pi x) \quad \text{(almost everywhere)},$$

this condition requires that the series obtained by differentiating term-by-term, $\Sigma_{n=1}^{\infty} 2c_n(n-\tfrac{1}{2})\pi \cos((n-\tfrac{1}{2})\pi x)$, converges in $L_2(0, 1)$ to some function (by the Riesz–Fischer Theorem). □

4.5 Applications to integral equations

At this stage we should take stock of what we have achieved. The various results above, mostly expressed in terms of pure operator theory, do not exhaust the possibilities in this direction, but the conclusions we draw are also expressed in functional analytic terms. For example, the various series expansions for a function, say ϕ, in $L_2(a, b)$ are known to converge in the Hilbert space $L_2(a, b)$, that is, they converge to ϕ in the mean. This is not the sort of convergence we would naturally seek were we to start from an integral equation and not express the equation in Hilbert space terms. In this section we shall devote some attention to results in which we can improve the convergence of the series arising to obtain stronger forms of convergence by using properties of the kernel which are not readily expressed in Hilbert space terms, together with some related results concerning the various series expansions we have encountered.

Theorem 4.19

Let k be a hermitian L_2-kernel on $[a, b] \times [a, b]$ and suppose that the sequences of eigenvalues and eigenvectors of the corresponding integral operator K on $L_2(a, b)$ appearing in the Spectral Theorem are (μ_n) and (ϕ_n). Then $\Sigma \mu_n^2$ converges and

$$\sum_{n=1}^{\infty} \mu_n^2 = \int_a^b \int_a^b |k(x, t)|^2 \, \mathrm{d}x \, \mathrm{d}t.$$

Moreover,

$$k(x, t) = \sum_{n=1}^{\infty} \mu_n \phi_n(x)\overline{\phi_n(t)}$$

in the sense that

$$\int_a^b \int_a^b |k(x, t) - \sum_{n=1}^{N} \mu_n \phi_n(x)\overline{\phi_n(t)}|^2 \, \mathrm{d}x \, \mathrm{d}t \to 0 \text{ as } N \to \infty.$$

As usual, if K has only finitely many eigenvalues the sums involved are finite sums.

Proof

The sequence (ϕ_n) is orthonormal and by Corollary 4.17 to the Spectral Theorem, we can extend it to a complete orthonormal set in $L_2(a, b)$. Call the resulting set $\{\phi_n\}\cup\{\chi_n\}$ where the set of vectors $\{\chi_n\}$ is at most countable (by Theorem A10) and may be finite or even empty. We know that $K\phi_n=\mu_n\phi_n$ and $K\chi_n=0$ for each relevant n.

Suppose that x is chosen so that $\int_a^b|k(x, t)|^2\,dt<\infty$. Then the function f_x given by $f_x(t)=k(x, t)$ is in $L_2(a, b)$, and if we set

$$a_n(x)=\int_a^b k(x, t)\phi_n(t)\,dt$$

and

$$b_n(x)=\int_a^b k(x, t)\chi_n(t)\,dt \quad (a\leqslant x\leqslant b)$$

these give the Fourier coefficients of f_x with respect to the complete orthonormal set $\{\bar{\phi}_n\}\cup\{\bar{\chi}_n\}$ in $L_2(a, b)$. By Parseval's Theorem, therefore,

$$\int_a^b|k(x, t)|^2\,dt=\sum|a_n(x)|^2+\sum|b_n(x)|^2.$$

For almost all x in $[a, b]$, $\int_a^b|k(x, t)|^2\,dt<\infty$, $a_n(x)=(K\phi_n)(x)=\mu_n\phi_n(x)$ and $b_n(x)=(K\chi_n)(x)=0$, whence $\int_a^b|k(x, t)|^2\,dt=\sum_{n=1}^\infty|\mu_n\phi_n(x)|^2$. Integrating from a to b yields

$$\int_a^b\int_a^b|k(x, t)|^2\,dt\,dx=\sum_{n=1}^\infty\mu_n^2\int_a^b|\phi_n(x)|^2\,dx$$

$$=\sum_{n=1}^\infty\mu_n^2, \tag{4.23}$$

the interchange of summation and integration being justified by the Monotone Convergence Theorem.

Noticing that, for all n,

$$\int_a^b\int_a^b k(x, t)\overline{\phi_n(x)}\phi_n(t)\,dt\,dx=\mu_n$$

and expanding the modulus term, we see that

$$\int_a^b\int_a^b|k(x, t)-\sum_{n=1}^N\mu_n\phi_n(x)\overline{\phi_n(t)}|^2\,dx\,dt=\int_a^b\int_a^b|k(x, t)|^2\,dx\,dt-\sum_{n=1}^N\mu_n^2$$

and (4.23) now finishes the proof. $\qquad\square$

Remarks

This result has the merit that the expression given for $\Sigma\mu_n^2$ is easily calculated. It shows, moreover, that the estimate $\{\int_a^b\int_a^b|k(x,t)|^2\,dx\,dt\}^{\frac{1}{2}}$ for $\|K\|$ is an overestimate except in the case where K has at most one non-zero eigenvalue with one corresponding eigenvector, since $\|K\|=\sup\{|\mu_n|:n\in\mathbb{N}\}$. Notice also that since compact operators can be found whose sequence of eigenvalues is any prescribed sequence which tends to zero (Problem 4.13), the L_2-kernels generate a proper subset of the set of all compact operators on $L_2(a,b)$.

The expression of $k(x,t)$ in the form $\Sigma_{n=1}^\infty\mu_n\phi_n(x)\overline{\phi_n(t)}$ can be viewed as a fairly obvious generalisation of a degenerate kernel. Notice, however, that we have only shown that the series converges in the mean; this does not of itself guarantee pointwise convergence.

Theorem 4.19 has an extension to L_2-kernels which are not necessarily hermitian, for which we need to consider the spectral expansion obtained in Theorem 4.18.

Corollary 4.20

Let k be an L_2-kernel on $[a,b]\times[a,b]$, and let K be the corresponding integral operator on $L_2(a,b)$. Suppose that the spectral expansion of K^*K is $K^*K\phi=\Sigma_{n=1}^\infty\mu_n(\phi,\phi_n)\phi_n$ where (ϕ_n) is orthonormal. Then, letting $v_n=\mu_n^{\frac{1}{2}}$ and $\psi_n=v_n^{-1}K\phi_n$, so that (ψ_n) is orthonormal,

$$k(x,t)=\sum_{n=1}^\infty v_n\psi_n(x)\overline{\phi_n(t)}$$

in the sense that

$$\int_a^b\int_a^b|k(x,t)-\sum_{n=1}^N v_n\psi_n(x)\overline{\phi_n(t)}|^2\,dx\,dt\to 0 \text{ as } N\to\infty$$

and

$$\sum_{n=1}^\infty v_n^2=\int_a^b\int_a^b|k(x,t)|^2\,dx\,dt.$$

Proof

The bulk of the work was done in the proof of Theorem 4.18. Since K^*K is self-adjoint, by Corollary 4.17 to the Spectral Theorem there is an orthonormal sequence (possibly finite) (χ_n) such that $K\chi_n=0$ for all n and $\{\phi_n\}\cup\{\chi_n\}$ is a complete orthonormal sequence. The Fourier coefficients of the function $t\mapsto k(x,t)$ with respect to the complete orthonormal sequence $\{\overline{\phi_n}\}\cup\{\overline{\chi_n}\}$

are $\int_a^b k(x, t)\phi_n(t)\, dt = (K\phi_n)(x) = v_n\psi_n(x)$ and $\int_a^b k(x, t)\chi_n(t)\, dt = \theta_n(x)$, say. By Parseval's Theorem, A8, then, if $\int_a^b |k(x, t)|^2\, dt < \infty$,

$$\sum_{n=1}^{\infty} |v_n\psi_n(x)|^2 + \sum_n |\theta_n(x)|^2 = \int_a^b |k(x, t)|^2\, dt. \tag{4.24}$$

Since the vectors χ_n were chosen so that $K\chi_n = 0$ as members of the Hilbert space $L_2(a, b)$, it follows that $\int_a^b k(x,t)\chi_n(t)\, dt = 0$ for almost all $x \in [a, b]$ and therefore $\theta_n(x) = 0$ for almost all $x \in [a, b]$. This is true of all $n \in \mathbb{N}$, so the set of all $x \in [a, b]$ for which at least one natural number n has $\theta_n(x) \neq 0$ also has measure zero. Therefore, for all $x \in [a, b]$ other than some set of measure zero, (4.24) holds and $\theta_n(x) = 0\ (n \in \mathbb{N})$ whence

$$\sum_{n=1}^{\infty} v_n^2 |\psi_n(x)|^2 = \int_a^b |k(x, t)|^2\, dt.$$

Then, integrating both sides with respect to x gives

$$\int_a^b \int_a^b |k(x, t)|^2\, dt\, dx = \int_a^b \sum_{n=1}^{\infty} v_n^2 |\psi_n(x)|^2\, dx = \sum_{n=1}^{\infty} v_n^2 \int_a^b |\psi(x)|^2\, dx$$

$$= \sum_{n=1}^{\infty} v_n^2,$$

the interchange of sum and integral being justified by the Monotone Convergence Theorem. Finally

$$\int_a^b \int_a^b |k(x, t) - \sum_{n=1}^{N} v_n\psi_n(x)\overline{\phi_n(t)}|^2\, dx\, dt = \int_a^b \int_a^b |k(x, t)|^2\, dt\, dx - \sum_{n=1}^{N} v_n^2$$

(since $\int_a^b \int_a^b k(x, t)\overline{v_n\psi_n(x)}\phi_n(t)\, dt\, dx = \int_a^b |v_n\psi_n(x)|^2\, dx = v_n^2$ with similar results from the other terms obtained by expanding the integral), which yields the desired conclusion. ☐

It normally turns out when dealing with integral equations that the kernels have rather better analytical properties than merely being L_2-kernels. One particularly important group is the set of *Schmidt kernels*. We shall define k to be a Schmidt kernel on $[a, b] \times [a, b]$ if it is an L_2-kernel with the additional property that there is a constant C such that

$$\text{for all } x \in [a, b] \int_a^b |k(x, t)|^2\, dt \leqslant C.$$

This property clearly holds if k is a bounded function (and, in particular, if k is continuous), and it remains true for certain unbounded kernels k, such as

$k(x, t) = |x - t|^{-\alpha} (0 < \alpha < \frac{1}{2})$. Notice that this condition is not symmetric in the two variables, so it is possible for k to be a Schmidt kernel while the kernel generating the adjoint operator is not.

The main use of the Schmidt condition is in the two results below.

Lemma 4.21

Let k be a Schmidt kernel on $[a, b] \times [a, b]$ and K be the integral operator in $L_2(a, b)$ which it generates. Then if (ϕ_n) is a sequence in $L_2(a, b)$ such that $\phi_n \to \phi$ in the L_2 norm, $K\phi_n \to K\phi$ uniformly as $n \to \infty$.

Proof

Suppose that for all $x \in [a, b]$ $\int_a^b |k(x, t)|^2 \, dt \leqslant C$. Then, for all $x \in [a, b]$,

$$|(K\phi_n)(x) - (K\phi)(x)| = |\int_a^b k(x, t)(\phi_n(t) - \phi(t)) \, dt|$$

$$\leqslant \left\{ \int_a^b |k(x, t)|^2 \, dt \right\}^{\frac{1}{2}} \left\{ \int_a^b |\phi_n(t) - \phi(t)|^2 \, dt \right\}^{\frac{1}{2}}$$

$$\leqslant C^{\frac{1}{2}} \|\phi_n - \phi\|. \qquad \square$$

Remark

We have actually proved that $\int_a^b k(x, t)\phi_n(t) \, dt \to \int_a^b k(x, t)\phi(t) \, dt$ uniformly, and have tacitly assumed that these functions are $(K\phi_n)(x)$ and $(K\phi)(x)$ respectively. Since $K\phi_n$ and $K\phi$ are members of $L_2(a, b)$, this is, strictly speaking, only known to be true for almost all x in $[a, b]$, unless we adopt the practice that $K\phi$ is the function given by the integral expression. Notice that even if k is an L_2-kernel, not known to be a Schmidt kernel, this shows that $K\phi \to K\phi_n$ pointwise on the set $\{x : \int_a^b |k(x, t)|^2 \, dt < \infty\}$.

Theorem 4.22 (Schmidt's Theorem)

Suppose that k is a hermitian Schmidt kernel on $[a, b] \times [a, b]$ and that the spectral expansion of the corresponding operator is $K\phi = \Sigma \mu_n(\phi, \phi_n)\phi_n$. Then the series $\Sigma \mu_n(\phi, \phi_n)\phi_n(x)$ converges both absolutely and uniformly in the sense that $\Sigma |\mu_n(\phi, \phi_n)\phi_n(x)|$ is uniformly convergent.

Proof

For all x, $t \mapsto k(x, t)$ is a L_2 function, so since $(\bar{\phi}_n)$ is an orthonormal sequence, and the Fourier coefficients of $t \mapsto k(x, t)$ with respect to this sequence are $(K\phi_n)(x) = \int_a^b k(x, t)\phi_n(t) \, dt$, we see that

$$\sum_{n=1}^{\infty} |\mu_n \phi_n(x)|^2 = \sum_{n=1}^{\infty} |(K\phi_n)(x)|^2 \leqslant \int_a^b |k(x, t)|^2 \, dt \leqslant C.$$

Let $\epsilon > 0$ and choose N so that $\Sigma_{n=N+1}^{\infty} |(\phi, \phi_n)|^2 < \epsilon^2$.
Then, for all $x \in [a, b]$

$$\sum_{n=N+1}^{\infty} |\mu_n(\phi, \phi_n)\phi_n(x)| \leqslant \left\{ \sum_{n=N+1}^{\infty} |(\phi, \phi_n)|^2 \right\}^{\frac{1}{2}} \left\{ \sum_{n=N+1}^{\infty} |\mu_n\phi_n(x)|^2 \right\}^{\frac{1}{2}}$$
$$< C^{\frac{1}{2}}\epsilon.$$

From this the result is immediate. ☐

Remark

Again, we have presumed that ϕ_n is chosen so that $\mu_n\phi_n(x) = \int_a^b k(x, t)\phi_n(t)\, dt$ for all x (rather than almost all x). The uniform convergence of $\Sigma\mu_n(\phi, \phi_n)$ $\phi_n(x)$ could be deduced from Lemma 4.21 immediately, but we shall occasionally need the stronger result above to allow the uniform convergence of related series to be deduced by comparison. A minor modification of the proof shows the expansion of Theorem 4.18 (in terms of two different orthonormal sequences) to be uniformly convergent in the same sense.

Example 4.8

Suppose that k is a hermitian Schmidt kernel on $[a, b] \times [a, b]$, that f is square-integrable and that $1/\lambda$ is not an eigenvalue of the integral operator K arising from k. Then we already know that the integral equation

$$\phi(x) - \lambda \int_a^b k(x, t)\phi(t)\, dt = f(x) \quad (a \leqslant x \leqslant b)$$

has a unique solution $\phi \in L_2(a, b)$. By taking inner products on both sides with ϕ_n, where the spectral expansion of K is $K\phi = \Sigma_{n=1}^{\infty} \mu_n(\phi, \phi_n)\phi_n$, we see that $(\phi, \phi_n) = (f, \phi_n)/(1 - \lambda\mu_n)$ and thus $\phi = f + \lambda K\phi = f + \Sigma_{n=1}^{\infty} \lambda\mu_n(1 - \lambda\mu_n)^{-1}(f, \phi_n)\phi_n$. So far, the series is only known to converge as a series in the Hilbert space $L_2(a, b)$. However, since $\mu_n \to 0$ as $n \to \infty$, $1/(1 - \lambda\mu_n)$ is a bounded sequence so, for some constant, M, we have for all $n \in \mathbb{N}$ $|\lambda\mu_n(1 - \lambda\mu_n)^{-1}| \leqslant M|\mu_n|$ and since $\Sigma|\mu_n(f, \phi_n)\phi_n(x)|$ converges uniformly, so does the series $\Sigma|\lambda\mu_n(1 - \lambda\mu_n)^{-1}(f, \phi_n)\phi_n(x)|$. Therefore we have

$$\phi(x) = f(x) + \sum_{n=1}^{\infty} \frac{\lambda\mu_n}{1 - \lambda\mu_n}(f, \phi_n)\phi_n(x)$$

where the series is (absolutely and) uniformly convergent. ☐

In the expansion of the form $\Sigma\mu_n(\phi, \phi_n)\phi_n(x)$ we know that, provided K is generated by a Schmidt kernel, the series is uniformly convergent in x. Writing $(\phi, \phi_n) = \int_a^b \phi(t)\overline{\phi_n(t)}\, dt$ allows us to express this series as

$\Sigma \int_a^b \mu_n \phi_n(x) \overline{\phi_n(t)} \phi(t) \, dt$ and it is tempting to seek conditions under which the sum and the integral can be interchanged. This we shall do, but we require some preparation.

The lemma below (Dini's Theorem) is a piece of pure analysis which is often omitted from standard treatments of uniform convergence. For this reason we give it here in detail, even though it may appear out of place.

Lemma 4.23 (Dini's Theorem)

Suppose that for each $n \in \mathbb{N}$ $f_n: [a, b] \to \mathbb{R}$ is a continuous function and that for each $x \in [a, b]$ $f_n(x)$ increases to $f(x)$ as $n \to \infty$. Then if f is continuous, $f_n \to f$ uniformly as $n \to \infty$.

Proof

Let $\epsilon > 0$. For each $x \in [a, b]$, $f_n(x)$ increases to $f(x)$ so there exists N_x such that for all $n \geq N_x$ $f(x) \geq f_{N_x}(x) > f(x) - \epsilon$. Now $f_{N_x} - f$ is continuous and its value at x is greater than $-\epsilon$, so

$$\text{there exists } \delta_x > 0 \text{ s.t. for all } y \in (x - \delta_x, x + \delta_x) \cap [a, b]$$
$$f_{N_x}(y) - f(y) > -\epsilon. \tag{4.25}$$

Since $[a, b]$ is contained in the union of the open intervals $(x - \delta_x, x + \delta_x)$ the compactness of $[a, b]$ now guarantees the existence of finitely many points $x_1, \ldots, x_k$ for which $[a, b] \subseteq (x_1 - \delta_{x_1}, x_1 + \delta_{x_1}) \cup \cdots \cup (x_k - \delta_{x_k}, x_k + \delta_{x_k})$. Let $N = \max(N_{x_1}, \ldots, N_{x_k})$. Then for all $n \geq N$ and for all $y \in [a, b]$, there is an $i \in \{1, 2, \ldots, k\}$ with $y \in (x_i - \delta_{x_i}, x_i + \delta_{x_i})$ so

$$f(y) \geq f_n(y) \geq f_{N_{x_i}}(y) > f(y) - \epsilon \quad \text{(by 4.25)}$$

whence

$$|f(y) - f_n(y)| < \epsilon. \qquad \square$$

For the next theorem we restrict our attention to kernels which generate *non-negative operators*, that is, self-adjoint operators K with the property that for all ϕ, $(K\phi, \phi) \geq 0$. An equivalent condition is that K be self-adjoint and have all its eigenvalues non-negative.

Theorem 4.24 (Mercer's Theorem)

Suppose that $k(x, t)$ is a continuous, hermitian kernel on $[a, b] \times [a, b]$ and that the integral operator K on $L_2(a, b)$ which it generates is non-negative. Then if the standard spectral expansion of K is $K\phi = \Sigma_{n=1}^{\infty} \mu_n(\phi, \phi_n)\phi_n$, we have

$$k(x, t) = \sum_{n=1}^{\infty} \mu_n \phi_n(x) \overline{\phi_n(t)},$$

the series converging absolutely and uniformly in the sense that

$$\sum |\mu_n \phi_n(x) \phi_n(t)|$$

converges uniformly with respect to both variables simultaneously.

Proof

Notice first that, for all $x \in [a, b]$, $k(x, x) \geq 0$. For if not, there is an $x_0 \in [a, b]$ with $k(x_0, x_0) < 0$ and, by the continuity of k, there exists $\delta > 0$ such that $|x - x_0| < \delta$ and $|t - t_0| < \delta$ implies $k(x, t) < 0$. Then let $\chi(t) = 1$ if $|t - t_0| < \delta$ and $t \in [a, b]$ and $\chi(t) = 0$ otherwise, so $\chi \in L_2(a, b)$, and $(K\chi, \chi) = \int_a^b \int_a^b k(x, t) \chi(t) \chi(x) \, dt \, dx < 0$ which contradicts the non-negativity of K.

Since $\mu_n \phi_n(x) = \int_a^b k(x, t) \phi_n(t) \, dt$ $(x \in [a, b])$ and k is continuous, ϕ_n is continuous if $\mu_n \neq 0$; $\mu_n \phi_n$ is at any rate continuous.

Also, for each $N \in \mathbb{N}$, $k(x, t) - \sum_{n=1}^{N} \mu_n \phi_n(x) \overline{\phi_n(t)}$ is a continuous kernel which generates a non-negative operator. It is easily seen that this kernel is hermitian, while if we call the corresponding operator K_N

$$(K_N \phi, \phi) = (K\phi, \phi) - \sum_{n=1}^{N} \mu_n |(\phi, \phi_n)|^2 = \sum_{n=N+1}^{\infty} \mu_n |(\phi, \phi_n)|^2 \geq 0,$$

since $\mu_n = (K\phi_n, \phi_n) \geq 0$ for each n. By the argument of the first paragraph, then, $k(x, x) - \sum_{n=1}^{N} \mu_n |\phi_n(x)|^2 \geq 0$ for all $x \in [a, b]$. Therefore, for all N and x, $\sum_{n=1}^{N} \mu_n |\phi_n(x)|^2 \leq k(x, x)$ and we see that $\sum \mu_n |\phi_n(x)|^2$ converges. Also if C is an upper bound for $\{k(x, x) : x \in [a, b]\}$, we have

$$\text{for all } x \in [a, b] \ \sum_{n=1}^{\infty} \mu_n |\phi_n(x)|^2 \leq C. \tag{4.26}$$

Choose $x \in [a, b]$ and keep it fixed for the time being. Let $\epsilon > 0$ and choose N so that $\sum_{n=N+1}^{\infty} \mu_n |\phi_n(x)|^2 < \epsilon^2$. Then, for all $t \in [a, b]$,

$$\sum_{n=N+1}^{\infty} |\mu_n \phi_n(x) \overline{\phi_n(t)}| \leq \left\{ \sum_{n=N+1}^{\infty} \mu_n |\phi_n(t)|^2 \right\}^{\frac{1}{2}} \left\{ \sum_{n=N+1}^{\infty} \mu_n |\phi_n(x)|^2 \right\}^{\frac{1}{2}} \tag{4.27}$$
$$< C^{\frac{1}{2}} \epsilon,$$

the first inequality from Cauchy's inequality, splitting $|\mu_n \phi_n(x) \overline{\phi_n(t)}|$ into $\mu_n^{\frac{1}{2}} |\phi_n(t)| \mu_n^{\frac{1}{2}} |\phi_n(x)|$. It follows that, for each fixed $x \in [a, b]$, the series $\sum |\mu_n \phi_n(x) \overline{\phi_n(t)}|$ converges uniformly with respect to t.

Let $k_0(x, t) = \sum_{n=1}^{\infty} \mu_n \phi_n(x) \overline{\phi_n(t)}$. By the uniform convergence we see that for each fixed x, $k_0(x, t)$ is continuous in t. Then, for fixed $x \in [a, b]$ and all $\phi \in L_2(a, b)$

$$\int_a^b (k_0(x, t) - k(x, t))\phi(t)\, dt = \int_a^b \sum_{n=1}^{\infty} \mu_n \phi_n(x)\overline{\phi_n(t)}\phi(t)\, dt - (K\phi)(x)$$

$$= \sum_{n=1}^{\infty} \mu_n \phi_n(x) \int_a^b \overline{\phi_n(t)}\phi(t)\, dt - \sum_{n=1}^{\infty} \mu_n(\phi, \phi_n)\phi_n(x)$$

$$= 0,$$

where the reversal of summation and integration in the first summand is justified by the uniform convergence of $\Sigma \mu_n \phi_n(x)\overline{\phi_n(t)}$ with respect to t. (For then, if N is sufficiently large, we have $\Sigma_{n=N+1}^{\infty}|\mu_n \phi_n(x)\overline{\phi_n(t)}| < \epsilon$ for all t, whence $|\int_a^b \Sigma_{n=N+1}^{\infty} \mu_n \phi_n(x)\overline{\phi_n(t)}\phi(t)\, dt| \leqslant \int_a^b \epsilon|\phi(t)|\, dt \leqslant \epsilon (b-a)^{\frac{1}{2}}\|\phi\|$.) Therefore, setting $g(t) = k_0(x, t) - k(x, t)$ we see that for all $\phi \in L_2(a, b) \int_a^b g(t)\phi(t)\, dt = 0$, whence $g(t) = 0$ for almost all $t \in [a, b]$. But g is continuous with respect to t, so $g(t) = 0$ for all $t \in [a, b]$. We now notice that the whole of this paragraph is true for all $x \in [a, b]$ so that, for all $x, t \in [a, b]$,

$$k(x, t) = k_0(x, t) = \sum_{n=1}^{\infty} \mu_n \phi_n(x)\overline{\phi_n(t)}.$$

In particular, then, $k(x, x) = \Sigma_{n=1}^{\infty} \mu_n |\phi_n(x)|^2$, for all $x \in [a, b]$. Since $|\phi_n|$ is continuous for each n and k is continuous, Dini's Theorem applied to the functions $f_n(x) = \Sigma_{j=1}^n \mu_j |\phi_j(x)|^2$ shows that $\Sigma \mu_n |\phi_n(x)|^2$ converges uniformly on $[a, b]$. Therefore if $\epsilon > 0$, we may choose N such that for all $x \in [a, b] \Sigma_{n=N+1}^{\infty} \mu_n |\phi_n(x)|^2 < \epsilon^2$. Then by (4.26) and (4.27), for all $x \in [a, b]$

$$\sum_{n=N+1}^{\infty} |\mu_n \phi_n(x)\phi_n(t)| \leqslant \left\{ \sum_{n=N+1}^{\infty} \mu_n |\phi_n(t)|^2 \right\}^{\frac{1}{2}} \left\{ \sum_{n=N+1}^{\infty} \mu_n |\phi_n(x)|^2 \right\}^{\frac{1}{2}}$$

$$< C^{\frac{1}{2}}\epsilon$$

giving the required uniform convergence.   □

We can, in fact, extend Mercer's Theorem slightly. Apart from the case where all the eigenvalues of K are non-positive (so that $-k$ satisfies the hypotheses), we may allow finitely many of the eigenvalues to have negative sign, all the rest being non-negative.

Corollary 4.25

Suppose that k is a continuous hermitian kernel on $[a, b] \times [a, b]$ all but finitely many of whose eigenvalues are non-negative. Then if (μ_n) and (ϕ_n)

are the usual sequences of eigenvalues and eigenvectors of the corresponding integral operator,

$$k(x, t) = \sum_{n=1}^{\infty} \mu_n \phi_n(x) \overline{\phi_n(t)},$$

the series $\Sigma |\mu_n \phi_n(x) \overline{\phi_n(t)}|$ being uniformly convergent with respect to both variables jointly.

Proof

Choose N so that for all $n \geq N$ $\mu_n \geq 0$. It is easily checked that if we set $k_N(x, t) = k(x, t) - \Sigma_{n=1}^{N-1} \mu_n \phi_n(x) \overline{\phi_n(t)}$, then the eigenvalues of the corresponding integral operator K_N are $\mu_N, \mu_{N+1}, \ldots$. Since these are all non-negative, for all $\phi \in L_2(a, b)$ $(K_N \phi, \phi) = \Sigma_{n=N}^{\infty} \mu_n |(\phi, \phi_n)|^2 \geq 0$ so K_N is non-negative. Since k_N is continuous and hermitian we may apply Mercer's Theorem to it, and $\Sigma_{n=N}^{\infty} |\mu_n \phi_n(x) \overline{\phi_n(t)}|$ converges uniformly in both variables. Adding the remaining finitely many terms does not affect the uniformity of the convergence. $\square$

Corollary 4.26

If k is a continuous, hermitian kernel on $[a, b] \times [a, b]$ all but finitely many of whose eigenvalues are non-negative, and if (μ_n) is the usual sequence of eigenvalues associated with the corresponding integral operator, then

$$\sum_{n=1}^{\infty} \mu_n = \int_a^b k(x, x)\, dx.$$

Proof

Since $k(x, x) = \Sigma_{n=1}^{\infty} \mu_n |\phi_n(x)|^2$ (in the usual notation) and the series is uniformly convergent

$$\int_a^b k(x, x)\, dx = \sum_{n=1}^{\infty} \mu_n \int_a^b |\phi_n(x)|^2\, dx = \sum_{n=1}^{\infty} \mu_n. \qquad \square$$

Example 4.9

Let us return to the problem of solving

$$\phi(x) - \lambda \int_a^b k(x, t)\phi(t)\, dt = f(x) \quad (a \leq x \leq b)$$

where this time k is a continuous, hermitian kernel and all but finitely many eigenvalues of the associated operator K are non-negative. Letting (μ_n) and

(ϕ_n) denote the usual sequences of eigenvalues and eigenvectors of K we see that, provided $1 - \lambda\mu_n \neq 0$ for all n, $(\phi, \phi_n) = (f, \phi_n)/(1 - \lambda\mu_n)$ and that

$$\phi = f + \sum_{n=1}^{\infty} \frac{\lambda\mu_n}{1 - \lambda\mu_n}(f, \phi_n)\phi_n. \tag{4.14}$$

This we have already deduced. However, since $\Sigma |\mu_n\phi_n(x)\overline{\phi_n(t)}|$ is uniformly convergent with respect to both variables by Mercer's Theorem, we may interchange the order of integration and summation, giving

$$\sum_{n=1}^{\infty} \frac{\lambda\mu_n}{1 - \lambda\mu_n} \int_a^b f(t)\overline{\phi_n(t)}\phi_n(x)\, dt = \int_a^b \sum_{n=1}^{\infty} \frac{\lambda\mu_n\phi_n(x)\overline{\phi_n(t)}}{1 - \lambda\mu_n} f(t)\, dt.$$

Letting $r_\lambda(x, t) = \Sigma_{n=1}^{\infty} \lambda\mu_n(1 - \lambda\mu_n)^{-1} \phi_n(x)\overline{\phi_n(t)}$, we see that, by the uniform convergence, r_λ is continuous and

$$\phi(x) = f(x) + \int_a^b r_\lambda(x, t)f(t)\, dt \quad (a \leqslant x \leqslant b).$$

If $\lambda\mu_n = 1$ for some n (possibly several), the solution of the integral equation will exist provided f is orthogonal to the corresponding eigenvectors ϕ_n; this solution is not unique. In this case the solutions are given by (4.16) and Mercer's Theorem again shows the required uniform convergence to allow summation and integration to be interchanged. $\square$

We can say a little more than the result of Example 4.9. Suppose that $k(x, t)$ is a continuous, hermitian kernel on $[a, b] \times [a, b]$. Then if K is the associated integral operator on $L_2(a, b)$, K^2 is a non-negative operator whose kernel is continuous. (The kernel of K^2 has value at (x, t) of $\int_a^b k(x, s)k(s, t)\, ds$.) If the eigenvalues and eigenvectors of K are (μ_n) and (ϕ_n), with the usual conventions, then the spectral expansion of K^2 is easily checked to be $K^2\phi = \Sigma_{n=1}^{\infty} \mu_n^2(\phi, \phi_n)\phi_n$. By Mercer's Theorem, then, the series $\Sigma |\mu_n^2\phi_n(x)\phi_n(t)|$ is uniformly convergent with respect to both variables jointly.

Now consider the equation

$$\phi(x) - \lambda \int_a^b k(x, t)\phi(t)\, dt = f(x) \quad (a \leqslant x \leqslant b),$$

with k as above, in the case where $1/\lambda$ is not an eigenvalue of K. Then by (4.14)

$$\phi = f + \sum_{n=1}^{\infty} \frac{\lambda \mu_n}{1 - \lambda \mu_n} (f, \phi_n)\phi_n$$

$$= f + \sum_{n=1}^{\infty} \lambda \mu_n (f, \phi_n)\phi_n + \sum_{n=1}^{\infty} \frac{(\lambda \mu_n)^2}{1 - \lambda \mu_n} (f, \phi_n)\phi_n, \qquad (4.28)$$

all of the series converging in the Hilbert space $L_2(a, b)$. The middle term on the right of (4.28) is the spectral expansion for $\lambda K f$, while the third term may be written as

$$\sum_{n=1}^{\infty} (\lambda \mu_n)^2 (1 - \lambda \mu_n)^{-1} \int_a^b f(t)\overline{\phi_n(t)} \, dt \phi_n(x).$$

Noticing that $\sum |(\lambda \mu_n)^2 (1 - \lambda \mu_n)^{-1} \phi_n(x)\phi_n(t)|$ converges uniformly with respect to both variables by comparison with $\sum |\mu_n^2 \phi_n(x)\phi_n(t)|$, and letting

$$s(x, t) = \sum_{n=1}^{\infty} \frac{(\lambda \mu_n)^2}{1 - \lambda \mu_n} \phi_n(x)\overline{\phi_n(t)} \quad (a \leqslant x, t \leqslant b),$$

we see that s is continuous (since each ϕ_n is). Moreover, the third term in (4.28) is $\int_a^b s(x, t)f(t) \, dt$ whence the solution ϕ is given by

$$\phi(x) = f(x) + \int_a^b r_\lambda(x, t)\phi(t) \, dt \quad (a \leqslant x \leqslant b)$$

where

$$r_\lambda(x, t) = \lambda k(x, t) + s(x, t) \quad (a \leqslant x, t \leqslant b).$$

Notice that r_λ is continuous.

At this point we should collect our latest results into a formal theorem.

Theorem 4.27

Suppose that k is a continuous hermitian kernel on $[a, b] \times [a, b]$ and that $1/\lambda$ is not an eigenvalue of the associated integral operator K. Then the unique solution of

$$\phi(x) - \lambda \int_a^b k(x, t)\phi(t) \, dt = f(x) \quad (a \leqslant x \leqslant b)$$

is given by

$$\phi(x) = f(x) + \int_a^b r_\lambda(x, t)f(t) \, dt \quad (a \leqslant x \leqslant b), \qquad (4.29)$$

where

$$r_\lambda(x,t) = \lambda k(x,t) + \sum_{n=1}^{\infty} \frac{(\lambda\mu_n)^2}{1-\lambda\mu_n} \phi_n(x)\overline{\phi_n(t)}$$

is continuous.   □

The corresponding result to Theorem 4.27 in the case where K is a hermitian L_2-kernel is that ϕ is given by (4.29) where r_λ is an L_2-kernel. This is much easier to prove, since the series $\Sigma\lambda\mu_n(1-\lambda\mu_n)^{-1}\phi_n(x)\overline{\phi_n(t)}$ converges in the mean on $[a,b]\times[a,b]$. Let $r_n(x,t)=\Sigma_{j=1}^n \lambda\mu_j(1-\lambda\mu_j)^{-1}\phi_j(x)\overline{\phi_j(t)}$, so that

$$\int_a^b \int_a^b |r_m(x,t)-r_n(x,t)|^2 \,\mathrm{d}x\,\mathrm{d}t = \sum_{j=n+1}^m |\lambda\mu_j(1-\lambda\mu_j)^{-1}|^2.$$

Since $\Sigma|\mu_j|^2$ converges and $(1-\lambda\mu_j)$ is bounded, we see that in the Hilbert space $L_2([a,b]\times[a,b])$, with norm given by

$$\|\psi\| = \left\{\int_a^b \int_a^b |\psi(x,t)|^2 \,\mathrm{d}x\,\mathrm{d}t\right\}^{\frac{1}{2}}$$

the sequence (r_n) is a Cauchy sequence. Therefore there is a function r such that $r_n \to r$ in this Hilbert space, i.e. $\int_a^b\int_a^b |r(x,t)-r_n(x,t)|^2 \,\mathrm{d}x\,\mathrm{d}t \to 0$ as $n\to\infty$. But this is exactly the condition that $\int_a^b r_n(x,t)\phi(t)\,\mathrm{d}t$ should tend to $\int_a^b r(x,t)\phi(t)\,\mathrm{d}t$ in the norm on $L_2(a,b)$ as $n\to\infty$, so

$$\int_a^b r(x,t)\phi(t)\,\mathrm{d}t = \lim_{n\to\infty} \int_a^b \sum_{j=1}^n \lambda\mu_j(1-\lambda\mu_j)^{-1}\phi_j(x)\overline{\phi_j(t)}f(t)\,\mathrm{d}t$$

$$= \lim_{n\to\infty} \sum_{j=1}^n \lambda\mu_j(1-\lambda\mu_j)^{-1}(f,\phi_j)\phi_j(x)$$

$$= \sum_{j=1}^{\infty} \lambda\mu_j(1-\lambda\mu_j)^{-1}(f,\phi_j)\phi_j(x),$$

the result we wish.

All of the integral operators we have dealt with in this chapter have the property of being bounded linear maps of $L_2(a,b)$ into itself. From this it is automatic that if $\phi_n \to \phi$ in the mean, that is $\|\phi_n - \phi\| \to 0$ as $n\to\infty$, then $K\phi_n \to K\phi$ in the same sense. Lemma 4.21 shows that if K arises from a Schmidt kernel, then the knowledge that $\phi_n \to \phi$ in the mean allows us to conclude that $K\phi_n \to K\phi$ uniformly. In other cases, one can deduce other forms of convergence, which may be useful in particular cases. For example, if k is a Schur kernel and $C = \sup_x \int_a^b |k(x,t)|\, dt$, then

$$|(K\phi_n)(x) - (K\phi)(x)| \leqslant \int_a^b |k(x,t)|\, |\phi_n(t) - \phi(t)|\, dt$$

$$\leqslant C \sup_t |\phi_n(t) - \phi(t)|$$

if the supremum on the right exists. From this we see that if $\phi_n \to \phi$ uniformly, then $K\phi_n \to K\phi$ uniformly.

Problems

4.1 Suppose that H is a Hilbert space and that $T \in B(H)$. Show that if T is invertible then T^{-1} commutes with every operator which commutes with T. Show also that if $K \in B(H)$ is compact and $I - \lambda K$ is invertible then $(I - \lambda K)^{-1} = I + \lambda K (I - \lambda K)^{-1}$ and so $(I - \lambda K)^{-1}$ is the sum of the identity operator and a compact operator.

4.2 Show that if H is a Hilbert space and $T, U \in B(H)$ where U is invertible, then $\sigma(T^*) = \{\mu : \bar{\mu} \in \sigma(T)\}$ and $\sigma(U^{-1}TU) = \sigma(T)$.

4.3 Let H be a Hilbert space and $S, T \in B(H)$ where S and T commute. Show that ST is invertible if and only if both S and T are invertible; to show the 'only if' part you should show that if $UST = STU = I$ then $TU = UT = S^{-1}$.

By writing $(T - \mu I)(T + \mu I) = T^2 - \mu^2 I$ show that $v \in \sigma(T^2) \Leftrightarrow v \in \{\mu^2 : \mu \in \sigma(T)\}$. Generalise this to show that if p is a polynomial with complex coefficients and H is a complex Hilbert space $\sigma(p(T)) = \{p(\mu) : \mu \in \sigma(T)\}$. (Factorise $p(z) - p(\mu)$ into linear factors.)

4.4 Let K be the operator from $L_2(0,1)$ to itself defined by $(K\phi)(x) = \int_0^1 \min(x,t)\, \phi(t)\, dt$. By showing that $\mu\phi = K\phi$ if and only if $\mu\phi''(x) + \phi(x) = 0 \ (0 < x < 1)$ and $\phi(0) = \phi'(1) = 0$, deduce that the sequences (μ_n) and (ϕ_n) of eigenvalues and eigenvectors of K predicted by the Spectral Theorem are $\mu_n = 1/((n - \frac{1}{2})^2 \pi^2)$ and $\phi_n(x) = 2^{\frac{1}{2}} \sin(n - \frac{1}{2})\pi x$.

4.5 Consider the operator from $L_2(0,1)$ to itself defined by $(K\phi)(x) = \int_0^{1-x} \phi(t)\, dt$ $(0 \leqslant x \leqslant 1)$; show that K is compact and self-adjoint and that its sequence of eigenvalues is $((-1)^{n-1}/((n-\frac{1}{2})\pi))$ and find the corresponding eigenvectors. (Notice that $K = UV$ where $(U\phi)(x) = \phi(1-x)$ and $(V\phi)(x) = \int_0^x \phi(t)\, dt$; since we know V has no eigenvalues and that $\sigma(V) = \{0\}$, this demonstrates that if U and V do not commute $\sigma(V)$ and $\sigma(UV)$ bear no discernible relation to each other.)

4.6 Find the sequences of eigenvalues and corresponding eigenvectors for the operators on $L_2(0,1)$ given by the formulae

$$\int_0^1 (1 - \max(x,t))\phi(t)\, dt; \quad \int_0^1 |x - t|\phi(t)\, dt; \quad \int_0^1 (2|x-t| - 1)\phi(t)\, dt.$$

(Beware that in the last example the set of eigenvectors corresponding to each eigenvalue has dimension 2. Notice, however, that in this case (and the first) Corollary 4.26 provides a check that no eigenvalues have been overlooked. Theorem 4.19 may also be used for this check.)

4.7 Use the expansion

$$-\log|\cos x - \cos t| = \log 2 + \sum_{n=1}^{\infty} \frac{2}{n} \cos nx \cos nt$$

(valid for $0 < x, t < \pi$, $x \neq t$) to find the eigenvalues and eigenvectors of the operator K where $(K\phi)(x) = -\int_0^\pi \log|\cos x - \cos t|\phi(t)\,dt$ $(0 \leqslant x \leqslant \pi)$. (Notice that the series for $-\log|\cos x - \cos t|$ converges in the mean.) Hence solve the integral equation

$$\phi(x) + \lambda \int_0^\pi \log|\cos x - \cos t|\phi(t)\,dt = x \quad (0 \leqslant x \leqslant \pi)$$

provided $\lambda \neq n/\pi$ $(n \in \mathbb{N})$ and $\lambda \neq 1/(\pi \log 2)$.

(The given expansion for $-\log|\cos x - \cos t|$ in terms of an expansion in orthonormal functions allows us to spot the eigenvalues and eigenvectors directly. Notice that since $|\cos x - \cos t| \leqslant |x - t|$ $(0 \leqslant x, t \leqslant \pi)$ we can see directly that $\log|\cos x - \cos t|$ is an L_2-kernel.)

4.8 Classical Fourier series theory establishes that if $0 \leqslant x, t \leqslant 2\pi$ and $0 \leqslant a < 1$, $-\log(1 - 2a\cos(x - t) + a^2) = \sum_{n=1}^{\infty} (2/n)a^n\cos n(x - t)$. Use this to find the eigenvalues and eigenvectors of the operator K on $L_2(0, 2\pi)$ given by

$$(K\phi)(x) = -\frac{1}{2\pi}\int_0^{2\pi} \log(1 - 2a\cos(x - t) + a^2)\phi(t)\,dt \quad (0 \leqslant x \leqslant 2\pi)$$

and solve

$$\phi(x) = x(2\pi - x) + \lambda(K\phi)(x) \quad (0 \leqslant x \leqslant 2\pi).$$

4.9 The Poisson kernel, $k(x, t)$, is given by the formula

$$k(x, t) = \frac{1}{2\pi}\frac{1 - a^2}{1 - 2a\cos(x - t) + a^2} \quad (0 \leqslant x, t \leqslant 2\pi, 0 < a < 1).$$

Show that

$$k(x, t) = \frac{1}{2\pi} + \frac{1}{\pi}\sum_{n=1}^{\infty} a^n \cos n(x - t).$$

Find the usual sequences (μ_n) and (ϕ_n) of eigenvalues and eigenvectors of the operator K on $L_2(0, 2\pi)$ generated by k, and verify that

$$\sum_{n=1}^{\infty} \mu_n = \int_0^{2\pi} k(x, x)\,dx.$$

Calculate the kernel $k_m(x, t)$ which generates K^m and verify that, for each $m \in \mathbb{N}$, $\Sigma_{n=1}^{\infty} \mu_n^m = \int_0^{2\pi} k_m(x, x) \, dx$.

Show that, provided $\lambda \mu_n \neq 1$ (for $n = 1, 2, \ldots$)

$$(I - \lambda K)^{-1} f = f + \int_0^{2\pi} r_\lambda(x, t) f(t) \, dt$$

where

$$r_\lambda(x, t) = \frac{1}{2\pi} \frac{\lambda}{1 - \lambda} + \frac{1}{\pi} \sum_{n=1}^{\infty} \frac{\lambda a^n \cos n(x - t)}{(1 - \lambda a^n)} \quad (0 \leqslant x, t \leqslant 2\pi).$$

4.10 Show that if μ_1 and μ_2 are scalars and K is a bounded operator then, provided $\mu_1, \mu_2 \notin \sigma(K)$,

$$(\mu_1 I - K)^{-1} - (\mu_2 I - K)^{-1} = -(\mu_1 - \mu_2)(\mu_1 I - K)^{-1}(\mu_2 I - K)^{-1}. \quad (4.30)$$

Verify this in the case of the operator K in Problem 4.9.

(This equation is a particular case of the identity $A^{-1} - B^{-1} = A^{-1}(B - A)B^{-1}$ valid for any pair A, B of invertible operators. The form (4.30) may be rewritten as

$$(f(\mu_1) - f(\mu_2))/(\mu_1 - \mu_2) = -f(\mu_1)f(\mu_2)$$

where $f(\mu) = (\mu I - K)^{-1}$, a form in which we see that the left hand side tends to a limit as $\mu_1 \to \mu_2$. When the Hilbert space is complex this allows us to consider f as a (vector-valued) analytic function; conventional (complex-valued) analytic functions may be obtained by considering $((\mu I - K)^{-1} \phi, \psi)$ for fixed ϕ, ψ.)

4.11 (i) Let H be a complex Hilbert space and $T \in B(H)$ be *skew-adjoint*, that is, let $T^* = -T$. Show that $\sigma(T) \subset \{ix : x \in \mathbb{R}\}$. Deduce that if $S \in B(H)$ then $I + S - S^*$ is invertible.

(ii) Let H be a complex Hilbert space and $S \in B(H)$. Show that, for all $\phi \in H$, $(S^*S\phi, \phi) \geqslant 0$ and adapt the ideas of Lemma 4.13 to prove that $I + S^*S$ is invertible.

4.12 Theorem 3.5 shows us that if T is compact then so is T^*T. Modify Theorem 4.18 to prove the converse, that if T is bounded and T^*T is compact then T is also compact.

4.13 Suppose that H is a Hilbert space, (ϕ_n) is a complete orthonormal sequence in H and (α_n) is a bounded sequence of scalars. Show that there is a unique $T \in B(H)$ for which $T\phi_n = \alpha_n \phi_n$ $(n \in \mathbb{N})$ and that this T is compact if and only if $\alpha_n \to 0$ as $n \to \infty$. Show further that if $\Sigma |\alpha_n|^2$ converges and $H = L_2(0, 1)$ then T is generated by an L_2-kernel.

4.14 The kernel $k(x, t) = (x + t)^{-\alpha}$ $(0 < \alpha < 1)$ is a Schur kernel and it therefore generates a bounded linear map K on $L_2(0, 1)$. Let $A_n(x) = \int_0^1 \min(n, k(x, t)) \, dt$

and use Dini's Theorem to show that $A_n(x) \to \int_0^1 k(x, t)\, dt$ uniformly on $[0, 1]$, to show that the operator K is compact.

4.15 Not all operators arising from Schur kernels are compact. To see this let $k(x, t) = n(n + 1)$ if $x, t \in (1/(n + 1), 1/n)$ with $k(x, t) = 0$ for other values of x and t in $[0, 1]$; k is a Schur kernel. Consider the orthonormal sequence (χ_n) where $\chi_n(t) = (n(n + 1))^{\frac{1}{2}}$ $(t \in [1/(n + 1), 1/n])$ with $\chi_n(t) = 0$ for other $t \in [0, 1]$, and show that $K\chi_n = \chi_n$ $(n \in \mathbb{N})$ to show that the operator K generated by k is not compact.

4.16 Consider the integral equation

$$\phi(x) - \lambda \int_0^1 (\min(x, t) - xt)\phi(t)\, dt = f(x) \quad (0 \leqslant x \leqslant 1)$$

where $f \in L_2(0, 1)$. By calculating the solution using the spectral expansion as in Example 4.4, show that if $\lambda \neq n^2\pi^2$ for all $n \in \mathbb{N}$ then $\phi = (I - \lambda K)^{-1} f$ where

$$\|(I - \lambda K)^{-1}\| = \sup\left\{ \left| \frac{n^2\pi^2}{n^2\pi^2 - \lambda} \right| : n = 1, 2, 3, \ldots \right\}.$$

(In other words, calculate the norm of the self-adjoint operator defined by the solution formula.) Deduce that if $\lambda = 1$ and $\tilde{f}$ is an approximation to f, then the solution $\tilde{\phi}$ to the equation $\tilde{\phi} - \lambda K \tilde{\phi} = \tilde{f}$ satisfies

$$\|\phi - \tilde{\phi}\| \leqslant \pi^2(\pi^2 - 1)^{-1}\|f - \tilde{f}\|.$$

4.17 Suppose that K is a compact operator on $L_2(0, 1)$ and that L is a bounded linear map on $L_2(0, 1)$. Show that if $I - K$ is invertible and $\|L\| < 1/\|(I - K)^{-1}\|$ then the operator equation $\tilde{\phi} - (K + L)\tilde{\phi} = f$ is soluble and that if $\tilde{\phi}$ is its solution and $\phi - K\phi = f$ then

$$\tilde{\phi} - \phi = \sum_{n=1}^{\infty} ((I - K)^{-1}L)^n (I - K)^{-1} f$$

and hence

$$\|\tilde{\phi} - \phi\| \leqslant \frac{\|(I - K)^{-1}\|^2 \|L\| \|f\|}{1 - \|(I - K)^{-1}\| \|L\|}.$$

4.18 (i) Show that the boundary value problem

$$\phi''(x) + (a^2 + \epsilon r(x))\phi(x) = h(x) \quad (0 < x < 1),$$
$$\phi(0) = \phi(1) = 0,$$

where r and h are continous, ϵ and a are constants and a is not a multiple of π, is equivalent to the integral equation

$$\phi(x) = \epsilon \int_0^1 g(x, t)r(t)\phi(t)\, dt + f(x) \quad (0 \leqslant x \leqslant 1)$$

where

$$g(x,t) = \begin{cases} \dfrac{\sin(ta)\sin((1-x)a)}{a \sin a} & (t \leqslant x), \\[2ex] \dfrac{\sin(xa)\sin((1-t)a)}{a \sin a} & (x \leqslant t), \end{cases}$$

and $f(x) = -\int_0^1 g(x,t)h(t)\,dt$. Show also that if $a=0$ the boundary value problem is equivalent to the same integral equation but with $g(x,t) = \min(x,t) - xt$.

(ii) Show that the boundary value problem

$$\phi''(x) + \epsilon x^2 \phi(x) = 1 \quad (0 < x < 1) \quad \phi(0) = \phi(1) = 0$$

is satisfied by $\phi(x) = x^{-1}\psi(x)$ where

$$\psi(x) = \epsilon \int_0^1 (\min(x,t) - xt)xt\psi(t)\,dt - \tfrac{1}{2}x^2(1-x) \quad (0 \leqslant x \leqslant 1),$$

that is, $\psi = \epsilon K\psi + f$ where

$$(K\psi)(x) = \int_0^1 (\min(x,t) - xt)xt\psi(t)\,dt \quad \text{and} \quad f(x) = -\tfrac{1}{2}x^2(1-x).$$

By showing that $\|K\| < 1/30$ show that $\|\psi\| < 0.049/(1-\epsilon/30)$ and therefore that, provided $\epsilon < 30$,

$$\phi(x) = -\tfrac{1}{2}x(1-x) - \frac{\epsilon}{120}(x - 3x^5 + 2x^6) + \epsilon^2 x^{-1} K^2 \psi(x).$$

By observing that $K(K\psi(x)) = x\int_0^1 (\min(x,t) - xt)t(K\psi)(t)\,dt$ and using the Schmidt condition, show that for all $x \neq 0$

$$x^{-1}(K^2\psi)(x) \leqslant (1/90^{\frac{1}{2}})\|K\psi\| < \frac{(0.000176)}{1-\epsilon(0.034)}.$$

4.19 (i) Suppose that K is a compact self-adjoint operator on $B(H)$ and consider the equation $\phi - \lambda K\phi = f$. If $|\lambda|$ is sufficiently small then $(I - \lambda K)^{-1} = \sum_{n=0}^{\infty}(\lambda K)^n$ from which it is clear that $(I - \lambda K)^{-1}$ is the limit of a sequence of polynomials in K. Suppose now that $\lambda^{-1} \notin \sigma(K)$ (but not necessarily with $|\lambda|$ small) and use Example 4.3 to prove that $(I - \lambda K)^{-1}f = f_0 + \sum_{n=1}^{\infty}(1 - \lambda\mu_n)^{-1}(f,\phi_n)\phi_n$ where $f_0 = f - \sum_{n=1}^{\infty}(f,\phi_n)\phi_n$. Show also that if p is a polynomial then $p(K)f = p(0)f_0 + \sum_{n=1}^{\infty} p(\mu_n)(f,\phi_n)\phi_n$. Choose a sequence (p_m) of polynomials for which $p_m(t) \to 1/(1 - \lambda t)$ uniformly for $t \in \sigma(K)$ and show that $p_m(K) \to (I - \lambda K)^{-1}$, with $\|p_m(K) - (I - \lambda K)^{-1}\| \leqslant \sup\{|p_m(t) - (1/(1 - \lambda t)|: t \in \sigma(K)\}$. (The existence of the sequence of polynomials may be seen by using the Weierstrass approximation theorem to find polynomials approximating the function equal to $(1 - \lambda t)^{-1}$ on the interval $[-\|K\|, \|K\|]$ except between the two eigenvalues closest to $1/\lambda$ where the function is chosen to be linear.)

(ii)   Consider the integral equation

$$\phi(x) - 10 \int_0^1 \min(x, t)\phi(t)\, dt = f(x) \quad (0 \leqslant x \leqslant 1)$$

and show that the eigenvalues of the corresponding integral operator K are $(4/\pi^2, 4/(9\pi^2), 4/(25\pi^2), \ldots)$, where $(K\phi)(x) = \int_0^1 \min(x, t)\phi(t)\, dt$. Deduce that if $\alpha = \pi^2/(\pi^2 - 40)$ so that $p(t) = 1 + 10t + 100\alpha t^2$ coincides with $(1 - 10t)^{-1}$ for $t = 4/\pi^2$, then

$$\|(I - 10K)^{-1} - (I + 10K + 100\alpha K^2)\| < 0.44.$$

4.20 Suppose that H is a Hilbert space, that $K \in B(H)$ is compact and self-adjoint and that μ_0 is an eigenvalue of K. Suppose also that $f \in H$ has the property that $(f, \psi) = 0$ for all ψ satisfying the equation $\mu_0 \psi - K\psi = 0$. Show that the equation $\mu_0 \phi - K\phi = f$ has a unique solution ϕ which also has the property that it is orthogonal to all solutions ψ of the equation $\mu_0 \psi - K\psi = 0$.

Let $E = \{\psi \in H : \mu_0 \psi - K\psi = 0\}$ $(= \mathcal{N}(\mu_0 I - K))$, and show that $\phi \in E \Rightarrow (\mu_0 I - K)\phi \in E$ and that $\phi \in E^{\perp} \Rightarrow (\mu_0 I - K)\phi \in E^{\perp}$. By observing that $E^{\perp}$ is itself a Hilbert space and that the restriction of the operator $\mu_0 I - K$ to $E^{\perp}$ is invertible, deduce that there is a constant M such that for all $f \in E^{\perp}$ the (unique) solution $\phi \in E^{\perp}$ of the equation $\mu_0 \phi - K\phi = f$ has $\|\phi\| \leqslant M\|f\|$. (Putting this more casually, if we impose the additional conditions on f and ϕ then the solution ϕ depends continuously on f.)

Now suppose that μ is not an eigenvalue of K and show that if $f \in E^{\perp}$ then the solution ϕ of the equation $\mu\phi - K\phi = f$ is also in $E^{\perp}$. (Decompose ϕ into the sum of a vector in E and one in $E^{\perp}$.) Let $T_{\mu} : E^{\perp} \to E^{\perp}$ be the restriction of $(\mu I - K)^{-1}$ and show that if $f \in E^{\perp}$ then $T_{\mu} f \to \phi_0$ as $\mu \to \mu_0$ where ϕ_0 is the unique solution of $\mu_0 \phi - K\phi = f$ which lies in $E^{\perp}$.

(Again, imposing the additional conditions on f allows us to recover the information we would have expected had μ_0 not been an eigenvalue, in this case the information that the solution depends continuously on μ.)

5

The spectrum of a compact self-adjoint operator

5.1 Introduction

Chapter 4 was devoted to the spectral theorem in which we characterised a compact self-adjoint operator in terms of its eigenvalues and eigenvectors, and from this we were able to deduce substantial qualitative results about integral equations. Far reaching though these may be, they have the drawback that they involve the eigenvalues and eigenvectors of the particular integral operator being investigated. In many cases the exact determination of eigenvalues and eigenvectors will itself be a difficult problem, so in this chapter we shall derive a body of results giving characterisations of the various eigenvalues of a particular operator, and relations between the eigenvalues of the sum and product of two operators and those of the summand operators. The techniques used here yield some results immediately on the approximation of one operator by another, and these can be used to estimate the eigenvalues of otherwise recalcitrant operators. The main emphasis of this chapter, however, is not on approximation techniques, which will be dealt with in Chapters 7 and 8, but on suitable characterisations of the quantities involved which will form the foundations of these approximation techniques.

5.2 The Rayleigh quotient

If H is a Hilbert space and T a bounded linear map from H to itself, then associated with each vector $\phi \in H$ there is a scalar quantity $(T\phi, \phi)$, which is real if T is self-adjoint, for then $(T\phi, \phi) = (\phi, T^*\phi) = (\phi, T\phi) = \overline{(T\phi, \phi)}$. If we normalise ϕ and consider only those $\phi \in H$ of norm 1, then the resulting set $\{(T\phi, \phi): \|\phi\| = 1\}$ is a useful indicator of some of the properties of T. The restriction that $\|\phi\| = 1$ may be removed by considering instead $\{(T\phi, \phi)/(\phi, \phi): \phi \neq 0\}$, and the quantity $(T\phi, \phi)/(\phi, \phi)$ is called the *Rayleigh quotient*. Notice that by the Cauchy–Schwarz inequality

$|(T\phi, \phi)| \le \|T\phi\| \|\phi\| \le \|T\| \|\phi\|^2$ so that every element of $\{(T\phi, \phi): \|\phi\| = 1\}$ is a scalar of modulus not exceeding $\|T\|$.

Since we shall apply the following results only to compact self-adjoint operators, for which the ensuing proofs are much simplified by the Spectral Theorem, we shall from now on restrict attention to such operators. Let K be a compact self-adjoint operator in $B(H)$. Then there are sequences (μ_n) and (ϕ_n) of non-zero eigenvalues and eigenvectors of K respectively, with (ϕ_n) orthonormal, for which $K\phi = \Sigma_{n=1}^{\infty} \mu_n(\phi, \phi_n)\phi_n$. (As usual, if K has only finitely many non-zero eigenvalues this has to be modified to a finite sum; the consequent alterations are minor.) Then if $\phi \in H$

$$(K\phi, \phi) = \sum_{n=1}^{\infty} \mu_n |(\phi, \phi_n)|^2. \tag{5.1}$$

Now if $\|\phi\| = 1$, Bessel's inequality shows that $\Sigma_{n=1}^{\infty} |(\phi, \phi_n)|^2 \le 1$ whence

$$(K\phi, \phi) \le (\sup_n \mu_n) \sum_{n=1}^{\infty} |(\phi, \phi_n)|^2 \le (\sup_n \mu_n)$$

and, similarly

$$(K\phi, \phi) \ge (\inf_n \mu_n).$$

All of the members of the set $\{(K\phi, \phi): \|\phi\| = 1\}$ therefore lie in the interval $[\inf_n \mu_n, \sup_n \mu_n]$. It is worth noticing that if (ϕ_n) is a complete orthonormal sequence, so that $\|\phi\| = 1$ implies that $\Sigma_{n=1}^{\infty} |(\phi, \phi_n)|^2 = 1$, then (5.1) expresses $(K\phi, \phi)$ as a weighted average of the eigenvalues (μ_n). If (ϕ_n) is not a complete orthornormal sequence then we can write $\phi = \phi_0 + \Sigma_{n=1}^{\infty} (\phi, \phi_n)\phi_n$ where $K\phi_0 = 0$ and $\Sigma_{n=1}^{\infty} |(\phi, \phi_n)|^2 = \|\phi\|^2 - \|\phi_0\|^2$, in which case $(K\phi, \phi)$ is a weighted average of all the eigenvalues of K (the μ_n together with 0), again provided $\|\phi\| = 1$. The result that if $\|\phi\| = 1$ then $\inf_n \mu_n \le (K\phi, \phi) \le \sup_n \mu_n$ can now be seen more intuitively by noticing that a weighted average of the μ_n must lie in the interval between the supremum and the infimum of the set $\{\mu_n : n \in \mathbb{N}\}$.

If K possesses at least one positive eigenvalue, then because $\mu_n \to 0$ as $n \to \infty$ by Theorem 4.7, there will be a largest eigenvalue, that is, $\sup\{\mu_n : n \in \mathbb{N}\}$ will be attained, and we shall denote it by $\max\{\mu_n : n \in \mathbb{N}\}$, using the word maximum and the symbol max when it is known that the set has a greatest member. Similarly if K has negative eigenvalues, $\inf\{\mu_n : n \in \mathbb{N}\}$ is attained and we shall write it as $\min\{\mu_n : n \in \mathbb{N}\}$. Notice that if $\mu_1 = \max\{\mu_n : n \in \mathbb{N}\}$ then $(K\phi_1, \phi_1) = \mu_1$ so μ_1 is the maximum value of $\{(K\phi, \phi): \|\phi\| = 1\}$, while if K has some negative eigenvalues then the infimum of $\{(K\phi, \phi): \|\phi\| = 1\}$ is also attained. The only cases, indeed, in which the supremum and infimum of $\{(K\phi, \phi): \|\phi\| = 1\}$ are not both attained are where all the eigenvalues are of the same sign and 0 is not an eigenvalue. For example, if $\mu_n > 0$ for all n and 0 is not an eigenvalue, $(K\phi, \phi) > 0$ for all $\phi \neq 0$ so $\inf\{\mu_n: n \in \mathbb{N}\} = 0$ is not attained as a value of $(K\phi, \phi)$ for some ϕ of norm 1.

At this point matters are clarified if we expand our notation.

Definition 5.1

Let H be a Hilbert space and $K \in B(H)$ be compact and self-adjoint. We shall denote the positive eigenvalues of K by $\mu_1^+, \mu_2^+, \ldots$, ordered so that $\mu_1^+ \geqslant \mu_2^+ \geqslant \mu_3^+ \geqslant \ldots$, an eigenvalue appearing k times in the sequence if the corresponding space of eigenvectors has dimension k. We adopt the convention that if there are only N positive eigenvalues (including the case $N = 0$) then $\mu_n^+ = 0$ if $n > N$. The negative eigenvalues of K are similarly denoted by μ_n^- with $\mu_1^- \leqslant \mu_2^- \leqslant \ldots$, with $\mu_n^- = 0$ for all $n > N$ if there are only N negative eigenvalues. If it is necessary to distinguish between different operators we shall write $\mu_n^+(K)$ etc.

The usual sequence of (positive or negative) eigenvalues (μ_n) will henceforth be assumed ordered so that $|\mu_1| \geqslant |\mu_2| \geqslant \ldots$. Let (ϕ_n) be the corresponding orthonormal sequence of eigenvectors guaranteed by the Spectral Theorem, and split this into two orthonormal sequences (ϕ_n^+) and (ϕ_n^-) according as the corresponding eigenvalues are positive or negative, so that $K\phi_n^+ = \mu_n^+ \phi_n^+$ and $K\phi_n^- = \mu_n^- \phi_n^-$ $(n = 1, 2, \ldots)$.

Finally, let $R_K(\phi) = (K\phi, \phi)/(\phi, \phi)$ denote the *Rayleigh quotient*. This is, of course, only defined for $\phi \neq 0$ and we shall adopt the convention that ϕ is restricted to be non-zero so that, for example, $\inf_{\phi \in E} R_K(\phi)$ will be used as an abbreviation for $\inf\{R_K(\phi): \phi \in E, \phi \neq 0\}$. □

Lemma 5.1

If K is a compact, self-adjoint operator on the (non-zero) Hilbert space H, then

$$\mu_1^+ = \sup\{R_K(\phi): \phi \neq 0\} \ (= \max\{R_K(\phi): \phi \neq 0\} \text{ if } \mu_1^+ \neq 0)$$

and

$$\mu_1^- = \inf\{R_K(\phi): \phi \neq 0\} \ (= \min\{R_K(\phi): \phi \neq 0\} \text{ if } \mu_1^- \neq 0).$$

Also,

$$\|K\| = \max\{|R_K(\phi)|: \phi \neq 0\}.$$

Proof

The equations for $\mu_1^{\pm}$ have been proved in the discussion above, since $\mu_1^+ = \sup\{\mu_n : n \in \mathbb{N}\}$ and $\mu_1^- = \inf\{\mu_n : n \in \mathbb{N}\}$. The last equation arises from Lemma 4.14 whence we have

$$\|K\| = \sup\{|\mu| : \mu \in \sigma(K)\} = \sup\{|\mu_n| : n \in \mathbb{N}\} = \max(\mu_1^+, -\mu_1^-). \qquad \square$$

By considering $(K\phi, \phi)$ for $\|\phi\| = 1$ or, equivalently, by considering $R_K(\phi)$, we may identify $\mu_1^{\pm}$. Consideration of (5.1) shows that we can do rather better than this by restricting attention to $(K\phi, \phi)$ when ϕ belongs to some subspace of H. For example, if ϕ is a linear combination of $\phi_m, \phi_{m+1}, \ldots,$ ϕ_M then, omitting the terms in (5.1) for which $(\phi, \phi_n) = 0$, we see that $(K\phi, \phi) = \sum_{n=m}^{M} \mu_n |(\phi, \phi_n)|^2$; if also $\|\phi\| = 1$ then $(K\phi, \phi)$ is a weighted average of $\mu_m, \ldots, \mu_M$ and therefore lies in the interval $[\min_{m \leqslant n \leqslant M} \mu_n, \max_{m \leqslant n \leqslant M} \mu_n]$, the end points of this interval being attained by choosing ϕ to be the appropriate ϕ_n. By a similar choice of subspace, if we restrict ϕ to being a linear combination of $\phi_m^+, \phi_{m+1}^+, \ldots, \phi_M^+$ then for $\|\phi\| = 1$, $(K\phi, \phi)$ lies in the interval $[\mu_M^+, \mu_m^+]$. By varying the choice of basis vectors for the subspace in this last case, but keeping the number of them constant, we may choose the upper end of the interval to be μ_1^+ while the lower end can be made as large as μ_{M-m+1}^+, but no larger. This is the motivating idea for a method of identifying μ_n^+ for values of n greater than 1.

Lemma 5.2

Let K be a compact, self-adjoint operator on an infinite-dimensional Hilbert space H. Then, with the usual conventions, and letting E denote a typical subspace of H,

$$\mu_n^+ = \sup_{\dim E = n} \left(\inf_{\phi \in E} R_K(\phi) \right) \qquad (5.2)$$

and

$$\mu_n^- = \inf_{\dim E = n} \left(\sup_{\phi \in E} R_K(\phi) \right), \qquad (5.3)$$

where $\sup_{\dim E = n}$ indicates that the supremum is to be taken over all subspaces of dimension n.

Remark

The requirement that the space H be of infinite dimension in this lemma is to avoid awkward conventions' being introduced. For example, if n is the dimension of H then the right hand side of (5.2) equals the nth largest eigenvalue of K, which may be negative, so the right hand side of (5.2) would have to be replaced by the maximum of the quantity shown and zero.

Proof

For (5.2), choose E to be the subspace spanned by $\phi_1^+, \ldots, \phi_n^+$, provided these exist. Then if $\phi \in E$ and $\|\phi\| = 1$, $(K\phi, \phi)$ is some weighted average of $\mu_1^+, \ldots, \mu_n^+$ whence $(K\phi, \phi) \geqslant \mu_n^+$. Thus for $\phi \neq 0$ and $\phi \in E$, $R_K(\phi) = (K(\phi/\|\phi\|), \phi/\|\phi\|) \geqslant \mu_n^+$, whence $\inf_{\phi \in E} R_K(\phi) = \mu_n^+$, the infimum being attained at $\phi = \phi_n^+$. Therefore there is a subspace E of dimension n with $\inf_{\phi \in E} R_K(\phi) = \mu_n^+$ and so $\sup_{\dim E = n}(\inf_{\phi \in E} R_K(\phi)) \geqslant \mu_n^+$. In this case the supremum is attained.

If $\phi_1^+, \ldots, \phi_n^+$ do not all exist, that is, if there are fewer than n positive eigenvalues of K, counting multiple eigenvalues in the usual way, then by convention $\mu_n^+ = 0$. Choose E in this case to be the subspace spanned by $\phi_m^-, \ldots, \phi_{m+n-1}^-$, and a similar calculation shows that $\inf_{\phi \in E} R_K(\phi) \geqslant \mu_m^-$. Since $\mu_m^- \to 0$ as $m \to \infty$ by Theorem 4.7 we see that

$$\sup_{\dim E = n} \left(\inf_{\phi \in E} R_K(\phi) \right) \geqslant 0 = \mu_n^+.$$

To establish the reverse inequality, let E be a subspace of dimension n. Then E contains a non-zero vector ϕ orthogonal to $\phi_1^+, \ldots, \phi_{n-1}^+$, (since ϕ is required to satisfy only $n - 1$ conditions $(\phi, \phi_j^+) = 0 \, (j = 1, 2, \ldots, n - 1)$). For this ϕ,

$$(K\phi, \phi) = \sum_{j=n}^{\infty} \mu_j^+ |(\phi, \phi_j^+)|^2 + \sum_{j=1}^{\infty} \mu_j^- |(\phi, \phi_j^-)|^2$$

$$\leqslant \mu_n^+ \left(\sum_{j=n}^{\infty} |(\phi, \phi_j^+)|^2 + \sum_{j=1}^{\infty} |(\phi, \phi_j^-)|^2 \right)$$

$$\leqslant \mu_n^+ \|\phi\|^2$$

by Bessel's inequality. Therefore, $\inf_{\phi \in E} R_K(\phi) \leqslant \mu_n^+$. Since E is a typical subspace of dimension n, this is true of all such subspaces and

$$\sup_{\dim E = n} \left(\inf_{\phi \in E} R_K(\phi) \right) \leqslant \mu_n^+,$$

which, together with the earlier inequality, establishes (5.2).

Equation (5.3) is proved in an analogous way, with μ_n^+ and ϕ_n^+ being replaced by μ_n^- and ϕ_n^- throughout. $\qquad\square$

In Lemma 5.2 the supremum and infimum can frequently be replaced by maximum and minimum. For a finite-dimensional subspace E, $\inf\{R_K(\phi): \phi \in E\} = \inf\{(K\phi, \phi): \|\phi\| = 1, \phi \in E\}$ and the infimum on the right is attained, since $(K\phi, \phi)$ varies continuously with ϕ and $\{\phi \in E: \|\phi\| = 1\}$ is compact by the Heine–Borel Theorem. Therefore, in (5.2) and (5.3) the inner infimum or supremum may always be replaced by minimum or maximum respectively. The outer supremum or infimum is also attained by choosing E to be the subspace spanned by $\phi_1^+, \ldots, \phi_n^+$ or $\phi_1^-, \ldots, \phi_n^-$, respectively, provided these exist, that is, provided $\mu_n^\pm \neq 0$. If we restate the theorem with these implicit restrictions that the $\mu_n^\pm$ are to be non-zero or else the outer maximum or minimum is actually a supremum or infimum we have

$$\mu_n^+ = \max_{\dim E = n} \left(\min_{\phi \in E} R_K(\phi) \right)$$

$$\mu_n^- = \min_{\dim E = n} \left(\max_{\phi \in E} R_K(\phi) \right),$$

the *minimax principles* for the nth positive and negative eigenvalues.

Returning to the equation

$$(K\phi, \phi) = \sum_{n=1}^{\infty} \mu_n |(\phi, \phi_n)|^2 \qquad (5.1)$$

which we shall rewrite as

$$(K\phi, \phi) = \sum_{n=1}^{\infty} \mu_n^+ |(\phi, \phi_n^+)|^2 + \sum_{n=1}^{\infty} \mu_n^- |(\phi, \phi_n^-)|^2, \qquad (5.4)$$

we can see that there is another plausible way to proceed. Instead of restricting ϕ to a subspace of dimension n and considering the resulting minimum value of $R_K(\phi)$, we could identify μ_n^+ by restricting ϕ to be orthogonal to $\phi_1^+, \ldots, \phi_{n-1}^+$, when the maximum value of $(K\phi, \phi)$ will be μ_n^+, this maximum being taken over all ϕ of norm 1 orthogonal to $\phi_1^+, \ldots, \phi_{n-1}^+$. In this case varying the set to which ϕ is to be orthogonal may allow the maximum to increase (e.g. if ϕ is to be orthogonal to $\phi_2^+, \ldots, \phi_n^+$), and we obtain a second minimax principle.

Lemma 5.3

Let K be a compact, self-adjoint operator on the infinite-dimensional space H. Then, with the usual conventions, and letting E denote a subspace of H,

$$\mu_n^+ = \inf_{\dim E = n-1} \left(\sup_{\phi \in E^\perp} R_K(\phi) \right) \qquad (5.5)$$

and

$$\mu_n^- = \sup_{\dim E = n-1} \left(\inf_{\phi \in E^\perp} R_K(\phi) \right). \qquad (5.6)$$

Proof

Let E be a subspace of H of dimension $n-1$. Then if K has at least n positive eigenvalues there is a non-zero linear combination ϕ of $\phi_1^+, \ldots, \phi_n^+$ which is orthogonal to E. For this ϕ,

$$(K\phi, \phi) = \sum_{j=1}^{n} \mu_j^+ |(\phi, \phi_j^+)|^2 \geqslant \mu_n^+ \sum_{j=1}^{n} |(\phi, \phi_j^+)|^2 = \mu_n^+ \|\phi\|^2,$$

whence $R_K(\phi) \geqslant \mu_n^+$ and $\sup_{\phi \in E^\perp} R_K(\phi) \geqslant \mu_n^+$. Since this is true for all subspaces E of dimension $n-1$,

$$\inf_{\dim E = n-1} \left(\sup_{\phi \in E^\perp} R_K(\phi) \right) \geqslant \mu_n^+.$$

If K does not have n positive eigenvalues, so that $\mu_n^+ = 0$ by convention, then we proceed slightly differently. Again let E be a subspace of dimension $n-1$ and choose $m \in \mathbb{N}$. There is a non-zero linear combination, ϕ, of $\phi_m^-, \phi_{m+1}^-, \ldots, \phi_{m+n-1}^-$ which belongs to $E^\perp$ and, for this ϕ, $(K\phi, \phi) = \sum_{j=m}^{m+n-1} \mu_j^- |(\phi, \phi_j^-)|^2 \geqslant \mu_m^- \|\phi\|^2$ whence $R_K(\phi) \geqslant \mu_m^-$. Since m was arbitrary, this is true for all $m \in \mathbb{N}$ and $\sup_{\phi \in E^\perp} R_K(\phi) \geqslant \sup_m \mu_m^- = 0$. We now notice that E was a typical $(n-1)$-dimensional subspace so, in this case too,

$$\inf_{\dim E = n-1} \left(\sup_{\phi \in E^\perp} R_K(\phi) \right) \geqslant \mu_n^+.$$

Finally, let E_0 be the subspace spanned by $\phi_1^+, \ldots, \phi_{n-1}^+$ (or be a subspace of dimension $n-1$ containing all the ϕ_j^+ if there are fewer than $n-1$). Then $\phi \in E_0^\perp \Rightarrow (\phi, \phi_j^+) = 0$ for $j = 1, 2, \ldots, n-1$ and

$$(K\phi, \phi) = \sum_{j=n}^{\infty} \mu_j^+ |(\phi, \phi_j^+)|^2 + \sum_{j=1}^{\infty} \mu_j^- |(\phi, \phi_j^-)|^2$$
$$\leqslant \mu_n^+ \|\phi\|^2.$$

Thus $\sup_{\phi \in E_0^\perp} R_K(\phi) \leqslant \mu_n^+$, and therefore

$$\inf_{\dim E = n-1} \left(\sup_{\phi \in E^\perp} R_K(\phi) \right) \leqslant \mu_n^+,$$

establishing (5.5).

The proof of (5.6) is virtually identical. $\qquad \square$

Example 5.1

We can use some of the ideas above to calculate the best approximation to an operator by one arising from a degenerate kernel. Suppose that k is a hermitian kernel on $[a, b] \times [a, b]$ which generates the compact operator K. Then among all the kernels of the form $\Sigma_{i=1}^{N} a_i(x)b_i(t)$, where $a_1, \ldots, a_N$, $b_1, \ldots, b_N$ belong to $L_2(a, b)$, the one which generates the operator L closest to K (in the sense that $\|K - L\|$ is a minimum) is $\Sigma_{i=1}^{N} \mu_i \phi_i(x)\overline{\phi_i(t)}$, with (μ_i) and (ϕ_i) the usual sequences of eigenvalues and eigenvectors of K and with $|\mu_{i+1}| \leqslant |\mu_i|$ for all i.

Notice first that a bounded operator L on $L_2(a, b)$ is of rank at most N if and only if it arises from a kernel of the form $\Sigma_{i=1}^{N} a_i(x)b_i(t)$, where $a_1, \ldots, a_N, b_1, \ldots, b_N$ are functions in $L_2(a, b)$ (Problem 3.21). Suppose that L arises from such a kernel, and hence that it has rank at most N. Then among all the linear combinations of $\phi_1, \ldots, \phi_{N+1}$ there is a non-zero vector ϕ for which $L\phi = 0$; suppose, without loss of generality, that $\|\phi\| = 1$. Then

$$\|K - L\|^2 \geqslant \|(K - L)\phi\|^2 = \|K\phi\|^2 = \sum_{n=1}^{N+1} \mu_n^2 |(\phi, \phi_n)|^2 \geqslant |\mu_{N+1}|^2,$$

whence $\|K - L\| \geqslant |\mu_{N+1}|$.

To see that the best approximation of the desired sort is that arising from $\Sigma_{i=1}^{N} \mu_i \phi_i(x)\overline{\phi_i(t)}$, we need only recall that if the operator generated by this kernel is denoted by K_N, then $K_N\phi = \Sigma_{i=1}^{N} \mu_i(\phi, \phi_i)\phi_i$ and we showed in the course of the proof of the Spectral Theorem that $\|K - K_N\| = |\mu_{N+1}|$, so K_N attains the lower bound found above.   □

5.3 Eigenvalue inequalities

Lemmas 5.2 and 5.3 are minimax principles for finding the nth positive (or negative) eigenvalue of a self-adjoint compact operator K. Despite their apparent similarity we shall use them in rather different contexts. Lemma 5.3 is suited to various theoretical results and is the basis for the results which follow in this section, allowing us to relate the eigenvalues of one operator to those of another. Lemma 5.2, in contrast, is less well suited to the development of such theoretical results, but offers a practical method for estimating eigenvalues which is exploited in Chapter 7.

Theorem 5.4

Let K_1 and K_2 be compact self-adjoint operators on the Hilbert space H, and denote their corresponding sequences of positive eigenvalues, negative eigenvalues, and non-zero eigenvalues of either sign by $(\mu_n^+(K_i))$, $(\mu_n^-(K_i))$, $(\mu_n(K_i))$ respectively $(i = 1, 2)$. Then for all natural numbers m and n we have

$$\mu_{m+n-1}^+(K_1 + K_2) \leqslant \mu_m^+(K_1) + \mu_n^+(K_2), \tag{5.7}$$

$$\mu_{m+n-1}^-(K_1 + K_2) \geqslant \mu_m^-(K_1) + \mu_n^-(K_2) \tag{5.8}$$

and

$$|\mu_{m+n-1}(K_1 + K_2)| \leqslant |\mu_m(K_1)| + |\mu_n(K_2)|. \tag{5.9}$$

Proof

We shall prove the result when H has infinite dimension. The changes needed to adapt to the situation when H has finite dimension are slight.

Let $m, n \in \mathbb{N}$ and choose $\epsilon > 0$. Then by Lemma 5.3 we may choose two subspaces E_1 and E_2, of dimensions $m - 1$ and $n - 1$ respectively, for which

$$\sup_{\phi \in E_1^\perp} R_{K_1}(\phi) < \mu_m^+(K_1) + \epsilon \tag{5.10}$$

and

$$\sup_{\phi \in E_2^\perp} R_{K_2}(\phi) < \mu_n^+(K_2) + \epsilon. \tag{5.11}$$

Then if $\phi \in (E_1 + E_2)^\perp$, so that $\phi \in E_1^\perp$ and $\phi \in E_2^\perp$,

$$R_{K_1 + K_2}(\phi) = R_{K_1}(\phi) + R_{K_2}(\phi)$$

$$\leqslant \sup_{\phi \in E_1^\perp} R_{K_1}(\phi) + \sup_{\phi \in E_2^\perp} R_{K_2}(\phi)$$

$$< \mu_m^+(K_1) + \mu_n^+(K_2) + 2\epsilon,$$

by (5.10) and (5.11). Now the vector space sum of $E_1 + E_2$ will have dimension at most $m + n - 2$, but possibly less, so let F be a subspace of H of dimension $m + n - 2$ which contains $E_1 + E_2$. Then $\phi \in F^\perp \Rightarrow \phi \in (E_1 + E_2)^\perp$ and $\sup_{\phi \in F^\perp} R_{K_1 + K_2}(\phi) < \mu_m^+(K_1) + \mu_n^+(K_2) + 2\epsilon$.

Therefore

$$\mu_{m+n-1}^+(K_1 + K_2) = \inf_{\dim E = m+n-2} \left(\sup_{\phi \in E^\perp} R_{K_1 + K_2}(\phi) \right)$$

$$\leqslant \sup_{\phi \in F^\perp} R_{K_1 + K_2}(\phi) < \mu_m^+(K_1) + \mu_n^+(K_2) + 2\epsilon.$$

Since ϵ was an arbitrary positive number,

$$\mu_{m+n-1}^+(K_1+K_2) \leqslant \mu_m^+(K_1)+\mu_n^+(K_2),$$

proving (5.7).

Inequality (5.8) may be proved similarly, or deduced by noticing that if K is compact and self-adjoint then $\mu_n^-(K) = -\mu_n^+(-K)$.

The inequality (5.9) can now be deduced from (5.7) and (5.8), as follows. Suppose that among $\mu_1(K_1), \ldots, \mu_{m-1}(K_1)$ there are n_1 negative entries and p_1 positive ones, and that among $\mu_1(K_2), \ldots, \mu_{n-1}(K_2)$ there are n_2 negative and p_2 positive entries, repetitions being counted the number of times they appear. Since only p_1 of the numbers $\mu_1(K_1), \ldots, \mu_{m-1}(K_1)$ are positive, $\mu_{p_1+1}^+(K_1)$ equals $\mu_j(K_1)$ for some $j \geqslant m$, so $\mu_{p_1+1}^+(K_1) \leqslant |\mu_m(K_1)|$. By the same argument $\mu_{p_2+1}^+(K_2) \leqslant |\mu_n(K_2)|$, so (5.7) shows that

$$\mu_{p_1+p_2+1}^+(K_1+K_2) \leqslant \mu_{p_1+1}^+(K_1)+\mu_{p_2+1}^+(K_2) \leqslant |\mu_m(K_1)|+|\mu_n(K_2)|.$$

Therefore at most p_1+p_2 of the eigenvalues of K_1+K_2 are positive and have modulus greater than $|\mu_m(K_1)|+|\mu_n(K_2)|$. Similarly, because

$$\mu_{n_1+n_2+1}^-(K_1+K_2) \geqslant \mu_{n_1+1}^-(K_1)+\mu_{n_2+1}^-(K_2) \geqslant -|\mu_m(K_1)|-|\mu_n(K_2)|,$$

K_1+K_2 has at most n_1+n_2 negative eigenvalues whose modulus exceeds $|\mu_m(K_1)|+|\mu_n(K_2)|$. Therefore, since $n_1+n_2+p_1+p_2 = m+n-2$, K_1+K_2 has at most $m+n-2$ eigenvalues whose modulus exceeds $|\mu_m(K_1)|+|\mu_n(K_2)|$ and so

$$|\mu_{m+n-1}(K_1+K_2)| \leqslant |\mu_m(K_1)|+|\mu_n(K_2)|. \qquad \square$$

An interesting special case of Theorem 5.4 occurs when one of the summands has only finitely many eigenvalues of a particular sign. This is true of an operator arising from a degenerate kernel, since an operator of rank N can have a linearly independent set of at most N eigenvectors corresponding to non-zero eigenvalues, so if K has rank N then $\mu_n(K)=0$ for all $n>N$.

Theorem 5.5

Let K_1 and K_2 be compact self-adjoint operators.

(i) If K_2 has only N^+ positive eigenvalues, then

$$\mu_{m+N^+}^+(K_1+K_2) \leqslant \mu_m^+(K_1) \quad (m \in \mathbb{N}).$$

(ii) If K_2 has only N^- negative eigenvalues, then

$$\mu_{m+N^-}^-(K_1+K_2) \geqslant \mu_m^-(K_1) \quad (m \in \mathbb{N}).$$

(iii) If K_2 has only N non-zero eigenvalues then

$$|\mu_{m+N}(K_1+K_2)| \leqslant |\mu_m(K_1)| \quad (m \in \mathbb{N}).$$

Proof

The only point worth noting is that when we say K_2 has only N^+ positive eigenvalues, we mean that the usual sequence of eigenvalues $(\mu_n(K_2))$ arising from the Spectral Theorem has only N^+ positive terms. These N^+ terms need not, of course, be distinct. With this understanding, we see in (i) that $\mu_{N^++1}^+(K_2)=0$ and the result follows immediately from Theorem 5.4.

$\square$

Example 5.2

Consider the boundary value problem

$$\left. \begin{aligned} \phi''(x) + \lambda\phi(x) &= 0 \quad (0 < x < 1), \\ a_0\phi(0) + b_0\phi'(0) &= 0, \\ a_1\phi(1) + b_1\phi'(1) &= 0. \end{aligned} \right\} \tag{5.12}$$

We have already seen in §2.4 that in the case where $a_0 = a_1 = 1$ and $b_0 = b_1 = 0$, this is equivalent to the integral equation

$$\phi(x) - \lambda \int_0^1 k(x, t)\phi(t)\,dt = 0 \quad (0 \leqslant x \leqslant 1)$$

where $k(x, t) = \min(x, t) - xt$. (Consider the case $r = 1$ in (2.19).) The values of λ for which a non-trivial solution ϕ exists are the reciprocals of the non-zero eigenvalues of the integral operator generated by k, and are $n^2\pi^2$ $(n = 1, 2, 3, \ldots)$. The integral operator K generated by k is therefore non-negative. We can now obtain information about the eigenvalues of the integral operator associated with (5.12).

Following the conversion procedure for boundary value problems in Chapter 2, we see that $k(x, t) = \psi_1(\min(x, t))\psi_2(\max(x, t))$ where ψ_1 and ψ_2 are solutions of the differential equation $\psi''(x) = 0$ $(0 < x < 1)$ which also satisfy the Wronskian condition $\psi_1(x)\psi_2'(x) - \psi_1'(x)\psi_2(x) = -1$ $(0 < x < 1)$ and for which $\psi_1(0) = \psi_2(1) = 0$. To construct the integral equation equivalent of (5.12) we require two functions $\hat{\psi}_1$ and $\hat{\psi}_2$ which satisfy the differential equation $\hat{\psi}''(x) = 0$ $(0 < x < 1)$, the condition $\hat{\psi}_1(x)\hat{\psi}_2'(x) - \hat{\psi}_1'(x)\hat{\psi}_2(x) = -1$ $(0 < x < 1)$ and for which $\hat{\psi}_1$ satisfies the boundary condition at 0 and $\hat{\psi}_2$ satisfies that at 1. The differential equation shows that there are constants $\alpha_1, \alpha_2, \beta_1, \beta_2$ for which $\hat{\psi}_1 = \alpha_1\psi_1 + \alpha_2\psi_2$ and $\hat{\psi}_2 = \beta_1\psi_1 + \beta_2\psi_2$, whence the Wronskian condition implies that

$\alpha_1 \beta_2 - \alpha_2 \beta_1 = 1$. Then (5.12) is equivalent to the integral equation

$$\phi(x) - \lambda \int_0^1 \hat{k}(x, t)\phi(t) \, dt = 0 \quad (0 \leqslant x \leqslant 1)$$

where $\hat{k}(x, t) = \hat{\psi}_1(\min(x, t))\hat{\psi}_2(\max(x, t))$. It is now easy to check that

$$\begin{aligned}
\hat{k}(x, t) - k(x, t) &= \alpha_1 \beta_1 \psi_1(x)\psi_1(t) + \alpha_2 \beta_2 \psi_2(x)\psi_2(t) \\
&\quad + \alpha_2 \beta_1 (\psi_1(x)\psi_2(t) + \psi_2(x)\psi_1(t)) \\
&= \beta_1 \psi_1(x)(\alpha_1 \psi_1(t) + \alpha_2 \psi_2(t)) + \alpha_2 \psi_2(x)(\beta_1 \psi_1(t) + \beta_2 \psi_2(t))
\end{aligned}$$

so that $\hat{k} - k$ is degenerate.

Let $\hat{K}$ be the integral operator arising from $\hat{k}$. Then since the image of $\hat{K} - K$ is spanned by ψ_1 and ψ_2, $\hat{K} - K$ has rank at most two and therefore has at most two negative eigenvalues. Therefore $\mu_3^-(\hat{K} - K) = 0$ and for all $n \in \mathbb{N}$

$$\mu_{n+2}^-(\hat{K}) \geqslant \mu_n^-(K) + \mu_3^-(\hat{K} - K) = \mu_n^-(K) = 0.$$

We now deduce that $\hat{K}$ has at most two negative eigenvalues. $\qquad \square$

The method of Example 5.2 can be adapted to any Sturm–Liouville problem, to show that if the integral operator arising from a particular choice of boundary conditions has all but finitely many eigenvalues of one sign, then the same is true for any other choice of boundary conditions.

Recall that an operator $K \in B(H)$ is said to be non-negative if and only if $(K\phi, \phi) \geqslant 0$ for all $\phi \in H$. From this condition it is clear that the eigenvalues of a non-negative operator are all non-negative.

Corollary 5.6

Suppose that K_1 and K_2 are compact self-adjoint operators on the Hilbert space H, and that $K_1 - K_2$ is a non-negative operator. Then, for all natural numbers n,

$$\mu_n^+(K_2) \leqslant \mu_n^+(K_1) \quad \text{and} \quad \mu_n^-(K_2) \leqslant \mu_n^-(K_1).$$

Proof

Since $((K_1 - K_2)\phi, \phi) \geqslant 0$ for all $\phi \in H$, we see that, for all non-zero $\phi \in H$, $R_{K_2}(\phi) \leqslant R_{K_1}(\phi)$. Therefore, if $n \in \mathbb{N}$,

$$\mu_n^+(K_2) = \sup_{\dim E = n} \left(\inf_{\phi \in E} R_{K_2}(\phi) \right) \leqslant \sup_{\dim E = n} \left(\inf_{\phi \in E} R_{K_1}(\phi) \right) = \mu_n^+(K_1)$$

while a similar calculation shows that $\mu_n^-(K_2) \leqslant \mu_n^-(K_1)$. $\qquad \square$

We can use Theorem 5.4 to estimate the difference between the corresponding eigenvalues of two operators.

Theorem 5.7

Let K and K_1 be two compact self-adjoint operators. Then for all $m \in \mathbb{N}$,

$$|\mu_m^{\pm}(K) - \mu_m^{\pm}(K_1)| \leqslant \|K - K_1\|.$$

Proof

By Theorem 5.4,

$$\mu_m^+(K) \leqslant \mu_m^+(K_1) + \mu_1^+(K - K_1) \leqslant \mu_m^+(K_1) + \|K - K_1\|,$$

and

$$\mu_m^+(K_1) \leqslant \mu_m^+(K) + \mu_1^+(K_1 - K) \leqslant \mu_m^+(K) + \|K_1 - K\|.$$

Combining these gives

$$|\mu_m^+(K) - \mu_m^+(K_1)| \leqslant \|K - K_1\|.$$

The result for μ_m^- is immediate on noticing that $\mu_m^-(K) = -\mu_m^+(-K)$, with the corresponding result for K_1. $\qquad\Box$

Theorem 5.7 essentially only gives us new information for small values of m, since we know that both $\mu_m(K)$ and $\mu_m(K_1)$ tend to zero as $m \to \infty$. It does, however, give us the useful result that if (K_N) is a sequence of compact self-adjoint operators approximating K, the eigenvalues of the approximating operators approximate those of K. We can state this more precisely as a corollary.

Corollary 5.8

Suppose that (K_N) is a sequence of compact self-adjoint operators in the Hilbert space H and that $\|K_N - K\| \to 0$ as $N \to \infty$. Then for each $m \in \mathbb{N}$,

$$\mu_m^{\pm}(K_N) \to \mu_m^{\pm}(K) \text{ as } N \to \infty.$$

Proof

By Theorem 5.7, $|\mu_m^{\pm}(K_N) - \mu_m^{\pm}(K)| \leqslant \|K_N - K\|$. $\qquad\Box$

We can, of course, apply Theorem 5.7 to the case where K_1 is a finite-rank approximation to K. Suppose that K_1 has N_1^+ positive eigenvalues and N_1^- negative ones. Then if m is a natural number $|\mu_{m+N_1^+}^+(K)| \leqslant \|K - K_1\|$ and $|\mu_{m+N_1^-}^-(K)| \leqslant \|K - K_1\|$, showing that at most N_1^+ positive and

N_1^- negative eigenvalues of K exceed $\|K - K_1\|$ in modulus. This is slightly more specific information about the eigenvalues of K than obtained by Example 5.1, but with the penalty that in this case the finite rank operator is assumed self-adjoint.

Results concerning the eigenvalues of products of operators are more complicated than those for sums. One reason for this is not hard to see: although the sum of two self-adjoint operators is self-adjoint, the product will only be so if the two operators commute. The simplest result in this direction is Theorem 5.9.

Theorem 5.9

Suppose that K is a compact self-adjoint operator on the Hilbert space H and that $T \in B(H)$. Then for $m \in \mathbb{N}$

$$\mu_m^+(T^*KT) \leqslant \mu_m^+(K)\|T\|^2,$$

$$\mu_m^-(T^*KT) \geqslant \mu_m^-(K)\|T\|^2,$$

and

$$|\mu_m(T^*KT)| \leqslant |\mu_m(K)|\|T\|^2.$$

Proof

By Lemma 5.3 $\quad \mu_m^+(T^*KT) = \inf_{\dim E = m-1} \left(\sup_{\phi \in E^\perp} R_{T^*KT}(\phi) \right).$

Let $\epsilon > 0$ and choose the subspace E_0 to have dimension $m-1$ and satisfy

$$\sup_{\phi \in E_0^\perp} R_K(\phi) < \mu_m^+(K) + \epsilon.$$

Now

$$T\phi \in E_0^\perp \Leftrightarrow (T\phi, \psi) = 0 \quad (\psi \in E_0)$$

$$\Leftrightarrow (\phi, T^*\psi) = 0 \quad (\psi \in E_0)$$

$$\Leftrightarrow \phi \in \{T^*\psi : \psi \in E_0\}^\perp.$$

Let $F_0 = \{T^*\psi : \psi \in E_0\}$; F_0 has dimension at most $m-1$. Then, for all $\phi \in F_0^\perp$, $T\phi \in E_0^\perp$ so

$$(T^*KT\phi, \phi) = (KT\phi, T\phi)$$

$$\leqslant \sup\{(K\psi, \psi) : \psi \in E_0^\perp, \|\psi\| = 1\}\|T\phi\|^2$$

$$\leqslant \|T\phi\|^2(\mu_m^+(K) + \epsilon).$$

Therefore

$$\sup_{\phi \in F_0^\perp} R_{T^*KT}(\phi) \le \|T\|^2 (\mu_m^+(K) + \epsilon).$$

Let F_1 be a subspace of dimension $m-1$ which contains F_0. Then $F_1^\perp \subset F_0^\perp$ so that

$$\sup_{\phi \in F_1^\perp} R_{T^*KT}(\phi) \le \|T\|^2 (\mu_m^+(K) + \epsilon)$$

and therefore

$$\mu_m^+(T^*KT) = \inf_{\dim E = m-1} \left(\sup_{\phi \in E^\perp} R_{T^*KT}(\phi) \right) \le (\mu_m^+(K) + \epsilon) \|T\|^2.$$

Since $\epsilon > 0$ was arbitrary $\mu_m^+(T^*KT) \le \mu_m^+(K) \|T\|^2$.

The proof for the result involving $\mu_m^-(T^*KT)$ may be deduced immediately by substituting $-K$ for K. For the result about $|\mu_m(T^*KT)|$, suppose that among the first $m-1$ eigenvalues of K (counted in the usual way) n are negative and p are positive. Then $\mu_{n+1}^-(T^*KT) \ge \mu_{n+1}^-(K) \|T\|^2 \ge -|\mu_m(K)| \|T\|^2$ (the last inequality since $\mu_{n+1}^-(K)$ is the $(n+1)$-th negative eigenvalue of K and only n negative eigenvalues occur in $\mu_1(K), \ldots, \mu_{m-1}(K)$). Similarly $\mu_{p+1}^+(T^*KT) \le |\mu_m(K)| \|T\|^2$. Therefore at most n negative and p positive eigenvalues of T^*KT have modulus exceeding $|\mu_m(K)| \|T\|^2$. Since $n + p = m - 1$, $|\mu_m(T^*KT)| \le |\mu_m(K)|$. □

One immediate application of Theorem 5.9 is to the special case where T is chosen to be a projection. Suppose we wish to consider a finite-dimensional operator related to K with the intention of obtaining information about the eigenvalues of K. Let E be a finite-dimensional subspace of H and P be the orthogonal projection of H onto E. The operator K need not necessarily map E into itself so we 'compress' it into E and consider the operator $\hat{K}: E \to E$ defined by $\hat{K}\phi = PK\phi$ ($\phi \in E$). $\hat{K}$ acts in the finite-dimensional space E so we can in principle find its eigenvalues by the usual techniques of matrix theory. Now if $\phi \in E$ is an eigenvector of $\hat{K}$ corresponding to the non-zero eigenvalue μ, $\hat{K}\phi = \mu\phi$ and hence $PK\phi = \mu\phi$ and (since $P\phi = \phi$) $PKP\phi = \mu\phi$ so μ is an eigenvalue and $\phi \in H$ a corresponding eigenvector of PKP. Conversely if $PKP\phi = \mu\phi$ with $\phi \ne 0$ and $\phi \in H$, then either $\mu = 0$ or $\phi = \mu^{-1}PKP\phi = P(KP\mu^{-1}\phi) \in E$; in the latter case $\mu\phi = PK\phi = \hat{K}\phi$ so μ is an eigenvalue of $\hat{K}$. Therefore the non-zero eigenvalues of $\hat{K}$ are the same as those of PKP. But $P = P^*$ and $\|P\| \le 1$ so we have, by Theorem 5.9, that $\mu_m^+(\hat{K}) = \mu_m^+(PKP) \le \mu_m^+(K) \|P\|^2 \le \mu_m^+(K)$ and $\mu_m^-(\hat{K}) \ge \mu_m^-(K)$, these inequalities being valid for all $m \in \mathbb{N}$. Although the motivation for this arose from the restriction of K to the finite-dimensional subspace E, we have only used the existence of an orthogonal projection onto E, which exists so long as E is a closed subspace (which is automatic if E has finite dimension). We may sum up what we have proved in the following corollary.

Corollary 5.10

Suppose that E is a closed subspace of the Hilbert space H and that P is the orthogonal projection of H onto E. Then if K is a compact self-adjoint operator on H, we have, for all $m \in \mathbb{N}$, $\mu_m^+(PKP) \leqslant \mu_m^+(K)$, $\mu_m^-(PKP) \geqslant \mu_m^-(K)$ and $|\mu_m(PKP)| \leqslant |\mu_m(K)|$. In particular if $\hat{K}: E \to E$ is the operator defined by $\hat{K}\phi = PK\phi (\phi \in E)$ then, for all $m \in \mathbb{N}$, $\mu_m^+(\hat{K}) \leqslant \mu_m^+(K)$, $\mu_m^-(\hat{K}) \geqslant \mu_m^-(K)$ and $|\mu_m(\hat{K})| \leqslant |\mu_m(K)|$. $\qquad\square$

The result of Corollary 5.10 will be exploited in Chapter 7; its main feature is that we can construct the 'compression' $\hat{K}$ of K into E, calculate its eigenvalues and we then know that the positive eigenvalues of $\hat{K}$ are underestimates of those of K, while the negative eigenvalues of $\hat{K}$ are overestimates of those of K.

We can relate Theorem 5.9 specifically to integral operators as follows.

Theorem 5.11

Let k be a kernel on $[a, b] \times [a, b]$ which generates a compact self-adjoint operator K on $L_2(a, b)$, and let m be a bounded measurable function on $[a, b]$. Then if L is defined by

$$(L\phi)(x) = \int_a^b k(x, t)m(x)\overline{m(t)}\phi(t)\,dt \quad (a \leqslant x \leqslant b),$$

L is compact and, for all $n \in \mathbb{N}$,

$$\mu_n^+(L) \leqslant M^2 \mu_n^+(K), \ \mu_n^-(L) \geqslant M^2 \mu_n^-(K) \text{ and } |\mu_n(L)| \leqslant M^2 |\mu_n(K)|,$$

where $M = \sup\{|m(x)|: a \leqslant x \leqslant b\}$.

If, in addition, there is a positive constant M_0 for which $|m(x)| \geqslant M_0$ for all $x \in [a, b]$, then for all $n \in \mathbb{N}$,

$$\mu_n^+(L) \geqslant M_0^2 \mu_n^+(K), \ \mu_n^-(L) \leqslant M_0^2 \mu_n^-(K) \text{ and }$$
$$|\mu_n(L)| \geqslant M_0^2 |\mu_n(K)|.$$

Proof

Define T by $(T\phi)(x) = \overline{m(x)}\phi(x) (a \leqslant x \leqslant b)$. Then $T\phi$ belongs to $L_2(a, b)$ and $\|T\phi\| \leqslant M\|\phi\|$. It is easily checked that T^* is given by $(T^*\phi)(x) = m(x)\phi(x)$

$(a \leqslant x \leqslant b)$, whence we see that $L = T^*KT$. The compactness of L is now immediate, while an application of Theorem 5.9 shows that for all $n \in \mathbb{N}$

$$\mu_n^+(L) \leqslant \mu_n^+(K)\|T\|^2 \leqslant M^2 \mu_n^+(K), \ \mu_n^-(L) \geqslant M^2 \mu_n^-(K)$$

and

$$|\mu_n(L)| \leqslant M^2 |\mu_n(K)|.$$

The remaining inequalities are proved by observing that if for all $x \in [a, b]$ $|m(x)| \geqslant M_0 > 0$, then $(T^{-1}\phi)(x) = \phi(x)/\overline{m(x)}$ and $\|T^{-1}\| \leqslant 1/M_0$. Noticing that $K = T^{*-1}LT^{-1}$ and applying Theorem 5.9 now proves the remainder. □

Remark

It is not necessary that the function m of Theorem 5.11 be bounded, only that it be essentially bounded, that is, equal almost everywhere to a bounded function. This minor strengthening of the result will not concern us in this book.

In the conversion of Sturm–Liouville boundary value problems into integral equations we encounter integral operators of the form $\int_0^1 k(x, t)r(t) \phi(t) \, dt$ where k and r are continuous and real-valued and k is symmetric. In the case where r is positive, we produce an alternative formulation by multiplying through by $(r(x))^{\frac{1}{2}}$ to obtain $\int_0^1 k(x, t) (r(x)r(t))^{\frac{1}{2}} \psi(t) \, dt$. It is easily checked that the operators K_1 and L on $L_2(0, 1)$ defined by $(K_1\phi)(x) = \int_0^1 k(x, t)r(t)\phi(t) \, dt$ and $(L\phi)(x) = \int_0^1 k(x, t)(r(x)r(t))^{\frac{1}{2}}\phi(t) \, dt$ have the same non-zero eigenvalues. (Notice, however, that K_1 is not generally self-adjoint.) This provides the background for the example below.

Example 5.3

The operators K_1 and K where

$$(K_1\phi)(x) = \int_0^1 k(x, t)r(t)\phi(t) \, dt$$

and

$$(K\phi)(x) = \int_0^1 k(x, t)\phi(t) \, dt,$$

where k and r are continuous and real-valued, r is non-negative and k is symmetric, are compact operators on $L_2(0, 1)$. We show that their eigenvalues are related by

$$\left. \begin{array}{l} \mu_n^+(K) \inf_{0 \leqslant x \leqslant 1} r(x) \leqslant \mu_n^+(K_1) \leqslant \mu_n^+(K) \sup_{0 \leqslant x \leqslant 1} r(x) \\ \\ \mu_n^-(K) \inf_{0 \leqslant x \leqslant 1} r(x) \geqslant \mu_n^-(K_1) \geqslant \mu_n^-(K) \sup_{0 \leqslant x \leqslant 1} r(x). \end{array} \right\} \quad (5.13)$$

and

With L as defined immediately before this example, we can easily check that $K_1\phi = \mu\phi$ implies that $L\psi = \mu\psi$ where $\psi = r^{\frac{1}{2}}\phi$, so that every eigenvalue of K_1 is also an eigenvalue of L. Moreover, if $\psi \in L_2(0,1)$ and $L\psi = \mu\psi$, then $\mu\psi(x) = (r(x))^{\frac{1}{2}}\int_0^1 k(x,t)(r(t))^{\frac{1}{2}}\psi(t)\,dt$, whence it follows that $\mu\psi/r^{\frac{1}{2}} \in L_2(0,1)$. If $\mu \neq 0$, set $\phi = \psi/r^{\frac{1}{2}}$ and we have $\phi \in L_2(0,1)$ and $K_1\phi = \mu\phi$. Therefore the non-zero eigenvalues of K_1 and L are identical. The rest of the example is just an application of Theorem 5.11 with $m(x) = (r(x))^{\frac{1}{2}}$, producing the right hand inequalities in (5.13) on noticing that $\mu_n^+(K_1) = \mu_n^+(L)$. The left hand inequalities also follow from Theorem 5.11 if $\inf r(x) > 0$ and are trivial if $\inf r(x) = 0$.   □

Theorem 5.12

Suppose that K is a compact self-adjoint operator on H and that T is a second compact operator on H. Then, for $n \in \mathbb{N}$,

$$\mu_n^+(T^*KT) \leqslant \min_{1 \leqslant k \leqslant n} \mu_k^+(K)\mu_{n-k+1}^+(T^*T) \quad (5.14)$$

and

$$\mu_n^-(T^*KT) \geqslant \max_{1 \leqslant k \leqslant n} \mu_k^-(K)\mu_{n-k+1}^+(T^*T). \quad (5.15)$$

Proof

Choose $n \in \mathbb{N}$ and fix it. Now choose k with $1 \leqslant k \leqslant n$ and let $\epsilon > 0$.

By Lemma 5.3 there are subspaces E of dimension $k-1$ and F of dimension $n-k$ for which

$$\sup_{\phi \in E^\perp} R_K(\phi) < \mu_k^+(K) + \epsilon$$

and

$$\sup_{\phi \in F^\perp} R_{T^*T}(\phi) < \mu_{n-k+1}^+(T^*T) + \epsilon.$$

Now

$$\phi \in \{T^*\psi : \psi \in E\}^\perp \Rightarrow \text{for all } \psi \in E \quad (\phi, T^*\psi) = 0$$
$$\Rightarrow \text{for all } \psi \in E \quad (T\phi, \psi) = 0$$
$$\Rightarrow T\phi \in E^\perp,$$

so let $E_0 = \{T^*\psi : \psi \in E\}$. E_0 has dimension at most $k-1$, and

$$\phi \in E_0^\perp \Rightarrow T\phi \in E^\perp.$$

Let $\phi \in F^\perp \cap E_0^\perp$. Then $T\phi \in E^\perp$ and for $\phi \neq 0$

$$(KT\phi, T\phi) \leqslant \|T\phi\|^2 \sup_{\psi \in E^\perp} R_K(\psi)$$
$$\leqslant (T^*T\phi, \phi)(\mu_k^+(K) + \epsilon),$$
$$< (\mu_{n-k+1}^+(T^*T) + \epsilon)(\mu_k^+(K) + \epsilon)\|\phi\|^2.$$

Therefore

$$\sup_{\phi \in F^\perp \cap E_0^\perp} R_{T^*KT}(\phi) < (\mu_{n-k+1}^+(T^*T) + \epsilon)(\mu_k^+(K) + \epsilon).$$

Now $F^\perp \cap E_0^\perp = (F + E_0)^\perp$, and $F + E_0$ is a subspace of dimension at most $n-1$. Let F_0 be a subspace of dimension $n-1$ containing $F + E_0$, so that $F_0^\perp \subset F^\perp \cap E_0^\perp$ and

$$\sup_{\phi \in F_0^\perp} R_{T^*KT}(\phi) < (\mu_{n-k+1}^+(T^*T) + \epsilon)(\mu_k^+(K) + \epsilon).$$

However, because F_0 has dimension $n-1$, and letting $\tilde{F}$ denote a subspace of H,

$$\mu_n^+(T^*KT) = \inf_{\dim \tilde{F} = n-1} \left(\sup_{\phi \in \tilde{F}^\perp} R_{T^*KT}(\phi) \right)$$
$$< (\mu_{n-k+1}^+(T^*T) + \epsilon)(\mu_k^+(K) + \epsilon).$$

Since $\epsilon > 0$ was arbitrary, we deduce that

$$\mu_n^+(T^*KT) \leqslant \mu_k^+(K)\mu_{n-k+1}^+(T^*T). \tag{5.16}$$

Because (5.16) holds for all k with $1 \leqslant k \leqslant n$, we deduce (5.14), and (5.15) follows on substituting $-K$ for K in (5.14). $\qquad \square$

Theorem 5.12 requires that T be compact, which is not in general true of the operator of Example 5.3; if $r(x) = x$ $(0 \leqslant x \leqslant 1)$, for example, and $(T\phi)(x) =$

$(r(x))^{\frac{1}{2}}\phi(x)$, then it is easily checked that for all $\mu \in [0,1]$, $T-\mu I$ is not invertible and μ is not an eigenvalue. In the case when T is compact, Theorem 5.12 contains the first two parts of the result of Theorem 5.9, for $\mu_1^+(T^*T)=\|T^*T\|=\|T\|^2$ and then $\mu_m^+(T^*KT) \leqslant \mu_m^+(K)\mu_1^+(T^*T)$.

Example 5.4

Suppose that k is a continuous hermitian kernel on $[a,b] \times [a,b]$ with the property that $\partial^2 k/\partial t\partial x$ exists and is a hermitian L_2-kernel. Then we shall show that if μ_n is the usual sequence of eigenvalues of the operator K generated by k, $\Sigma (n^2\mu_n)^2$ converges.

Let $k_1(x,t)=(\partial^2 k/\partial t\partial x)(x,t)$ which we know to be an L_2-kernel; let K_1 be the operator it generates on $L_2(a,b)$. For all x for which $\int_a^b |k_1(x,t)|^2 \, \mathrm{d}t < \infty$, $t \mapsto k_1(x,t)$ is an L_1 function (by Schwarz's inequality) and so if we set $F(x,t)=\int_a^t k_1(x,v) \, \mathrm{d}v$, $(\partial F/\partial t)(x,t)=k_1(x,t)$ for almost all $t \in [a,b]$ by Theorem B6. Then

$$\int_a^b -F(x,t)\phi(t) \, \mathrm{d}t + \int_a^b k_1(x,t)\Phi(t) \, \mathrm{d}t = 0 \qquad (5.17)$$

where $\phi \in L_2(a,b)$ and

$$\Phi(t) = \int_t^b \phi(u) \, \mathrm{d}u = (V^*\phi)(t) \text{ with } (V\psi)(x) = \int_a^x \psi(u) \, \mathrm{d}u.$$

Then, since (5.17) is true for almost all $x \in [a,b]$,

$$(VK_1V^*\phi)(x) = \int_a^x \mathrm{d}u \int_a^b k_1(u,t)\Phi(t) \, \mathrm{d}t = \int_a^x \mathrm{d}u \int_a^b F(u,t)\phi(t) \, \mathrm{d}t$$

$$= \int_a^b k_0(x,t)\phi(t) \, \mathrm{d}t \qquad (5.18)$$

where

$$k_0(x,t) = \int_a^x \mathrm{d}u \int_a^t k_1(u,v) \, \mathrm{d}v.$$

Let K_0 be the operator generated by k_0, so that $K_0=VK_1V^*$. Now $(VV^*\phi)(x)=\int_a^b (\min(x,t)-a)\phi(t) \, \mathrm{d}t$ and $\mu_n^+(VV^*)=(b-a)^2/((n-\frac{1}{2})^2\pi^2)$, by extending Problem 4.4. Then for each $m \in \mathbb{N}$, by Theorem 5.12,

$$\mu_m^+(K_0) \leqslant \min_{1 \leqslant k \leqslant m} \mu_k^+(K_1)\mu_{m-k+1}^+(VV^*).$$

In particular

$$\mu_{2m}^+(K_0) \leqslant \mu_m^+(K_1)\mu_{m+1}^+(VV^*) = (b-a)^2 \ \mu_m^+(K_1)/((m+\tfrac{1}{2})^2\pi^2),$$

and

$$\mu_{2m-1}^+(K_0) \leqslant \mu_m^+(K_1)\mu_m^+(VV^*) = (b-a)^2 \ \mu_m^+(K_1)/((m-\tfrac{1}{2})^2\pi^2).$$

From these two inequalities, and the knowledge that $\Sigma\,(\mu_m^+(K_1))^2$ converges, because k_1 is an L_2-kernel, we see that $\Sigma\,(m^2\mu_m^+(K_0))^2$ converges. A similar calculation with $\mu_m^-(K_0)$ shows that $\Sigma\,(m^2\mu_m^-(K_0))^2$ converges.

Now $k(x,t) = k_0(x,t) + k(a,t) + k(x,a) - k(a,a)$ so that $k - k_0$ is a degenerate kernel, of rank at most two. (Regard $k(x,a) - k(a,a)$ as a single function of x.) Then, by Theorem 5.5, $|\mu_{m+2}(K)| \leqslant |\mu_m(K_0)|$ $(m \in \mathbb{N})$ and we see that $\Sigma\,(m^2\mu_m(K))^2$ converges.

This technique is capable of variation to produce results on the eigenvalue distribution of kernels of appropriate smoothness. □

Problems

5.1   Suppose that k is a hermitian kernel on $[a,b] \times [a,b]$ which generates a compact integral operator K, and that for all $x,t \in [a,b]$ $k(x,t) \geqslant 0$. By observing that $\int_a^b \int_a^b k(x,t)\,dt\,dx > 0$, show that, unless $k(x,t) = 0$ almost everywhere, K has at least one positive eigenvalue. By considering $|\phi_1^+|$ show that the eigenvector ϕ_1^+ corresponding to the positive eigenvalue μ_1^+ may be assumed to have non-negative values.

(To see that this is the best possible such result consider the kernel $k(x,t) = \max(x,t)$ which has one positive and infinitely many negative eigenvalues.)

5.2   Let K be a compact self-adjoint operator on the Hilbert space H. Show that if $\omega \in \mathbb{R}$ then $((K - \omega I)^2\phi, \phi) \geqslant \inf_n |\mu_n - \omega|^2 \|\phi\|^2$ for all $\phi \in H$. (Use the Spectral Theorem and its corollaries.) Deduce that

$$\inf_{\phi \neq 0} R_{(K-\omega I)^2}(\phi) = \inf_{n \in \mathbb{N}} |\mu_n - \omega|^2.$$

Let $(K\phi)(x) = \int_0^1 \max(x,t)\phi(t)\,dt$ $(0 \leqslant x \leqslant 1)$. (This K has one positive and infinitely many negative eigenvalues, the positive one being $1/\alpha^2$ where α satisfies $\alpha = \coth\alpha$.) By putting $\phi = 1$, and choosing the value of ω to minimise $R_{(K-\omega I)^2}(\phi)$ show that there is an eigenvalue μ of K satisfying $|\mu - \tfrac{2}{3}| < 0.149$. (More systematic and accurate methods of estimating eigenvalues are discussed in Chapter 7.)

5.3   Let H be a Hilbert space and $T \in B(H)$. Denote by $W(T)$ the set $\{(T\phi, \phi): \phi \in H, \|\phi\| = 1\} = \{R_T(\phi): \phi \neq 0\}$, and for a scalar μ let $d_\mu = \inf\{|\mu - \alpha|: \alpha \in W(T)\}$.

Show that if $\|\phi\|=1$ then $\|(T-\mu I)\phi\|\geq d_\mu$ and that $\|(T-\mu I)^*\phi\|\geq d_\mu$. Deduce that if $d_\mu>0$ then $T-\mu I$ is invertible and $\|(T-\mu I)^{-1}\|\leq d_\mu^{-1}$.

5.4   Consider the operator K from $L_2(0,1)$ to itself defined by $(K\phi)(x)=\int_0^1 (x+t)^{-\alpha}\phi(t)\,dt$ where $0<\alpha<1$. By observing that this has a Schur kernel and approximating K by operators whose kernels are bounded functions, prove that K is compact and self-adjoint with $\|K\|\leq 1/(1-\alpha)$. Show that

$$\frac{2(2^{1-\alpha}-1)}{(1-\alpha)(2-\alpha)}\leq\mu_1^+\leq\frac{1}{1-\alpha}.$$

5.5   Show that the operator L defined by $(L\phi)(x)=\int_0^1 (x+t)\phi(t)\,dt$ $(0\leq x\leq 1)$ has one positive and one negative eigenvalue. Let K_1 and K_2 be defined on $L_2(0,1)$ by

$$(K_1\phi)(x)=\int_0^1 \min(x,t)\phi(t)\,dt, \quad (K_2\phi)(x)=\int_0^1 \max(x,t)\phi(t)\,dt$$

(both for $0\leq x\leq 1$). (The eigenvalues of K_1 are calculated in Problem 4.4.) Hence show that for $n\in\mathbb{N}$

$$\mu_{n+1}^+(K_1)\leq\mu_n^+(-K_2)=-\mu_n^-(K_2)$$

and $|\mu_{n+1}^-(K_2)|\leq\mu_n^+(K_1)$, to deduce that $\mu_n^-(K_2)$ lies in the interval $(-1/((n-\frac{3}{2})^2\pi^2), -1/((n+\frac{1}{2})^2\pi^2))$ if $n\geq 2$.

5.6   It is very tempting to consider whether Theorem 5.4 could be strengthened to compare $\mu_n^+(K_1+K_2)$ with $\mu_n^+(K_1)+\mu_n^+(K_2)$. To see that the former is not in general less than or equal to the latter, consider the 2×2 matrices $\left(\begin{smallmatrix}2&0\\0&0\end{smallmatrix}\right)$ and $\left(\begin{smallmatrix}1&0\\0&2\end{smallmatrix}\right)$.

5.7   Suppose that k_α is a hermitian L_2-kernel on $[0,1]\times[0,1]$ which generates the operator K_α, this being true for $0\leq\alpha<1$. Show that if $\int_0^1\int_0^1 |k_\alpha(x,t)-k_0(x,t)|^2\,dx\,dt\to 0$ as $\alpha\to 0+$ then $\mu_n^+(K_\alpha)\to\mu_n^+(K_0)$ and $\mu_n^-(K_\alpha)\to\mu_n^-(K_0)$ as $\alpha\to 0+$.

Let $k_\alpha(x,t)=(x+t)^{-\alpha}$; this generates a compact self-adjoint operator on $L_2(0,1)$ for $0\leq\alpha<1$. (For $\alpha\geq\frac{1}{2}$ the compactness follows from Problem 3.8.) Show that $\|K_{\frac{1}{4}}-K_0\|<0.17$ and deduce that $K_{\frac{1}{4}}$ has one eigenvalue in the interval $(0.83, 1.17)$, all the others lying in $(-0.17, +0.17)$.

5.8   Suppose that K is a compact self-adjoint operator on the Hilbert space H and that P is a projection. Show that if K is non-negative then all the eigenvalues of PKP are non-negative but that the converse is false (that is, there may exist projections P for which PKP is non-negative even if K is not non-negative). Deduce that if PKP has eigenvalues of both signs then K also has eigenvalues of both signs.

5.9   Suppose that the kernel k on $[a,b]\times[a,b]$ generates a compact self-adjoint operator K on $L_2(a,b)$. Let $[c,d]\subset[a,b]$ and let $\ell(x,t)=k(x,t)$ for $x,t\in[c,d]$ with $\ell(x,t)=0$ otherwise. Show that if L is the operator on $L_2(a,b)$ generated by the kernel ℓ, then $|\mu_n(L)|\leq|\mu_n(K)|$ $(n=1,2,\ldots)$.

5.10   Suppose that $K=S^*S$ for some bounded linear map S and that T is a compact operator on the same Hilbert space. By observing that $\mu_n^+(T^*KT) = \mu_n^+(STT^*S^*)$ show that $\mu_n^+(T^*KT) \leqslant \|S\|^2 \mu_n^+(TT^*) = \|K\| \mu_n^+(T^*T)$. (This would be a consequence of Theorem 5.12 if S were known to be compact.) Use this in connection with the operator K on $L_2(0,1)$ defined by $(K\phi)(x) = e^x \phi(x)$ to show that the kernel $e^{\min(x,t)} - 1$ generates a non-negative operator on $L_2(0,1)$, say L, with $\mu_n^+(L) \leqslant e/((n-\tfrac{1}{2})^2 \pi^2)$.

5.11 Let $k(x,t)$ and $\ell(x,t)$ be hermitian L_2-kernels on $[a,b] \times [a,b]$, where $\ell(x,t)$ is degenerate of rank N (that is, $\ell(x,t)$ is of the form $\Sigma_{i=1}^N a_i(x) b_i(t)$). Letting K and L be the operators on $L_2(a,b)$ generated by k and ℓ respectively, show that $|\mu_{n+N}(K)| \leqslant |\mu_n(K-L)|$ $(n=1,2,3,\ldots)$ and hence that

$$\sum_{n=N+1}^{\infty} |\mu_n(K)|^2 \leqslant \int_a^b \int_a^b |k(x,t) - \ell(x,t)|^2 \, dt \, dx.$$

5.12 Suppose that $k(x,t)$ is a hermitian kernel on $[a,b] \times [a,b]$ which is continuous and for which the partial derivatives $\partial k/\partial x$ and $\partial k/\partial t$ are continuous. Choose $\ell(x,t)$ by dividing $[a,b] \times [a,b]$ into N^2 squares of side $(b-a)/N$ and letting $\ell(x,t)$ be constant on each square with value equal to the value of $k(x,t)$ at the centre of the square. Deduce that there is a constant C such that, for all natural numbers N, $\Sigma_{n=N+1}^{\infty} |\mu_n(K)|^2 \leqslant CN^{-2}$. Using the fact that $(|\mu_n(K)|)$ is, by definition, a decreasing sequence, deduce that for some constant D, $|\mu_n(K)| \leqslant Dn^{-\frac{3}{2}}$ $(n=1,2,\ldots)$.

(This result, proved by Weyl in 1912, is the prototype for others relating the smoothness of the kernel to the rate at which $|\mu_n|$ tends to zero.)

5.13 In Problem 2.4 we related the boundary value problem for the buckling of an elastic column under a compressive load

$$(I(x)\phi''(x))'' + \lambda \phi''(x) = 0 \quad (0 < x < 1),$$
$$\phi(0) = \phi''(0) = \phi(1) = \phi''(1) = 0,$$

where $I(x) > 0$ $(0 \leqslant x \leqslant 1)$ to an integral equation with kernel

$$(\min(x,t) - xt)/(I(x)I(t))^{\frac{1}{2}}.$$

Given that $0 < I_0 \leqslant I(x) \leqslant I_1$ $(0 \leqslant x \leqslant 1)$ show that the eigenvalues (μ_n) of the integral operator satisfy $1/(n^2\pi^2 I_1) \leqslant \mu_n \leqslant 1/(n^2\pi^2 I_0)$ and hence that the eigenvalues (λ_n) of the boundary value problem satisfy $n^2\pi^2 I_0 \leqslant \lambda_n \leqslant n^2\pi^2 I_1$ $(n \in \mathbb{N})$.

5.14 The regular Sturm–Liouville boundary value problem

$$(p(x)\phi'(x))' - q(x)\phi(x) + \lambda r(x)\phi(x) = 0 \quad (0 < x < 1),$$
$$\phi(0) = \phi(1) = 0,$$

is equivalent to the integral equation

$$\phi(x) = \lambda \int_0^1 k_0(x,t) r(t) \phi(t) \, dt \quad (0 \leqslant x \leqslant 1)$$

where $k_0(x, t) = \psi_1(\min(x, t))\psi_2(\max(x, t))$ for appropriately defined functions ψ_1 and ψ_2, subject to there being no non-trivial solution of the boundary value problem corresponding to $\lambda = 0$. Use the result of Problem 2.8 to show that, provided $\cos \omega_0 \psi_2(0) - \sin \omega_0 \psi_2'(0) \neq 0$, the same differential equation together with the boundary conditions

$$\cos \omega_0 \phi(0) - \sin \omega_0 \phi'(0) = 0, \quad \phi(1) = 0, \tag{5.18}$$

is equivalent to

$$\phi(x) = \lambda \int_0^1 k_{\omega_0}(x, t) r(t) \phi(t) \, dt \quad (0 \leq x \leq 1)$$

where

$$k_{\omega_0}(x, t) = k_0(x, t) + A_{\omega_0} \psi_2(x) \psi_2(t).$$

By noticing that $A_0 = A_\pi = 0$ show that if α is the least positive value of ω for which $\cos \omega \psi_2(0) - \sin \omega \psi_2'(0) = 0$ then $A_{\omega_0} > 0$ for $\omega_0 \in (0, \alpha)$ and $A_{\omega_0} < 0$ for $\omega_0 \in (\alpha, \pi)$. Deduce that the eigenvalues of the operator generated by k_{ω_0} increase as ω_0 increases from 0 to α.

(A similar increase is evident as ω_1 increases if the boundary conditions are replaced by $\phi(0) = 0$, $\cos \omega_1 \phi(1) + \sin \omega_1 \phi'(1) = 0$.)

5.15 (i) Let E be a finite-dimensional subspace of a Hilbert space, $T: E \to E$ be a self-adjoint linear map and $(\phi_1, \ldots, \phi_N)$ and $(\psi_1, \ldots, \psi_N)$ be two orthonormal bases for E. Show that

$$\sum_{n=1}^N (T\phi_n, \phi_n) = \sum_{m,n=1}^N (T\phi_n, \psi_m)(\psi_m, \phi_n) = \sum_{m=1}^N (T\psi_m, \psi_m)$$

and deduce that $\sum_{n=1}^N (T\phi_n, \phi_n)$ equals the sum of the eigenvalues of T.

(ii) Let K be a compact self-adjoint operator on the Hilbert space H and let $(\phi_1, \ldots, \phi_N)$ be an orthonormal set in H. Prove that

$$\sum_{n=1}^N \mu_n^-(K) \leq \sum_{n=1}^N (K\phi_n, \phi_n) \leq \sum_{n=1}^N \mu_n^+(K).$$

By considering separately the terms for which $(K\phi_n, \phi_n) \geq 0$ and those with $(K\phi_n, \phi_n) < 0$, deduce that

$$\sum_{n=1}^N |(K\phi_n, \phi_n)| \leq \sum_{n=1}^N |\mu_n(K)|.$$

6

Positive operators

6.1 Introduction

We have already defined a non-negative operator K to be one which is self-adjoint and such that $(K\phi, \phi) \geqslant 0$ for all ϕ. A positive operator is self-adjoint and satisfies the stronger condition $(K\phi, \phi) > 0$ for all $\phi \neq 0$. In Mercer's Theorem we saw that if the operator generated by a continuous kernel $k(x, t)$ $(0 \leqslant x, t \leqslant 1)$ is non-negative, then $k(x, t)$ can be expanded in a uniformly convergent series and the eigenvalues of the operator are such that $\Sigma_{n=1}^{\infty} \mu_n = \int_0^1 k(x, x) \, dx$. If we know that all of these eigenvalues are positive then each of them does not exceed the value of the integral, which is therefore an upper bound for the largest eigenvalue.

Given an arbitrary operator T in a Hilbert space, we can construct the operator T^*T which is non-negative since $(T^*T\phi, \phi) = \|T\phi\|^2 \geqslant 0$. We are therefore well supplied with non-negative operators. In practice, however, we usually need to determine whether a given operator is non-negative and this is a difficult issue to resolve. We give it some attention in this chapter, because there are many techniques, especially in approximation theory, which apply only for non-negative or positive operators. The eigenvalue bound referred to above is a simple example.

As we intend to apply our results to integral equations, we shall restrict attention to compact operators and those closely related to them. More generality is possible for most of the purely operator-theoretic results than we need to give here.

A particular application of positive integral operators is to those arising from Sturm–Liouville boundary value problems, and in the final section we conclude our discussion of this class of operators, drawing together the information we have gathered so far.

6.2 General properties

Definition 6.1

Let H be a Hilbert space and let $K \in B(H)$ be self-adjoint. K is said to be *non-negative* if and only if $(K\phi, \phi) \geqslant 0$ for all $\phi \in H$. It is said to be *positive* if and only if $(K\phi, \phi) > 0$ for all $\phi \neq 0$ in H. If there is a positive constant m for which $(K\phi, \phi) \geqslant m\|\phi\|^2$ for all $\phi \in H$, then K is said to be *positive and bounded below*. □

The terms non-negative and positive are usually replaced by positive semi-definite and positive definite in the context of matrix theory. The analogous definitions of non-positive and negative operators follow immediately. Note that if H is a complex Hilbert space then the stipulation in Definition 6.1 that K be self-adjoint is superfluous, in the sense that if $(K\phi, \phi)$ is real for all $\phi \in H$, then K is necessarily self-adjoint. If H is a real Hilbert space the requirement that K is self-adjoint is necessary. (See Problem 6.1.)

Lemma 6.1

A compact self-adjoint operator is non-negative (positive) if and only if all of its eigenvalues are non-negative (positive).

Proof

Let $K \in B(H)$ be the operator and (μ_n) and (ϕ_n) be the sequences of its eigenvalues and eigenvectors respectively. Then for $\phi \in H$

$$(K\phi, \phi) = \sum_{n=1}^{\infty} \mu_n |(\phi, \phi_n)|^2$$

from which the result is immediate. □

Lemma 6.2

The sum of two non-negative operators is non-negative and is positive if one of the summands is positive. If an operator is positive and bounded below, then it is invertible and its inverse is positive and bounded below.

Proof

Suppose that K_1 and K_2 are non-negative and operate in the Hilbert space H. Then for all $\phi \in H$, $((K_1 + K_2)\phi, \phi) = (K_1\phi, \phi) + (K_2\phi, \phi) \geqslant 0$, the inequality being strict if at least one of the operators K_1, K_2 is positive.

Let K be positive and bounded below, with $(K\phi, \phi) \geqslant m\|\phi\|^2$ for all $\phi \in H$, where m is a positive constant. Then, by the Cauchy–Schwarz inequality, $\|K\phi\| \|\phi\| \geqslant (K\phi, \phi) \geqslant m\|\phi\|^2$, so that $\|K\phi\| \geqslant m\|\phi\|$. Since this holds for all ϕ, K is injective. By Theorem A13 $\overline{\mathscr{R}(K^*)} = \overline{\mathscr{R}(K)} = H$, so if $\psi \in H$ there is a sequence (ϕ_n) for which $K\phi_n \to \psi$ as $n \to \infty$. Since $\|\phi_n - \phi_{n'}\| \leqslant m^{-1}\|K\phi_n - K\phi_{n'}\|$, for all $n, n' \in \mathbb{N}$, (ϕ_n) is a Cauchy sequence in H and is therefore convergent. Let $\phi_n \to \phi$ as $n \to \infty$, then $\psi = K\phi$ and so

$\psi \in \mathcal{R}(K)$. Since ψ is a typical element of H, $\mathcal{R}(K)=H$. It follows that K is bijective and has an inverse, K^{-1} say, as a linear map. Now $\|K\phi\| \geqslant m\|\phi\|$ for all $\phi \in H$ and therefore, for all $\psi \in H$, $m^{-1}\|\psi\| \geqslant \|K^{-1}\psi\|$, so that K^{-1} is bounded. To show that K^{-1} is positive and bounded below, we notice that, if $\phi \in H$,

$$(K^{-1}\phi, \phi) = (K^{-1}\phi, K(K^{-1}\phi)) \geqslant m\|K^{-1}\phi\|^2$$

and $\|\phi\| \leqslant \|K\| \|K^{-1}\phi\|$, and therefore

$$(K^{-1}\phi, \phi) \geqslant (m/\|K\|^2)\|\phi\|^2$$

for all $\phi \in H$. $\qquad\qquad\qquad\qquad\qquad\qquad\qquad\qquad\qquad\qquad\square$

It is worth pointing out here that while a compact self-adjoint operator K on an infinite-dimensional space may or may not be positive, it cannot be positive and bounded below. This is a consequence of Lemma 4.5, since if we choose (ϕ_n) to be an orthonormal sequence in H, then $(K\phi_n, \phi_n) \to 0$ as $n \to \infty$. However, an operator of the form $I - \lambda K$ may be positive and bounded below – indeed, it will be if $|\lambda|$ is sufficiently small.

Example 6.1

Let $(K\phi)(x) = \int_0^1 \min(x, t)\phi(t)\,dt$ $(0 \leqslant x \leqslant 1)$ define the operator K on $L_2(0, 1)$. From previous examples we know that this operator is compact and it is clearly self-adjoint. By reducing the integral equation $\mu\phi = K\phi$ to a boundary value problem it is easy to show that the eigenvalues of K are $\{1/(n-\frac{1}{2})^2\pi^2 : n \in \mathbb{N}\}$ and therefore, according to Lemma 6.1, K is a positive operator. The largest eigenvalue of K is $\mu_1^+ = 4/\pi^2$, and by Lemma 5.1 we have $(K\phi, \phi) \leqslant 4\pi^{-2}\|\phi\|^2$. If $\lambda \geqslant 0$ then, $((I - \lambda K)\phi, \phi) \geqslant (1 - 4\lambda/\pi^2)\|\phi\|^2$, so that $I - \lambda K$ is positive and bounded below for $0 \leqslant \lambda < \pi^2/4$. With $\lambda < 0$, $((I - \lambda K)\phi, \phi) \geqslant \|\phi\|^2$, since K is a positive operator. Therefore $I - \lambda K$ is positive and bounded below for $\lambda < \pi^2/4$. $\qquad\qquad\qquad\square$

Of course it is not often possible to find all (or even any) of the eigenvalues of an operator exactly, so the method of demonstrating positivity presented in the last example is of little practical value. In fact the converse viewpoint is potentially more useful: if we can show by some means that a given operator is positive, we can infer that it has only positive eigenvalues. A direct proof that the operator K in the last example is positive involves showing that, for all non-zero $\phi \in L_2(0, 1)$,

$$\int_0^1 \int_0^1 \min(x, t)\phi(t)\overline{\phi(x)}\,dt\,dx > 0$$

and this is not easily achieved. A more oblique approach is required and we have hinted at one way forward in the introduction to this chapter.

We noted that if T is an operator on H, then T^*T is a non-negative operator; it is positive if $T\phi = 0$ only for $\phi = 0$. Therefore one prospective method of determining whether a given operator in H is non-negative, or positive, is to seek to express it in the form T^*T for some $T \in B(H)$. Before we explore this idea, which turns out to be quite fruitful, a few basic results are required. For example, we need to assure ourselves at the outset that for every non-negative operator $K \in B(H)$ there is an operator $T \in B(H)$ which satisfies the equation $K = T^*T$.

Theorem 6.3

Every non-negative compact operator K in a Hilbert space H has a unique non-negative square root L; that is, if K is non-negative and compact, there is a unique non-negative bounded linear map L such that $L^2 = K$. L is compact and commutes with every bounded operator which commutes with K.

Proof

The Spectral Theorem makes this easy. For the sequence of eigenvalues (μ_n), all non-negative, and eigenfunctions (ϕ_n), $K\phi = \sum_{n=1}^{\infty} \mu_n(\phi, \phi_n)\phi_n$ for all $\phi \in H$. Define L by $L\phi = \sum_{n=1}^{\infty} v_n(\phi, \phi_n)\phi_n$ where $v_n = \mu_n^{\frac{1}{2}}$. L is non-negative since $v_n \geqslant 0$ for all n, and compact since $v_n \to 0$ as $n \to \infty$, by Problem 4.13. Also $L^2 = K$, so we have shown that K has at least one non-negative square root.

We now show that L commutes with every bounded operator which commutes with K. Let $TK = KT$. Then for $n \in \mathbb{N}$, $KT\phi_n = TK\phi_n = T\mu_n\phi_n = \mu_n T\phi_n$, so that $T\phi_n$ is an eigenvector of K corresponding to the eigenvalue μ_n. $T\phi_n$ is therefore a linear combination of $\{\phi_k : \mu_k = \mu_n\}$. Hence $\mu_k = \mu_n \Rightarrow L\phi_k = v_k\phi_k = v_n\phi_k$ and $LT\phi_n = v_n T\phi_n = TL\phi_n$. By Corollary 4.16 applied to K, for each $\phi \in H$, there is a ϕ_0 such that $K\phi_0 = 0$ and $\phi = \phi_0 + \sum_{n=1}^{\infty}(\phi, \phi_n)\phi_n$. Thus $T\phi = T\phi_0 + \sum_{n=1}^{\infty}(\phi, \phi_n)T\phi_n$ and $K(T\phi_0) = T(K\phi_0) = 0$ whence $L(T\phi_0) = 0$ and

$$LT\phi = LT\phi_0 + \sum_{n=1}^{\infty}(\phi, \phi_n)LT\phi_n$$

$$= \sum_{n=1}^{\infty} v_n(\phi, \phi_n)T\phi_n = TL\phi.$$

Since ϕ is a typical element of H, $LT = TL$.

To show that L is unique, suppose that $M \in B(H)$ is non-negative and

$M^2 = K$. Then M commutes with K and therefore with L, so that $0 = L^2 - M^2 = (L+M)(L-M)$. Let $\phi \in H$. Then

$$((L+M)(L-M)\phi, (L-M)\phi) = (0, (L-M)\phi) = 0. \qquad (6.1)$$

Now for $\psi \in H$ $((L+M)\psi, \psi) = (L\psi, \psi) + (M\psi, \psi)$ so $((L+M)\psi, \psi) = 0$ implies that $(L\psi, \psi) = 0$ and $(M\psi, \psi) = 0$ because both L and M are non-negative. Moreover $(L\psi, \psi) = \sum_{n=1}^{\infty} \nu_n |(\psi, \phi_n)|^2$ so $(L\psi, \psi) = 0$ implies that $L\psi = 0$. Therefore $\|M\psi\|^2 = (M^2\psi, \psi) = (L^2\psi, \psi) = \|L\psi\|^2 = 0$ so that $M\psi = 0$ also. Applying this to $\psi = (L-M)\phi$, we deduce from (6.1) that $(L-M)^2\phi = 0$ and therefore that $\|(L-M)\phi\|^2 = ((L-M)^2\phi, \phi) = 0$. Since ϕ is arbitrary this shows that $L\phi = M\phi$ for all $\phi \in H$, and therefore $L = M$. $\qquad\square$

We adopt the convention of writing the non-negative square root L of K as $K^{\frac{1}{2}}$.

Theorem 6.3 shows that for every non-negative compact operator K there is at least one operator T such that $K = T^*T$, namely $T = K^{\frac{1}{2}}(=T^*)$. In fact, by using the basic idea in the first part of the proof, we can demonstrate the existence of a profusion of such operators for each given non-negative compact K. Since $K\phi = \sum_{n=1}^{\infty} \mu_n(\phi, \phi_n)\phi_n$, then $K = T^2 = T^*T$ where T is now defined by $T\phi = \sum_{n=1}^{\infty} \nu_n(\phi, \phi_n)\phi_n$ with $\nu_n = \pm\mu_n^{\frac{1}{2}}$. If infinitely many of the μ_n's are non-zero we therefore have infinitely many distinct, self-adjoint square roots of K. Nor have we yet located every T such that $T^*T = K$. If we presume that H has infinite dimension, then there is an infinite orthonormal sequence (ϕ_n) with $K\phi = \sum_{n=1}^{\infty} \mu_n(\phi, \phi_n)\phi_n$ (setting $\mu_n = 0$ for $n > N$ if K has only N non-zero eigenvalues). Now let α_n be a complex number of modulus $\mu_n^{\frac{1}{2}}$, and define T by $T\phi = \sum_{n=1}^{\infty} \alpha_n(\phi, \phi_n)\phi_{n+1}$, then $T^*T = K$ but this T is not self-adjoint and it has no non-zero eigenvalues. (To check this last statement notice that $\|T^m\| \leq |\alpha_1\alpha_2 \dots \alpha_m|$ so, since $\alpha_m \to 0$ as $m \to \infty$, $\rho(T) = 0$.)

We now see that if K is a non-negative compact operator, it may be expressed in the form T^*T for many bounded linear maps T. In principle this considerably eases our task of demonstrating that a compact operator K is non-negative by writing it in the form $K = T^*T$. Even if K is an integral operator, not all of the operators T satisfying $T^*T = K$ need be integral operators.

An operator of the form T^*T has already occurred in Chapter 4, in particular in Theorem 4.18. There we used the fact that if K is compact, K^*K is compact and self-adjoint and therefore the Spectral Theorem can be applied to it. If we pass lightly over the same ground again, suppose now that T is a bounded linear map and that T^*T is compact. Then for some sequence (μ_n) of positive eigenvalues and a corresponding orthonormal sequence (ϕ_n) of eigenvectors,

$$(T^*T)\phi = \Sigma_{n=1}^{\infty} \mu_n(\phi, \phi_n)\phi_n \quad (\phi \in H)$$

(with the usual possibility that the sum may be finite). In the proof of Theorem 4.18 we noticed that the sequence (ψ_n) defined by $\psi_n = v_n^{-1} T\phi_n$, with $v_n = \mu_n^{\frac{1}{2}}$, is orthonormal and that $T\phi = \Sigma_{n=1}^{\infty} v_n(\phi, \phi_n)\psi_n$. This last equation shows us that T is compact (since $\|T - T_N\| \to 0$ as $N \to \infty$ where $T_N\phi = \Sigma_{n=1}^{N} v_n(\phi, \phi_n)\psi_n$ for all $\phi \in H$) so we have shown that T is compact if T^*T is.

If we denote the positive square root of T^*T by S, so that $S\phi = \Sigma_{n=1}^{\infty} v_n$ $(\phi, \phi_n)\phi_n$, and assume that the nullspace of T^*T is zero, then (ϕ_n) is a complete orthonormal set and we can define a linear map on H by $U\phi_n = \psi_n$ $(n \in \mathbb{N})$. In this case $\|U\phi\| = \|\phi\|$ for all ϕ and $U^*U = I$, and $T = US$, giving a decomposition of T into the product of a norm-preserving operator and a positive operator. Something similar may be done even if T^*T is not injective, but at the expense of altering the type of the operator U.

Notice that $T^*T\phi = 0 \Leftrightarrow T\phi = 0$. In one direction this is obvious, since $T\phi = 0 \Rightarrow T^*T\phi = 0$. The converse follows because $T^*T\phi = 0 \Rightarrow 0 = (T^*T\phi, \phi) = \|T\phi\|^2 \Rightarrow T\phi = 0$.

Against this background it may appear that finding a T^*T decomposition of a given non-negative compact operator K is likely to be straightforward, because of the profusion of suitable operators T. This impression is misleading; it is true that in theory there is a lot of scope, but in practice it is often difficult to find an explicit decomposition $K = T^*T$ if the eigenvalues and eigenvectors of K are not known.

Because it is useful to know whether a compact operator is non-negative, as we have already indicated, we now give a series of examples, using different methods to investigate particular integral operators and producing T^*T forms where appropriate. We do not necessarily offer the most direct approach to non-negativity or positivity in each case, but rather concentrate on giving a range of prospective techniques. The examples are interspersed with statements of more general results, where these suggest themselves.

Example 6.2

Let $(K\phi)(x) = \int_0^1 \min(x, t)\phi(t)\,dt$ $(0 \leq x \leq 1)$ define the operator K on $L_2(0, 1)$. We have already seen that K is positive, because all of its eigenvalues are positive, but we now show that it is positive independently of a knowledge of the eigenvalues.

(7.36) and gives

$$(A_{2m+2}/A_{2m})^{\frac{1}{2}} \leqslant |\mu_1| \leqslant (A_{2m}/M)^{1/(2m)} \quad (m \in \mathbb{N}), \qquad (7.43)$$

where the ordering $|\mu_1| \geqslant |\mu_2| \geqslant \cdots \geqslant 0$ has been assumed and M now denotes the number of eigenvectors associated with $|\mu_1|$. Cauchy's inequality (7.37) can be used to show that $A_{2m}^2 \leqslant A_{2m-2}A_{2m+2}$, so the lower bound sequence is non-decreasing. Reference to (7.38) shows that the lower bounds in (7.43) form a subsequence of the lower bounds derived earlier.

Example 7.7

Let

$$(K\phi)(x) = \int_0^1 \max(x, t)\phi(t)\, dt \quad (0 \leqslant x \leqslant 1)$$

define the operator K on $L_2(0, 1)$. We can use (7.39) to give $A_2 = 1/2$. Also

$$k_2(x, t) = \int_0^1 \max(x, s)\max(s, t)\, ds = \tfrac{1}{6}(2 + x^3) + \tfrac{1}{2}xt^2,$$

and, from (7.39) again, we find that $A_4 = 7/30$. The tightest bounds available are $(A_4/A_2)^{\frac{1}{4}} = (7/15)^{\frac{1}{4}} \leqslant |\mu_1|$ and $|\mu_1| \leqslant (A_4/M)^{\frac{1}{4}} = (7/(30M))^{\frac{1}{4}}$. It follows that $M = 1$ and hence that $0.6831 \leqslant |\mu_1| \leqslant 0.6950$ (to four decimal places).

We have derived greater lower bounds for the largest eigenvalue of this operator by other methods, but the upper bound is significantly less than the value 0.7297 obtained in Example 7.2. Combining the results of Example 7.3 with the present calculation, we know that the largest eigenvalue of K is positive and such that $0.6944 \leqslant \mu_1^+ \leqslant 0.6950$. $\qquad \square$

7.5 Comparison and related methods

It is sometimes possible to infer properties of the solution of an operator equation indirectly, by relating the equation to another one which can be solved explicitly. The idea of dealing with more or less intractable problems by devising easier ones of a suitable kind is quite general and can be implemented in a variety of ways.

Here our only concern is with eigenvalue estimation and the theoretical background required has been anticipated and dealt with in Chapter 5. In fact Theorem 5.11 establishes inequalities which directly provide comparison results for integral operators arising from Sturm–Liouville boundary value problems, as Example 5.3 illustrates. Our applications are not limited to operators of this particular type, however, and we give examples which convey more general aspects of comparison methods.

Example 7.8

Let

$$(L\phi)(x) = \frac{1}{\pi} \int_0^1 \log\left|\frac{x^{\frac{1}{2}} + t^{\frac{1}{2}}}{x^{\frac{1}{2}} - t^{\frac{1}{2}}}\right| \phi(t)\, dt \quad (0 \leqslant x \leqslant 1)$$

define the operator L on $L_2(0, 1)$. We know that L is a positive operator (see Example 6.7) although this fact will emerge anyway during the process of bounding the eigenvalues of L, which is our objective here.

The Fourier series

$$\log\left|\frac{\cos\theta + \cos\sigma}{\cos\theta - \cos\sigma}\right| = \sum_{n=1}^{\infty} \frac{4}{2n-1} \cos\{(2n-1)\theta\} \cos\{(2n-1)\sigma\}$$

$$(0 \leqslant \theta, \sigma \leqslant \pi, \theta \neq \sigma)$$

(derived in Appendix C) allows us to deduce that the eigenvalues and eigenvectors of the operator defined by

$$(K_2\phi)(\sigma) = \frac{1}{\pi} \int_0^{\pi/2} \log\left|\frac{\cos\theta + \cos\sigma}{\cos\theta - \cos\sigma}\right| \phi(\theta)\, d\theta \quad (0 \leqslant \sigma \leqslant \pi/2)$$

are respectively $\mu_n^+ = 1/(2n-1)$ and $\phi_n^+(\sigma) = 2\cos\{(2n-1)\sigma\}/\pi^{\frac{1}{2}}$ $(n \in \mathbb{N})$. It follows that the operator K_1 obtained by making the variable changes $x = \cos^2\sigma$, $t = \cos^2\theta$, so that

$$(K_1\phi)(x) = \frac{1}{\pi} \int_0^1 \log\left|\frac{x^{\frac{1}{2}} + t^{\frac{1}{2}}}{x^{\frac{1}{2}} - t^{\frac{1}{2}}}\right| \frac{\phi(t)\, dt}{2(t(1-t))^{\frac{1}{2}}} \quad (0 \leqslant x \leqslant 1),$$

also has eigenvalues $\mu_1^+ = 1/(2n-1)$ $(n \in \mathbb{N})$.

It is easy to check that the eigenvalues of the self-adjoint operator

$$(K\phi)(x) = \frac{1}{\pi} \int_0^1 \log\left|\frac{x^{\frac{1}{2}} + t^{\frac{1}{2}}}{x^{\frac{1}{2}} - t^{\frac{1}{2}}}\right| \frac{\phi(t)\, dt}{m(x)m(t)} \quad (0 \leqslant x \leqslant 1),$$

where $m(x) = 2^{\frac{1}{2}} x^{\frac{1}{2}} (1-x)^{\frac{1}{4}}$ $(0 \leqslant x \leqslant 1)$ are therefore $\mu_n^+(K) = 1/(2n-1)$ $(n \in \mathbb{N})$. We now have a structure allowing us to use Theorem 5.11, according to which the positive eigenvalues of the two operators

$$(K\phi)(x) = \int_a^b k(x, t)\phi(t)\, dt, \quad (L\phi)(x) = \int_a^b k(x, t)m(x)\overline{m(t)}\phi(t)\, dt \quad (a \leqslant x \leqslant b)$$

defined on $L_2(a, b)$ are related by $\mu_n^+(L) \leqslant M^2 \mu_n^+(K)$, where $M =$

$\sup\{|m(x)|:a\leqslant x\leqslant b\}$. Since $M=1$ in the present application, we have

$$\mu_n^+(L)\leqslant 1/(2n-1)\quad(n\in\mathbb{N}).\qquad(7.44)$$

The complementary inequalities provided by Theorem 5.11 are not applicable in this case as $m(x)$ is not bounded below away from zero. In practice the production of only one-sided bounds is a fairly common outcome of this comparison technique. To obtain underestimates of the eigenvalues of L requires a different comparison operator, or a fresh approach.

Note from Example 6.7 that $L=TT^*$ where

$$(T\phi)(x)=\frac{1}{\pi^{\frac{1}{2}}}\int_0^x\frac{\phi(t)\,dt}{(x-t)^{\frac{1}{2}}}\quad(0\leqslant x\leqslant 1).$$

It is easy to show that $(T^2\phi)(x)=\int_0^x\phi(t)\,dt$ and hence that $(T^{*2}\phi)(x)=\int_x^1\phi(t)\,dt$. Therefore

$$(TLT^*\phi)(x)=\int_0^x ds\int_s^1\phi(t)\,dt=\int_0^1\min(x,t)\phi(t)\,dt,$$

the eigenvalues of this operator being $\mu_n^+(TLT^*)=4/(\pi^2(2n-1)^2)\,(n\in\mathbb{N})$, as follows from the equivalent Sturm–Liouville boundary value problem. Now $\|T\|\leqslant 1$ and, by Theorem 5.9, $\mu_n^+(TLT^*)\leqslant\mu_n^+(L)\|T\|^2$ giving $\mu_n^+(L)\geqslant 4/(\pi^2(2n-1)^2)\,(n\in\mathbb{N})$. These lower bounds are a good deal less satisfactory than the upper bounds (7.44) as more has been given away to obtain them. In fact the eigenvalues of TLT^* are likely to be more comparable in magnitude with the eigenvalues of L^2 than with those of L, and this suspicion leads us to consider the eigenvalue inequalities given in Theorem 5.12. In the present notation, (5.14) implies that

$$\mu_{2n-1}^+(TLT^*)\leqslant\min_{1\leqslant k\leqslant 2n-1}\mu_k^+(L)\mu_{2n-k}^+(TT^*)$$

$$\leqslant\mu_n^+(L)\mu_n^+(TT^*)=\{\mu_n^+(L)\}^2\quad(n\in\mathbb{N}).$$

Therefore $\mu_n^+(L)\geqslant 2/\{\pi(4n-3)\}$ for $n\in\mathbb{N}$, which significantly improves our previous lower bounds, and we conclude that

$$2/\{\pi(4n-3)\}\leqslant\mu_n^+(L)\leqslant 1/(2n-1)\quad(n\in\mathbb{N}).$$

Note that the comparison method, as we have used it here, does not produce the type of information we have sought previously in this chapter. We have not attempted to find, and could not have expected to obtain, sharp bounds for the dominant eigenvalue. $\qquad\square$

Bounds of the type (7.44) can be found for other practically important operators by expanding their kernels in terms of orthogonal functions. In many cases these are necessarily so-called special functions, rather than the elementary functions which sufficed in the previous example.

Example 7.9

Let $0 < \alpha < 1$, $m(x) = (1 - x^2)^{(1-\alpha)/4}$ $(-1 \leqslant x \leqslant 1)$ and

$$v_n = \pi \Gamma(n + \alpha - 1)/(\cos(\tfrac{1}{2}\pi\alpha)\Gamma(\alpha)\Gamma(n))$$

$$\psi_n(x) = \frac{1}{\pi^{\frac{1}{2}}} \frac{\Gamma(\alpha/2)}{2^{1-\alpha/2}} \left\{ \frac{\Gamma(n)(2n+\alpha-2)}{\Gamma(n+\alpha-1)} \right\}^{\frac{1}{2}} \frac{C_{n-1}^{\alpha/2}(x)}{m(x)} \qquad (n \in \mathbb{N}).$$

The Gegenbauer polynomial C_n^ν is an eigenfunction of a Sturm–Liouville problem for Gegenbauer's equation, which we have referred to in Problem 2.12. The associated eigenvalue is $\lambda_n = n(n + 2\nu)$. The orthogonality of the functions ψ_n follows from Gegenbauer's equation, written in the form

$$\{(1 - x^2)^{\nu + \frac{1}{2}}\phi'(x)\}' + \lambda(1 - x^2)^{\nu - \frac{1}{2}}\phi(x) = 0 \quad (|x| < 1),$$

and the information we have derived about the solution of Sturm–Liouville problems in Chapter 6. (An account of the properties of the Gegenbauer polynomials is given in, for example, the *Handbook of Mathematical Functions* by M. Abramowitz and I. A. Stegun, Dover Publications Inc., New York, 1965.) The functions ψ_n form a complete orthonormal set on $[-1, 1]$ and

$$\{|x - t|^\alpha m(x)\overline{m(t)}\}^{-1} = \sum_{n=1}^{\infty} v_n \psi_n(x)\psi_n(t) \quad (-1 \leqslant x, t \leqslant 1).$$

Therefore the operator

$$(K\phi)(x) = \int_{-1}^{1} \frac{\phi(t)\,\mathrm{d}t}{|x - t|^\alpha m(x)\overline{m(t)}} \quad (-1 \leqslant x \leqslant 1)$$

has eigenvalues $\mu_n^+(K) = v_n$ $(n \in \mathbb{N})$ and, by Theorem 5.11, the eigenvalues of

$$(L\phi)(x) = \int_{-1}^{1} \frac{\phi(t)\,\mathrm{d}t}{|x - t|^\alpha} \quad (-1 \leqslant x \leqslant 1)$$

are such that

$$\mu_n^+(L) \leqslant v_n \quad (n \in \mathbb{N}).$$

From the asymptotic properties of the gamma function we can deduce that $v_n \sim \pi n^{\alpha - 1}(\cos(\tfrac{1}{2}\pi\alpha)\Gamma(\alpha))^{-1}$ as $n \to \infty$.

Note that by a simple translation of variables, it can be shown that the eigenvalues of L_1, where

$$(L_1\phi)(x) = \int_0^1 \frac{\phi(t)\,dt}{|x-t|^\alpha} \quad (0 \leqslant x \leqslant 1),$$

are related to those of L by $\mu_n(L_1) = \mu_n(L)2^{\alpha-1}$ ($n \in \mathbb{N}$) and therefore $\mu_n(L_1) \leqslant 2^{\alpha-1}v_n$ ($n \in \mathbb{N}$). $\qquad\square$

Example 7.10

Let

$$(K\phi)(x) = \int_{-1}^1 \frac{\sin(x-t)}{x-t}\phi(t)\,dt \quad (-1 \leqslant x \leqslant 1)$$

define the operator K on $L_2(-1, 1)$. We showed in Example 6.6 that K is a positive operator. A comparison operator which suggests itself in this case is obtained by replacing $\sin(x-t)$ by a truncated version of its Taylor series. Therefore we consider the finite rank operator K_1 defined by

$$(K_1\phi)(x) = \int_{-1}^1 \{1 - \tfrac{1}{6}(x-t)^2\}\phi(t)\,dt \quad (-1 \leqslant x \leqslant 1).$$

By reducing $\mu\phi = K_1\phi$ to a system of (three) algebraic equations we find that the non-zero eigenvalues of K_1 are given by $\mu_1^+(K_1) = \tfrac{8}{9} + \tfrac{2}{\sqrt5} = 1.7833\ldots$, $\mu_2^+(K_1) = \tfrac{2}{9} = 0.2222\ldots$ and $\mu_1^-(K_1) = \tfrac{8}{9} - \tfrac{2}{\sqrt5} = -0.0055\ldots$. Obviously K_1 is merely an indefinite operator and only its two positive eigenvalues can be related to those of K. The relationship is provided by Theorem 5.7 and requires us to estimate $\|K - K_1\|$. From the fact that

$$|\{(K - K_1)\phi\}(x)| \leqslant \frac{1}{5!}\int_{-1}^1 |x-t|^4|\phi(t)|\,dt \quad (-1 \leqslant x \leqslant 1),$$

by Taylor's series, and using the Schwarz inequality, it is not difficult to show that $\|K - K_1\| \leqslant 0.03975$. Therefore $|\mu_1^+(K) - \tfrac{8}{9} - \tfrac{2}{\sqrt5}| \leqslant 0.03975$ and $|\mu_2^+(K) - \tfrac{2}{9}| \leqslant 0.03975$. It also follows (see the discussion after Corollary 5.8) that $\mu_3^+(K) \leqslant 0.03975$. $\qquad\square$

Problems

7.1   Let $K \in B(H)$ be a compact self-adjoint operator and let the sequences of eigenvalues and eigenvectors of K be (μ_n) and (ϕ_n) respectively. Use variational calculus to show that the Rayleigh quotient $R_K(p) = (Kp, p)/(p, p)$ ($p \in H: p \neq 0$) is stationary at $c_n\phi_n$ ($n \in \mathbb{N}$), where c_n is a non-zero constant, and that the corresponding stationary value is μ_n. Show that the Rayleigh–Ritz method used in conjunction with $R_K(p)$ is equivalent to the process in §7.2.2.

7.2   Evaluate the Rayleigh quotient $R_K(p) = (Kp, p)/\|p\|^2$ with $p(x) = x^\alpha$ $(0 \leqslant x \leqslant 1)$ where $\alpha > -\frac{1}{2}$ and where K is the operator defined on $L_2(0, 1)$ by

$$(K\phi)(x) = \int_0^1 \max(x, t)\phi(t)\, dt \quad (0 \leqslant x \leqslant 1).$$

Show that $R_K(p)$ is maximised by taking $\alpha = (\sqrt{2} - 1)/2$ and deduce that the largest eigenvalue of K exceeds $4(3 - 2\sqrt{2})$.

7.3   Let K be a positive compact operator on H with eigenvalues μ_n $(n \in \mathbb{N})$ and eigenvectors ϕ_n $(n \in \mathbb{N})$. Let the functional $R_{m,j}: H \to \mathbb{R}$ be defined by

$$R_{m,j}(p) = (K^m p, p)/(K^j p, p) \quad (p \in H, p \neq 0),$$

where $m \in \mathbb{N}$ and $j \geqslant 0$ is such that $m - j \in \mathbb{N}$. Show that $R_{m,j}$ is stationary at $c_n \phi_n$ $(n \in \mathbb{N})$, where c_n is an arbitrary, non-zero constant, and that its stationary values are μ_n^{m-j}. Show also that

$$R_{m,m-j}(p) \leqslant R_{m+j,m}(p) \leqslant \mu_1^j \quad (p \in H, p \neq 0),$$

where μ_1 is the largest eigenvalue of K, for any $m, j \in \mathbb{N}$ such that $m \geqslant j$. Use $R_{1,0}(p)$ with the trial function $p(x) = x$ $(0 \leqslant x \leqslant 1)$ to determine an underestimate for the largest eigenvalue μ_1 of the operator generated on $L_2(0, 1)$ by the kernel $k(x, t) = \min(x, t)$ $(0 \leqslant x, t \leqslant 1)$. Use $R_{2,1}(p)$ to obtain an improved underestimate of μ_1.

7.4   Let the operator K on $L_2(-1, 1)$ be defined by

$$(K\phi)(x) = \int_{-1}^1 |x - t|\phi(t)\, dt \quad (-1 \leqslant x \leqslant 1).$$

(i)   Let $(K_1\phi)(x) = \int_{-1}^1 \phi(t)\, dt$ $(-1 \leqslant x \leqslant 1)$ and use Lemma 6.4 to show that $K_1 - K$ is a positive operator. Deduce, using Corollary 5.6, that K has one positive eigenvalue μ_1^+ and infinitely many eigenvalues μ_n^- $(n \in \mathbb{N})$ where $\mu_1^- \leqslant \mu_2^- \leqslant \cdots < 0$. Show that $\mu_1^+ \leqslant 2$ and $\mu_1^- \geqslant -2$.

(ii)   Use the Rayleigh–Ritz method (as illustrated in Example 7.1) with the test functions $\chi_n(x) = x^{n-1}$ $(-1 \leqslant x \leqslant 1)$ to show that

$$2(2 + (41/5)^{\frac{1}{2}})/7 \leqslant \mu_1^+, \; \mu_1^- \leqslant -4/5, \; \mu_2^- \leqslant 2(2 - (41/5)^{\frac{1}{2}})/7.$$

Calculate the trace A_2 of K and deduce that μ_1^+ is the largest eigenvalue of K in magnitude.

(iii)   Let $\psi_0(x) = 1$ $(-1 \leqslant x \leqslant 1)$ and evaluate $\Psi_m = (\psi_0, K^m \psi_0)$ for $m = 0, 1, \ldots, 4$. Deduce that

$$\mu_1^+ \leqslant 2(102 + 5(2/3)^{\frac{1}{2}})/147.$$

7.5   Let K be the integral operator on $L_2(0, 1)$ generated by the L_2-kernel $k(x, t)$ defined on $[0, 1] \times [0, 1]$. Let the eigenvalues μ_n $(n \in \mathbb{N})$ of K be ordered so that $|\mu_1| > |\mu_2| > \cdots > 0$, and let A_{2m} denote the $2m$th trace of K. Prove that

$$(A_{2m+2}/A_{2m})^{\frac{1}{2}} \leqslant |\mu_1| \leqslant (A_{2m}/M)^{1/(2m)} \quad (m \in \mathbb{N}),$$

where M denotes the dimension of the set of eigenvectors of K associated with its eigenvalues $\pm \mu_1$. If the eigenvalues of K are all simple, prove that $(B_{2m+2}/B_{2m})^{\frac{1}{2}} \leqslant |\mu_1 \mu_2| \leqslant (B_{2m}/2)^{1/(2m)}$ $(m \in \mathbb{N})$, where $B_{2m} = A_{2m}^2 - A_{4m}$ $(m \in \mathbb{N})$.

7.6   Let K be the operator on $L_2(0, 1)$ generated by the symmetric kernel k, where $k(x, t) = \frac{1}{2} xt^2 - \frac{1}{6} t^3$ $(0 \leqslant t \leqslant x \leqslant 1)$. By referring to Problem 1.10, show that K is a positive operator. Evaluate the traces A_1 and A_2 of K and deduce that the largest eigenvalue, μ_1^+, of K is simple and such that

$$\tfrac{11}{140} \leqslant \mu_1^+ \leqslant \tfrac{1}{4}(\tfrac{11}{105})^{\frac{1}{2}},$$

and that the second largest eigenvalue of K is no larger than $1/396$.

(The bounds for μ_1^+ can be used to bound the fundamental frequency of flexural oscillation of a rod having one free end and one clamped end. Example 1.3 and Problem 1.10 provide some background to this interpretation.)

7.7   Let $K \in L_2(0, 1)$ be the integral operator generated by the continuous hermitian kernel $k(x, t)$ $(0 \leqslant x, t \leqslant 1)$. If K is a positive operator with simple eigenvalues μ_n^+ $(n \in \mathbb{N})$, ordered so that $\mu_1^+ > \mu_2^+ > \cdots > 0$, show that

$$B_{m+1}/B_m \leqslant \mu_1^+ \mu_2^+ \leqslant (B_m/2)^{1/m} \quad (m \in \mathbb{N}),$$

where $B_m = A_m^2 - A_{2m}$ $(m \in \mathbb{N})$ and A_m denotes the mth trace of K.

7.8   The kernel k is defined on $[0, 1] \times [0, 1]$ by

$$k(x, t) = \begin{cases} 1 & (x + t \leqslant 1), \\ -1 & (x + t > 1). \end{cases}$$

Show that

$$\left.\begin{aligned}
k_2(x, t) &= 1 - 2|x - t|, \\
k_3(x, t) &= 2k(x, t)(|1 - x - t| - (1 - x - t)^2), \\
k_4(x, t) &= (1 - 6(x - t)^2 + 4|x - t|^3)/3, \\
k_5(x, t) &= 2k(x, t)(|1 - x - t| - 2|1 - x - t|^3 + (1 - x - t)^4)/3.
\end{aligned}\right\} \quad (0 \leqslant x, t \leqslant 1).$$

Deduce that, for the operator K generated by k, $A_2 = 1$, $A_4 = \frac{1}{3}$, $A_6 = 2/15$, $A_8 = 17/315$ and $A_{10} = 62/2835$. Let the largest eigenvalue of K be μ_1. Use the trace bounds to show that $|\mu_1|$ has multiplicity one or two, and that if the multiplicity is two then

$$0.63657 \leqslant |\mu_1| \leqslant 0.63663.$$

7.9   Let the operator K be defined on $L_2(0, 1)$ by

$$(K\phi)(x) = \int_0^1 \{\beta^2 + \alpha(\alpha - \beta)\min(x, t) + \beta(\alpha - \beta)\max(x, t)\}\phi(t)\,dt \quad (0 \leqslant x \leqslant 1),$$

where α and β are real, distinct, non-zero constants. Show that the traces A_1 and A_2 of K are given by $A_1 = (\alpha^2 + \beta^2)/2$ and $A_2 = A_1(2A_1 + \alpha\beta)/3$. Deduce that the largest eigenvalue, μ_1, of K is simple if $\alpha^2 + \beta^2 + 4\alpha\beta > 0$ and that $\mu_1 > 1$ if $\alpha^2 + \beta^2 + \alpha\beta > 3$.

In the case where $\beta = (3^{\frac{1}{2}} - 1)^{-1}$ and $\alpha = 3^{\frac{1}{2}}\beta$ show that $3.565 < \mu_1 < 3.648$. (The operator K is positive, by Lemma 6.4, and it is related to a mixed boundary value problem, as Problem 2.14 shows.)

7.10 (i)   Let K be a compact self-adjoint operator on a Hilbert space H. Let μ_m be a simple eigenvalue of K and let $p \in H$ be an approximation to the corresponding eigenvector ϕ_m, such that (Kp, p) is nearer to μ_m than to any other eigenvalue of K. Write $p = (p, \phi_m)\phi_m + \psi$, where $(\phi_m, \psi) = 0$ and let $S = K - (Kp, p)I$. Show that $\|Sp\| \geq \|S\psi\|$ and deduce that

$$\|\psi\| \leq \|Sp\|/\alpha,$$

where $\alpha = \inf\{|\mu_j - (Kp, p)|: j \neq m\}$.

 (ii)   Let the positive operator K be defined on $L_2(0, 1)$ by

$$(K\phi)(x) = \int_0^1 (1 - \max(x, t))\phi(t)\, dt \quad (0 \leq x \leq 1).$$

Let $\psi_0(x) = 1$ $(0 \leq x \leq 1)$ and evaluate the Kellogg approximation $p = \psi_1/\|\psi_1\|$ to the eigenvector ϕ_1^+ of K belonging to its largest eigenvalue μ_1^+, where $\psi_1 = K\psi_0$. Deduce that $p = (p, \phi_1^+)\phi_1^+ + \psi$ where $\|\psi\| \leq 0.0444$.

(Hint: for the given approximation, (Kp, p) supplies an underestimate of μ_1^+. The trace A_1 of K may be used to show that μ_1^+ is a simple eigenvalue, and to provide an overestimate of the second largest eigenvalue of K. This knowledge allows the upper bound for $\|\psi\|$ derived in (i) to be used.)

7.11   Let $K \in B(H)$ be a compact self-adjoint operator whose largest eigenvalue, μ_1^+, is positive.

 (i)   Let $\psi_0 \in H$ be such that $K\psi_0 \neq 0$ and let $\Psi_m = (\psi_0, \psi_m)$ $(m + 1 \in \mathbb{N})$ where $\psi_m = K\psi_{m-1}$ $(m \in \mathbb{N})$. Prove that

$$\frac{\Psi_{2m-1}}{\Psi_{2m-2}} \leq \mu_1^+ \quad (m \in \mathbb{N}).$$

 (ii)   Let A_m denote the mth trace of K. Show that

$$\frac{A_{2m+1}}{A_{2m}} \leq \mu_1^+ \quad (m \in \mathbb{N}).$$

7.12   Let

$$(K\phi)(x) = \int_{-1}^1 \frac{\{1 - \cos(x - t)\}}{|x - t|}\phi(t)\, dt \quad (-1 \leq x \leq 1)$$

and

$$(K_1\phi)(x) = \int_{-1}^{1} \tfrac{1}{2}|x-t|\phi(t)\,dt \quad (-1 \leqslant x \leqslant 1)$$

define the operators K and K_1 on $L_2(-1, 1)$. Show that $\|K - K_1\| \leqslant 1/3\sqrt{7}$ and deduce, using the information in Problem 7.4, that the eigenvalues of the operator K are such that $0.56 < \mu_1^+ < 0.85$, $\mu_1^- < -0.27$ and $\mu_2^+ < 0.13$.

7.13 Show that the operator K_1 defined by

$$(K_1\phi)(x) = \int_{-1}^{1} \{1 - \tfrac{1}{6}\alpha^2(x-t)^2\}\phi(t)\,dt \quad (-1 \leqslant x \leqslant 1)$$

has two positive eigenvalues and one negative eigenvalue for all real, non-zero α. Deduce bounds for the two largest eigenvalues of the operator K where

$$(K\phi)(x) = \int_{-1}^{1} \frac{\sin\{\alpha(x-t)\}}{\alpha(x-t)}\phi(t)\,dt \quad (-1 \leqslant x \leqslant 1),$$

α being real and non-zero. By translating the variables, deduce bounds for the two largest eigenvalues of the operator defined by

$$(K_2\phi)(x) = \int_{0}^{1} \frac{\sin(x-t)}{x-t}\phi(t)\,dt \quad (0 \leqslant x \leqslant 1).$$

7.14 Let

$$(K\phi)(x) = \int_{0}^{1} -(xt)^{\frac{1}{4}}\log\{\max(x, t)\}\phi(t)\,dt \quad (0 \leqslant x \leqslant 1)$$

define the operator K on $L_2(0, 1)$. Note that K is a positive operator by Example 6.14. Evaluate the traces A_1, A_2, A_3 and A_4 of K and deduce the modified traces B_1 and B_2, where $B_m = A_m^2 - A_{2m}$ ($m \in \mathbb{N}$). Hence show that if the eigenvalues of K are ordered so that $\mu_1^+ \geqslant \mu_2^+ \geqslant \cdots > 0$ then

$$\frac{11}{64} < \mu_1^+ < \frac{1}{8}\left(\frac{11}{13}\right)^{\frac{1}{4}}, \quad \frac{1}{48}\left(\frac{3}{11}\right)^{\frac{1}{4}} < \mu_2^+ < \frac{1}{11\sqrt{6}}$$

(The integrals $\int_0^1 x^n \log x\,dx = -(n+1)^{-2}$, $\int_0^1 x^n(\log x)^2\,dx = 2(n+1)^{-3}$, where $n \in \mathbb{N}$, will prove to be useful. Reference to Example 2.3 shows that the operator K is associated with Bessel's equation of order zero. The eigenvalues μ_n^+ of K and the eigenvalues λ_n^+ of the boundary value problem

$$(x\phi'(x))' + \lambda x\phi(x) = 0 \quad (0 < x < 1),$$
$$\phi(x) \text{ bounded as } x \to 0, \quad \phi(1) = 0,$$

are related by $\mu_n^+ \lambda_n^+ = 1$. The eigenfunctions of the boundary value problem are $\phi_n(x) = c_n J_0(x(\lambda_n^+)^{\frac{1}{2}})$ $(0 \leqslant x \leqslant 1)$ and the eigenvalues λ_n^+ are the roots of $J_0(\lambda^{\frac{1}{2}}) = 0$. The Bessel function J_0 was defined by its Taylor series in Problem 3.16. It follows that $(\mu_n^+)^{-\frac{1}{2}}$ are the zeros of J_0. Using the bounds given for μ_1^+, we deduce that the smallest zero of J_0 lies in the interval (2.404, 2.412).)

7.15   Use the eigenvalue bounds derived in Example 7.8 together with Theorem 5.11 to show that $1/(4n-3) \leqslant \mu_n^+(K)$ where

$$(K\phi)(x) = \int_0^1 \log\left|\frac{x+t}{x-t}\right| \phi(t)\, dt \quad (0 \leqslant x \leqslant 1).$$

(The operator K is of importance because of its association with elliptic boundary value problems. A simple illustration of the relationship is given in §1.3.3. In more complicated cases the relevant operator can often be written as $K + K_1$, where K_1 is generated by a continuous kernel, and inequalities such as those derived in Theorem 5.4 are then of importance in obtaining eigenvalue bounds.)

7.16   Let the operators K and T be defined on $L_2(0, 1)$ by

$$(K\phi)(x) = \int_0^1 \frac{\phi(t)\, dt}{|x-t|^{\frac{1}{2}}}, \quad (T\phi)(x) = \int_0^x \frac{\phi(t)\, dt}{(x-t)^{\frac{1}{2}}} \quad (0 \leqslant x \leqslant 1).$$

Note that K is a positive operator by Example 6.8 and that $K = T + T^*$.

(i)   Show that $K_1 = T^2 + (T^*)^2$ is the rank-one operator given by $(K_1\phi)(x) = \pi \int_0^1 \phi(t)\, dt$ $(0 \leqslant x \leqslant 1)$. Using the result in Problem 6.5 show that the kernel of K^2 is

$$k_2(x, t) = \pi + \log\left|\frac{x^{\frac{1}{2}} + t^{\frac{1}{2}}}{x^{\frac{1}{2}} - t^{\frac{1}{2}}}\right| + \log\left|\frac{(1-x)^{\frac{1}{2}} + (1-t)^{\frac{1}{2}}}{(1-x)^{\frac{1}{2}} - (1-t)^{\frac{1}{2}}}\right| \quad (0 \leqslant x, t \leqslant 1, x \neq t).$$

Deduce that the trace A_3 of K is given by

$$A_3 = \frac{8\pi}{3} + 4\int_0^1 dx \int_0^x \log\left(\frac{x^{\frac{1}{2}} + t^{\frac{1}{2}}}{x^{\frac{1}{2}} - t^{\frac{1}{2}}}\right) \frac{dt}{(x-t)^{\frac{1}{2}}},$$

and, by putting $t = x \cos^2 \theta$, show that $A_3 = 8\pi$.

(ii)   Let $\psi_0(x) = 1$ $(0 \leqslant x \leqslant 1)$ and evaluate Ψ_0, Ψ_1 and Ψ_2 where $\Psi_m = (\psi_0, K^m \psi_0)$ $(m \in \mathbb{N})$. Show that the largest eigenvalue μ_1^+ of K satisfies $\mu_1^+ > 3(4+\pi)/8$ and, using A_3, deduce that $\mu_1^+ < \frac{8}{3} + (\pi - 28/9)^{\frac{1}{2}}$. Deduce also that the second largest eigenvalue of K is such that $\mu_2^+ < 1.81$.

(iii)   Obtain upper bounds for the eigenvalues $\mu_{n+1}^+(K)$ $(n \in \mathbb{N})$ as follows. Use Theorem 5.5 to show that $\mu_{m+1}^+(K_1 + TT^*) \leqslant \mu_m^+(TT^*)$ $(m \in \mathbb{N})$, where K_1 is the operator defined in (i). Use Theorem 5.4 and Lemma 3.7 to show that

$$\mu_{m+n}^+(K^2) \leqslant \mu_m^+(TT^*) + \mu_n^+(TT^*) \quad (m, n \in \mathbb{N}).$$

Finally, make use of the bounds evaluated in Example 7.8 (*cf.* (7.44)) to deduce that

$$\mu_{2n}^+(K)\leqslant\left(\frac{2\pi}{2n-1}\right)^{\frac{1}{2}},\quad \mu_{2n+1}^+(K)\leqslant2\left(\frac{n\pi}{4n^2-1}\right)^{\frac{1}{2}}\quad(n\in\mathbb{N}).$$

(iv) Obtain lower bounds for the eigenvalues $\mu_n^+(K)$ $(n\in\mathbb{N})$ by observing that K^2-TT^* is a positive operator and using Corollary 5.6 together with the lower bounds derived in Example 7.8 to show that

$$\mu_n^+(K)\geqslant\left(\frac{2}{4n-3}\right)^{\frac{1}{2}}\quad(n\in\mathbb{N}).$$

(The upper bounds in (iii) are weaker than those derived in Example 7.9 with $\alpha=\frac{1}{2}$ there. We have of course used two comparison methods in succession to obtain the bounds in (iii), and cannot expect a sharp result. The lower bounds in (iv) have been achieved without devising a comparison operator for K, which is a less than straightforward exercise. We can conclude that the eigenvalues $\mu_n^+(K)$ behave asymptotically like $n^{-\frac{1}{2}}$ for large n; this deduction does not of course follow from Example 7.9 alone.)

7.17 Let $(K\phi)(x)=\int_0^1\phi(t)|x-t|^{-\frac{1}{2}}\,dt$ $(0\leqslant x\leqslant1)$ and $(T\phi)(x)=\int_0^x\phi(t)(x-t)^{-\frac{1}{2}}\,dt$ $(0\leqslant x\leqslant1)$.

(i) By using Theorem 5.12 in conjunction with the bounds deduced in Examples 7.8 and 7.9 show that

$$\mu_{2m-1}^+(TKT^*)\leqslant\frac{\pi^{\frac{3}{2}}\Gamma(m-\frac{1}{2})}{(2m-1)\Gamma(m)},\quad \mu_{2m}^+(TKT^*)\leqslant\frac{\pi^{\frac{3}{2}}\Gamma(m-\frac{1}{2})}{(2m+1)\Gamma(m)}\quad(m\in\mathbb{N}).$$

(ii) Using the fact that $(T^2\phi)(x)=\pi\int_0^x\phi(t)\,dt$, show that the kernel of $TKT^*=(T)^2T^*+T(T^*)^2$ is

$$2\pi(t^{\frac{1}{2}}+x^{\frac{1}{2}}-|x-t|^{\frac{1}{2}})\quad(0\leqslant x,t\leqslant1).$$

Determine the eigenvalues of the rank-two operator K_1 where

$$(K_1\phi)=\int_0^1(t^{\frac{1}{2}}+x^{\frac{1}{2}})\phi(t)\,dt\quad(0\leqslant x\leqslant1).$$

(iii) Show that the operator L defined by $(L\phi)(x)=\int_0^1|x-t|^{\frac{1}{2}}\phi(t)\,dt$ $(0\leqslant x\leqslant1)$ has one positive eigenvalue and infinitely many negative eigenvalues. Deduce that the negative eigenvalues are such that $\mu_1^-(L)\geqslant-(2+3/2^{\frac{1}{2}})/3$,

$$\mu_{2n}^-(L)\geqslant-\frac{\pi^{\frac{1}{2}}\Gamma(n-\frac{1}{2})}{2(2n-1)\Gamma(n)},\quad \mu_{2n+1}^-(L)\geqslant-\frac{\pi^{\frac{1}{2}}\Gamma(n-\frac{1}{2})}{2(2n+1)\Gamma(n)}\quad(n\in\mathbb{N}).$$

(As one would expect, a sharp bound is not obtained for $\mu_1^-(L)$ by this means. The lower bounds for $\mu_n^-(L)$ behave asymptotically like $\pi^{\frac{1}{2}}n^{-\frac{3}{2}}/2^{\frac{1}{2}}$ for large n.)

7.18 Determine underestimates for the two largest eigenvalues of the operator K on $L_2(0, 1)$, where

$$(K\phi)(x) = \int_0^1 \log\left|\frac{x^{\frac{1}{2}} + t^{\frac{1}{2}}}{x^{\frac{1}{2}} - t^{\frac{1}{2}}}\right| \phi(t)\, dt,$$

as follows. Recall from Example 6.7 that $K = TT^*$ where T is the operator on $L_2(0, 1)$ defined by $(T\phi)(x) = \int_0^x (x - t)^{-\frac{1}{2}} \phi(t)\, dt$ $(0 \leqslant x \leqslant 1)$. Show that with the test functions $\chi_n(x) = (1 - x)^{n - \frac{1}{2}}$ $(0 \leqslant x \leqslant 1; n \in \mathbb{N})$, $(T^*\chi_n)(x) = B(n + \frac{1}{2}, \frac{1}{2})(1 - x)^n$ $(0 \leqslant x \leqslant 1)$, where B is the beta function. Hence evaluate $(K\chi_n, \chi_m) = (T^*\chi_n, T^*\chi_m)$ and use the Rayleigh–Ritz method to show that the eigenvalues of K restricted to the space spanned by χ_1 and χ_2 are $3\pi^2(7 \pm 19^{\frac{1}{2}})/160$. The required underestimates follow.

(As is evident in this problem, an explicit TT^* factorisation of an operator can greatly facilitate the calculation of the quantity $(K\chi_n, \chi_m)$ occurring in the Rayleigh–Ritz method.)

7.19 Define the operator T on $L_2(0, 1)$ by

$$(T\phi)(x) = \int_0^x \frac{x^{\frac{1}{2}} \phi(t)\, dt}{t^{\frac{1}{4}}(x - t)^{\frac{3}{4}}},$$

let $\chi_n(x) = x^{n - \frac{3}{4}}$ $(0 \leqslant x \leqslant 1; n \in \mathbb{N})$ and show that $(T\chi_n)(x) = B(n, \frac{1}{4})x^{n - \frac{1}{4}}$ $(0 \leqslant x \leqslant 1)$ where B denotes the beta function defined in Appendix C. Evaluate $(T^*T\chi_n, \chi_m) = (T\chi_n, T\chi_m)$ and hence show that the Rayleigh–Ritz method used with the trial function $p = \alpha_1\chi_1 + \alpha_2\chi_2$ produces the underestimates $(57 + 561^{\frac{1}{2}})/6$ and $(57 - 561^{\frac{1}{2}})/6$ for the two largest eigenvalues of the operator T^*T.

Use Example 6.8 to show that

$$(TT^*\phi)(x) = B(\tfrac{1}{2}, \tfrac{1}{4}) \int_0^1 \frac{\phi(t)\, dt}{|x - t|^{\frac{1}{2}}} \quad (0 \leqslant x \leqslant 1)$$

and recall that T^*T and TT^* have the same eigenvalues. Deduce that the largest eigenvalue μ_1^+ of the operator K on $L_2(0, 1)$ given by $(K\phi)(x) = \int_0^1 |x - t|^{-\frac{1}{2}} \phi(t)\, dt$ $(0 \leqslant x \leqslant 1)$ is such that $\mu_1^+ > 2.56$.

(Note that $B(\frac{1}{2}, \frac{1}{4}) = 5.2441$ to four decimal places. It is more straightforward to apply the Rayleigh–Ritz method to T^*T than to TT^* and for eigenvalue estimation the more convenient operator can be used. Of course the sets of eigenvectors of TT^* and T^*T are different; note however that if ϕ_n is an eigenvector of T^*T then $T\phi_n$ is an eigenvector of TT^* belonging to the same eigenvalue.)

7.20 Use the Rayleigh–Ritz method to obtain lower bounds for the largest eigenvalue of the operator K on $L_2(0, 1)$ defined by

$$(K\phi)(x) = \int_0^1 \min(x, t)\phi(t)\, dt.$$

(i) Show that with the trial function $p = \alpha_1 \chi_1$ where $\chi_1(x) = x$ $(0 \leqslant x \leqslant 1)$ the lower bound $2/5$ is obtained.

(ii) Show that with the piecewise constant trial function $p = \alpha_1 \chi_1 + \alpha_2 \chi_2$, where

$$\left.\begin{array}{l} \chi_1(x) = 1 \\ \chi_2(x) = 0 \end{array}\right\} \quad (0 \leqslant x < \tfrac{1}{2}), \qquad \left.\begin{array}{l} \chi_1(x) = 0 \\ \chi_2(x) = 1 \end{array}\right\} \quad (\tfrac{1}{2} < x \leqslant 1),$$

the lower bound $(5 + 3\sqrt{2})/24 = 0.385\ldots$ is obtained.

(This simple example substantiates the cautionary remark made after Lemma 7.4. The example can be amplified by adding the test function χ_2, where $\chi_2(x) = x^2$ $(0 \leqslant x \leqslant 1)$, to the trial space in (i) and by using four piecewise constant test functions, defined in an obvious way, in place of the two given in (ii). Piecewise constant and piecewise linear trial functions are widely used in numerical analysis; one obvious advantage is that the test functions are orthogonal.)

7.21 (i) Deduce from Problem 6.9 that the operator defined by

$$(K\phi)(x) = \tfrac{1}{2} B(\tfrac{1}{4}, \tfrac{1}{4}) \int_0^1 \{|x^{\frac{1}{2}} - t^{\frac{1}{2}}|^{-\frac{1}{2}} - (x^{\frac{1}{2}} + t^{\frac{1}{2}})^{-\frac{1}{2}}\} \phi(t) \, dt \quad (0 \leqslant x \leqslant 1)$$

can be written as $K = TT^*$ where

$$(T\phi)(x) = \int_0^x t^{\frac{1}{4}} (x - t)^{-\frac{1}{2}} \phi(t) \, dt \quad (0 \leqslant x \leqslant 1).$$

Let $\chi_n(x) = x^{n - 11/8}$ $(0 \leqslant x \leqslant 1)$ and show that $(T\chi_n)(x) = B(n - \tfrac{1}{4}, \tfrac{1}{4}) x^{n-1}$ $(n \in \mathbb{N})$ where B is the beta function. Use the Rayleigh–Ritz method with the operator T^*T and the trial function $\alpha_1 \chi_1 + \alpha_2 \chi_2$ to produce underestimates for the two largest eigenvalues of K. By making an appropriate change of variables and using Theorem 5.11 deduce that $\mu_1^+(K_1) > 1.60$ and $\mu_2^+(K_1) > 0.36$ where

$$(K_1\phi)(x) = \int_0^1 \{|x - t|^{-\frac{1}{2}} - (x + t)^{-\frac{1}{2}}\} \phi(t) \, dt \quad (0 \leqslant x \leqslant 1).$$

(Note that $B(\tfrac{1}{4}, \tfrac{1}{4}) = 7.4163$ to four decimal places.)

(ii) Use Kellogg's method for the operator K_1 above with $\psi_0(x) = 1$ $(0 \leqslant x \leqslant 1)$ and show (by evaluating Ψ_1 and Ψ_2, in the usual notation) that $\mu_1^+(K_1) > 1.63$.

(In §1.3.3 a relationship is established between the operator K_1 and Tricomi's equation.)

7.22 Let K be the operator on $L_2(0, 1)$ defined by $(K\phi)(x) = \int_0^1 |x - t|^{-\frac{1}{2}} \phi(t) \, dt$ $(0 \leqslant x \leqslant 1)$ and let $p_1(x) = 1$ $(0 \leqslant x \leqslant 1)$ and $p_2(x) = 3^{\frac{1}{2}}(2x - 1)$ $(0 \leqslant x \leqslant 1)$. Note from Example 7.5 that $(Kp_1, p_1) = 32/21$ and show that $(Kp_2, p_2) = 32/105$.

Use the result in Problem 5.15(ii) and the information in Example 7.5 to deduce that $\mu_2^+(K) \geq 0.242$.

7.23 Let L be the operator of Example 7.9 defined by

$$(L\phi)(x) = \int_{-1}^{1} \frac{\phi(t)\,dt}{|x-t|^\alpha} \quad (-1 \leq x \leq 1),$$

where $0 < \alpha < 1$. By setting $\phi_j(x) = (n/2)^{\frac{1}{2}}$ for $x \in (-1 + 2(j-1)/n, -1 + 2j/n)$ and $\phi_j(x) = 0$ elsewhere in $[-1, 1]$, use Problem 5.15 to show that

$$\mu_1^+(L) + \cdots + \mu_n^+(L) \geq \frac{2^{2-\alpha}}{(1-\alpha)(2-\alpha)} n^\alpha.$$

7.24 (i)   Let $k: [0, 1] \times [0, 1] \to \mathbb{R}$ be a (real-valued) kernel which generates a compact operator K on $L_2(0, 1)$. Show that μ is an eigenvalue of K if and only if $\bar{\mu}$ is also an eigenvalue of K. Deduce (using the result of Problem 4.11) that if k is also skew-symmetric (that is $k(x, t) = -k(t, x)$) then the non-zero eigenvalues of K^2 are all negative and the corresponding sets of eigenvectors have even dimension.

(ii) Let the operators K_1 and K_2 be defined on $L_2(0, 1)$ by $K_1 = T - T^*$ and $K_2 = T + T^*$ where

$$(T\phi)(x) = \int_0^x (x-t)^{-\frac{1}{2}} \phi(t)\,dt.$$

Show that $K_1^2 + K_2^2 = 2\pi P$ where $(P\phi)(x) = \int_0^1 \phi(t)\,dt$ $(0 \leq x \leq 1)$ and that $-K_1^2$ is a positive operator. Use Theorem 5.4 and Corollary 5.6 to deduce that

$$\mu_{n+1}^+(K_2^2) \leq \mu_n^+(-K_1^2) \leq \mu_n^+(K_2^2) \quad (n \in \mathbb{N}).$$

Hence show that the eigenvalues of K_1 are $\pm i\mu_{2n}^+(K_2)$ $(n \in \mathbb{N})$.

8

Approximation methods for inhomogeneous integral equations

8.1 Introduction

In this chapter we tackle the all too common issue of finding approximations to the solution of an inhomogeneous integral equation whose exact solution we are unable to find explicitly. For definiteness, let us suppose the equation is

$$\phi(x) - \lambda \int_a^b k(x,t)\phi(t)\,\mathrm{d}t = f(x) \quad (a \leqslant x \leqslant b), \tag{8.1}$$

where k, f and λ are known, but ϕ is unknown. We wish to find ϕ, and in our presumed situation of being unable to determine an explicit expression for ϕ, we may choose to find functions which are close to ϕ in some sense. The sense in which we measure this closeness is likely to affect the outcome: do we, for example, require an approximate solution $\hat{\phi}$ such that, for all $x \in [a, b]$, $|\hat{\phi}(x) - \phi(x)| < \epsilon$ so that $\hat{\phi}$ approximates ϕ uniformly? Or is it sufficient to know that $\{\int_a^b |\hat{\phi} - \phi|^2\}^{\frac{1}{2}} < \epsilon$, the measure of difference being the L_2-norm? The choice between these two, or others that may be available, depends on the nature of the problem in hand. In some cases it may be sufficient to calculate $\phi(c)$ to some prescribed accuracy, where c is a particular point of $[a, b]$, while in a number of application areas it is not ϕ itself which is ultimately sought but its inner product (ϕ, g) with a given function g. The last two cases, where we ultimately seek not a function but a number, are at least potentially simpler if only because the object we seek is simpler.

We have already derived some methods for calculating approximations to the solution of (8.1). In Chapter 3 we showed that, under suitable circumstances, the solution of (8.1) could be expressed as the sum of the Neumann series $\phi = f + \sum_{n=1}^{\infty} \lambda^n K^n f$, the convergence of this series being in the sense of the norm on $L_2(a, b)$. This expression for ϕ, while quite specific, is slightly deceptive in that a general formula for $K^n f$ may not be available, nor may we be able to find an exact expression for the sum of the series.

241

Nevertheless, we can calculate $Kf, K^2f, \ldots, K^Nf$ directly for any finite value of N and hence find, explicitly, an expression for $f + \sum_{n=1}^{N} \lambda^n K^n f$. The difference between this and ϕ can be made as small as desired by a suitably large choice of N, and in this sense we can find an approximation to ϕ. The exact conditions under which the Neumann series converges depend on the particular kernel and have been tackled in Chapter 3, while some of the results at the end of Chapter 4 have been used to find circumstances under which the Neumann series converges uniformly. In these cases we have calculated simple upper bounds on the difference between the exact solution and the approximation to it.

Chapter 4 also contains some results in which the solution ϕ of (8.1) is expressed as a convergent series involving the eigenvalues μ_n and eigenvectors ϕ_n of the operator generated by k or some associated operator. Similar remarks apply to this as to the Neumann series if we know μ_n and ϕ_n exactly for all n, which, however, is seldom the case. Some worthwhile results can still be recovered from this series in cases where the first few values μ_n and functions ϕ_n are known sufficiently accurately.

The series approximations just mentioned are restricted in their applicability, either to sufficiently small values of λ in the case of the Neumann series or to cases where μ_n and ϕ_n can be found sufficiently accurately. We need methods of more general applicability, and it is the development of these which is the basis of this chapter. There are essentially three ideas to be pursued: we may replace (8.1) by another which 'approximates' it in some sense, and which we can solve, in the expectation that the solution to the altered problem is close to that of the original; we may restrict our attention to functions ϕ in some finite-dimensional subspace and solve (8.1) as best we can in that subspace, for example, by finding the member of the subspace which is closest to the exact solution. In this case we hope that we may obtain whatever accuracy is desired by choosing a sufficiently large subspace. The third method is to adapt the variational methods deployed in Chapter 7 to the current problem.

Addressing now the question of replacing (8.1) by another equation, we are faced with two issues; what sort of replacement equation would be easier to solve and what right have we to expect that the solution of this replacement equation is close to that of the original? There are relatively few integral equations for which we have methods of finding exact solutions, so we shall not attempt to replace (8.1) by another equation of the same form with a non-degenerate kernel, but rather to substitute a degenerate kernel for k, yielding a problem in a finite-dimensional space, that is, essentially a matrix problem. This substitute problem, of course, may not be trivial to solve but there is a wealth of literature on the approximate solution of matrix equations, so progress can be made. The issue of how to select a suitable degenerate kernel is connected with the second question of relating the solution of the altered problem to that of the original. Put in other terms, we wish to make a 'small' change in the problem and be certain that this results in only a 'small' change in the solution; in some sense the solution *process* must be continuous. This is the idea of the problem's being 'well-posed' in the traditional terminology, and which we considered briefly at the end of Chapter 6.

The second method, that of restriction of ϕ to some finite-dimensional subspace, is different in spirit from the first, although there are cases where the two amount to much the same thing. We shall need to pay attention to the choice of subspace, the 'error' involved in the approximate solution and the convergence of the approximate solutions as a sequence of subspaces is chosen. This will, once again, reduce the problem to one of finite dimension which has to be solved to produce an approximation to the exact solution.

The two foregoing methods will require some attention to be given to problems in finite-dimensional spaces, for which variational methods are a valuable technique. Indeed the variational methods, in practice, are applied by choosing 'test' functions with a finite number of parameters, that is, functions in some finite-dimensional space, and finding the stationary points in this situation. The theory then produces estimates relating to the true solution. We shall proceed by investigating the well-posedness issue and then we shall construct variational and other related methods. Finally, against the background of these methods, we shall examine the techniques of relating true and approximate solutions.

8.2 Well-posed problems

We wish to solve the integral equation

$$\phi(x) - \lambda \int_a^b k(x,t)\phi(t)\,\mathrm{d}t = f(x) \quad (a \leqslant x \leqslant b), \tag{8.1}$$

where λ, k and f are given and ϕ is unknown, by replacing it with another equation which approximates to (8.1), say,

$$\phi_1(x) - \lambda_1 \int_a^b k_1(x,t)\phi_1(t)\,\mathrm{d}t = f_1(x) \quad (a \leqslant x \leqslant b). \tag{8.2}$$

If λ_1, k_1 and f_1 are 'close' to λ, k and f respectively we may hope that the solution of (8.2) will resemble that of (8.1) and that the two solutions will be 'close' in some sense. The aim of this section is to turn this hope into certainty. Notice that we require some result of this sort before implementing any process of finding numerical solutions to (8.1), since in the course of such a process we are bound, whether through the action of rounding errors or more subtle causes, to replace the quantities involved with others close to them but not exactly equal.

Suppose that k is a kernel which gives rise to a bounded linear map K on $L_2(a, b)$, and that $f \in L_2(a, b)$. Then (8.1) may be recast as the equation $(I - \lambda K)\phi = f$ in $L_2(a, b)$. Supposing that k_1 gives rise to the bounded linear map K_1, (8.2) may be similarly recast as $(I - \lambda_1 K_1)\phi = f_1$. If $I - \lambda K$ is invertible as a bounded linear map, then we know from Corollary A3 that if T is a bounded linear map and $\|T\| < 1/\|(I - \lambda K)^{-1}\|$, $I - (I - \lambda K)^{-1}T$ and therefore $(I - \lambda K) - T$ are invertible. Therefore, if $\|\lambda K - \lambda_1 K_1\| < 1/\|(I - \lambda K)^{-1}\|$ the operator $I - \lambda_1 K_1 = I - \lambda K + (\lambda K - \lambda_1 K_1)$ is invertible and so (8.2) will have a unique solution $\phi_1 = (I - \lambda_1 K_1)^{-1}f_1$. Under these conditions

$$\phi - \phi_1 = (I - \lambda K)^{-1}(f - f_1) - (I - \lambda_1 K_1)^{-1}(\lambda_1 K_1 - \lambda K)(I - \lambda K)^{-1}f_1$$

so

$$\|\phi - \phi_1\| \leqslant C\|f - f_1\| + \frac{C^2\|\lambda_1 K_1 - \lambda K\|}{1 - C\|\lambda_1 K_1 - \lambda K\|}\|f_1\| \tag{8.3}$$

where $C = \|(I - \lambda K)^{-1}\|$; this makes use of the estimate

$$\|(I - \lambda_1 K_1)^{-1}\| = \|(I - \lambda K)^{-1}(I - (\lambda_1 K_1 - \lambda K)(I - \lambda K)^{-1})^{-1}\|$$

$$\leqslant C \sum_{n=0}^{\infty} \|(\lambda_1 K_1 - \lambda K)(I - \lambda K)^{-1}\|^n$$

$$\leqslant C \sum_{n=0}^{\infty} C^n \|\lambda_1 K_1 - \lambda K\|^n$$

in the final term. The principal result is that $\|\phi - \phi_1\|$ may be made as small as desired by choosing $\|f - f_1\|, |\lambda - \lambda_1|$ and $\|K - K_1\|$ sufficiently small, all of this on the presumption that $I - \lambda K$ is invertible.

The result just calculated shows us that $\|\phi - \phi_1\|$ will be small if we choose λ_1, K_1 and f_1 close to λ, K and f respectively. To give a more specific example, suppose that k is an L_2-kernel on $[a, b] \times [a, b]$ and that $I - \lambda K$ is invertible; the invertibility can in this case be checked by showing that the corresponding homogeneous equation has only the trivial solution. If we choose λ_1, k_1 and f_1 so that $|\lambda - \lambda_1|, \int_a^b \int_a^b |k(x, t) - k_1(x, t)|^2 \, dx \, dt$ and $\int_a^b |f(t) - f_1(t)|^2 \, dt$ are sufficiently small, then $\int_a^b |\phi(t) - \phi_1(t)|^2 \, dt$ will be as small as desired; by using (8.3) we can obtain an upper bound for $\int_a^b |\phi(t) - \phi_1(t)|^2 \, dt$.

The information that $\| \phi - \phi_1 \|$ is small is not always as specific as one might wish, since it does not indicate anything about $\phi(x) - \phi_1(x)$ for any particular value of x. However, by observing that $(I - \lambda K)^{-1}$ may be written in the form $I + \lambda K (I - \lambda K)^{-1}$ and using the result of Lemma 4.21, we may show that if k and k_1 are Schmidt kernels, then if $|\lambda - \lambda_1|$, $\int_a^b \int_a^b |k(x, t) - k_1(x, t)|^2 \, dx \, dt$ and $\sup_t |f(t) - f_1(t)|$ are all chosen sufficiently small, then $\sup_x |\phi(x) - \phi_1(x)|$ can be made as small as desired. In this case we may obtain a uniform approximation to ϕ, that is, one where we have an upper bound for $|\phi(x) - \phi_1(x)|$ which holds for all x. The remarks at the very end of Chapter 4 may be used to obtain a similar result in the case when k is a Schur kernel.

We have shown, therefore, that if the operator $I - \lambda K$ associated with the integral equation (8.1) is invertible as a map on $L_2(a, b)$, then the problem is 'well-posed' in the sense that its solution exists, is unique and varies only slightly if the data in the problem are slightly perturbed. This, of course, does not yet form a practical programme for finding approximate solutions.

The means by which we choose another kernel k_1 which is close to k and for which (8.2) may be solved depends on the nature of the kernel k. If k is, for example, continuous, then by dividing the square $[a, b] \times [a, b]$ into n^2 squares of side $(b - a)/n$, where n is sufficiently large, we may ensure that the value of $k(x, t)$ at any point in one of the squares differs from that at the centre by less than ϵ. Choosing k_1 to be the function whose value on each square is the value k takes at the centre of the square, we see that for all $x, t \in [a, b]$ $|k(x, t) - k_1(x, t)| < \epsilon$. Then, writing K and K_1 for the corresponding operators, $\| K - K_1 \| < \epsilon(b - a)$. Now k_1 is degenerate, so in this case solving (8.2) is equivalent to solving a system of n linear equations in n unknowns, which we can, in principle, carry out. (The operator $I - \lambda K_1$, and hence the matrix of the system of equations, is invertible if ϵ is chosen sufficiently small since we have presumed $I - \lambda K$ is invertible.) Similar approximations of k by a degenerate kernel are possible for other types of kernel; Theorem 3.4 is proved by carrying out such an approximation for an L_2-kernel k.

This approach, of finding a degenerate kernel k_1 for which $\| K - K_1 \|$ may be made as small as desired is, however, only possible for kernels which generate a compact operator. (It is possible that in a particular problem there is a degenerate kernel k_1 such that $\| K - K_1 \| < 1/\| (I - \lambda K)^{-1} \|$, and therefore that some information can be gained from this method. If K is not compact, it is not possible to choose $\| K - K_1 \|$ arbitrarily small, so there is an intrinsic limit to the attainable accuracy in such cases.) In this situation we need to resort to other methods, such as Galerkin's, which we shall consider later.

The foregoing remarks on the variation of the solution of (8.1) consequent on small changes in λ, k and f have presumed that $I - \lambda K$ is invertible and they are false if this is relaxed. If $I - \lambda_0 K$ is not invertible, and $I - \lambda K$ is invertible for $0 < |\lambda - \lambda_0| < \delta$, then $(I - \lambda K)^{-1} f$ need not tend to a solution of $(I - \lambda_0 K)\phi = f$ as $\lambda \to \lambda_0$, even if the latter equation is soluble. To see this consider the analogue in $\mathbb{R}^2$ where $\lambda_0 = 1$, $A = \begin{pmatrix} 1 & 1 \\ 0 & 1 \end{pmatrix}$, $\mathbf{f} = \begin{pmatrix} 1 \\ 0 \end{pmatrix}$ and we wish to solve $(I - \lambda A)\boldsymbol{\phi} = \mathbf{f}$.

8.3 Methods based on variational principles

In Chapter 7 we showed that the eigenvectors and eigenvalues of a compact self-adjoint operator K on a Hilbert space H can be characterised by stationary principles. For example, the Rayleigh quotient $R_K(p) = (Kp, p)/(p, p)$ $(p \in H, p \neq 0)$ is stationary at the eigenvectors of K and its stationary values are the eigenvalues of K. This property can be used to produce approximations to eigenvectors and eigenvalues of K by the Rayleigh–Ritz method, in which approximations to the stationary points of $R_K(p)$, and consequently to the eigenvectors, are sought in a finite-dimensional subspace of H.

Our aim here is to develop a corresponding method to approximate solutions of inhomogeneous operator equations in a Hilbert space H. The starting point is to identify a stationary principle, stating that a given functional, $J(p)$ say, is stationary at, and only at, the solution of a given equation. Locating the stationary point of $J(p)$ is therefore an indirect way of solving the equation, which can be exploited by the Rayleigh–Ritz procedure. Seeking an approximation $p_N \in E_N$ to the stationary point reduces the problem to one in matrix theory, E_N being a subspace of H of dimension N.

One of the virtues of the variational method, which we drew attention to in Chapter 7, is that approximations to the stationary value of a functional may be described as second-order accurate, relative to a first-order accuracy in the stationary point approximation. In the case of an inhomogeneous operator equation in H, whose exact solution is ϕ say, the stationary value of an associated functional is typically (ϕ, g), where $g \in H$ is known. The Rayleigh–Ritz method allied to the stationary principle produces the 'second order accurate' stationary value (p_N, g) where $p_N \in E_N$ is a 'first order accurate' approximation to ϕ. Obviously the method is at its most effective if only (ϕ, g) is required and there are numerous instances in application areas where this is the case. To give a simple illustration, consider the boundary value problem

$$\begin{cases} \phi''''(x) + \lambda\phi(x) = h(x) & (0 < x < 1), \\ \phi(0) = \phi(1) = \phi''(0) = \phi''(1) = 0. \end{cases}$$

One interpretation of this problem is that $\phi(x)$ represents the transverse displacement at x of the centroid of the cross-section of a uniform elastic beam. The beam is simply supported (hinged) at each end and is subjected to a transverse distributed load h. A restoring force also acts which is proportional to the local displacement, $\lambda > 0$ being a given constant. By using a method described in Chapter 1 (*cf.* Example 1.3, which is related to this illustration), the boundary value problem can be reduced to the equation $\phi = -\lambda K\phi + Kh$, where K is a positive compact operator. The equation therefore has a unique solution for any $\lambda > 0$. As we shall see shortly, one stationary principle which can be associated with an equation of the form in question has as its stationary value the real number $(\phi, Kh) = \lambda^{-1}\{(Kh, h) - (\phi, h)\}$. Since h is assigned, an approximation to this stationary value is effectively an approximation to (ϕ, h), the total work done by the load h in deflecting the beam from its $\phi = 0$ configuration. Thus if we only need an estimate of the work done, rather than the pointwise displacement $\phi(x)$ $(0 < x < 1)$, the variational approach is particularly well-suited to the problem.

If, in the last example or more generally, the unknown function ϕ itself is required, the variational method appears to be less satisfactory. Suppose that, by taking a sequence of trial spaces E_N, we obtain a sequence of increasingly accurate approximations (p_N, g) to the stationary value (ϕ, g) of a given functional. Then we may expect that the sequence of approximations p_N will also represent ϕ with increasing accuracy, at least in an overall sense (if (p_{N+1}, g) is closer to (ϕ, g) than (p_N, g), we cannot conclude that p_{N+1} represents ϕ more accurately than p_N). We would hope of course that if $(p_N, g) \to (\phi, g)$ as $N \to \infty$ then $p_N \to \phi$ as $N \to \infty$.

There is another way of obtaining an approximation to ϕ however, by concocting a variational principle whose stationary value is ϕ itself. This particular application of variational methods, and the others we shall discuss, are considerably strengthened if the associated variational principle is actually an extremum (maximum or minimum) principle or, failing that, if upper and lower bounds for the stationary value can be derived independently. In the case of a maximum principle, an approximation to the maximum value of the functional is of course a lower bound for the actual maximum. Likewise a minimum principle provides upper bounds for the true minimum. However upper and lower bounds need not be related to extremum principles and, unlike the latter, they need not offer a means of generating approximations. Instead they can be used to refine the estimate of a stationary value, using an approximation of the stationary point obtained independently. In this sense upper and lower bounds can be employed whether the stationary value is a maximum, a minimum or neither, but of course they serve most purpose in the last case where the variational principle does not give a bound directly.

In §8.3.1 to §8.3.6 inclusive we deal only with self-adjoint operators. The extensions of some of the ideas to the case of operators which are not self-adjoint are straightforward and are examined in §8.3.7 and §8.3.8.

8.3.1 A maximum principle

A suitable starting point from which to develop a collection of stationary principles and bounds is given by considering the equation

$$A\phi = F \tag{8.4}$$

where $A \in B(H)$ is a positive (self-adjoint) operator and $F \in H$ is regarded as given. The non-committal notation used here allows us to choose A and F in various ways later without any danger of confusion.

We suppose that A has a bounded inverse. The solution of (8.4) is unique, by virtue of the positivity of A, from which it also follows that, for any $p \in H$,

$$(A\phi - Ap, \phi - p) \geqslant 0, \tag{8.5}$$

with equality only if $p = \phi$. It follows that

$$(A\phi, \phi) \geqslant (A\phi, p) + (Ap, \phi) - (Ap, p)$$

and hence, using $A\phi = F$, that

$$(\phi, F) \geqslant (F, p) + (p, F) - (Ap, p). \tag{8.6}$$

If we now define the functional $J: H \to \mathbb{R}$ by

$$J(p) = (F, p) + (p, F) - (Ap, p) \quad (p \in H) \tag{8.7}$$

we have the maximum principle that the maximum value of $J(p)$ is equal to (ϕ, F), where $A\phi = F$, and the maximum is attained at, and only at, $p = \phi$. As we have already indicated, $J(p)$ takes only real values and in particular $(F, \phi) = (A\phi, \phi) = (\phi, A\phi) = (\phi, F)$ is a real number.

An alternative derivation of the maximum principle follows using variational calculus. If we now take $J(p)$ as the starting point and put

$p = \phi + \delta\phi$, a straightforward expansion yields

$$J(\phi + \delta\phi) = J(\phi) + (F - A\phi, \delta\phi) + (\delta\phi, F - A\phi) - (A\delta\phi, \delta\phi).$$

Thus $J(p)$ is stationary at $p = \phi$ where $A\phi = F$, the stationary value of $J(p)$ is $J(\phi) = (\phi, F)$ and $J(p) = (\phi, F) - (Ap - A\phi, p - \phi) \leqslant (\phi, F)$ because A is positive. Notice that for a particular p chosen to approximate ϕ, the difference between the corresponding approximation $J(p)$ to the maximum value of J and the actual maximum value is $O(\|p - \phi\|^2)$.

Suppose that we set $p = c\hat{p}$ where c is a constant, $\hat{p} \in H$ and $\hat{p} \neq 0$. If the resulting functional $J(c\hat{p}) = \bar{c}(F, \hat{p}) + c(\hat{p}, F) - c\bar{c}(A\hat{p}, \hat{p})$ is regarded as a function of c and $\bar{c}$, one can easily show that it is stationary with $c = (F, \hat{p})/(A\hat{p}, \hat{p})$. Using this value of c we can define the new functional $\hat{J}(\hat{p}) = J(c\hat{p})$, and

$$\hat{J}(\hat{p}) = (F, \hat{p})(\hat{p}, F)/(A\hat{p}, \hat{p}) \quad (\hat{p} \in H, \hat{p} \neq 0). \tag{8.8}$$

Since $\hat{J}(p)$ is just $J(p)$ optimised with respect to a constant multiplier in p, it is invariant under a scaling of $\hat{p}$, it is maximised by $\hat{p} = a\phi$ where $A\phi = F$ and a is any non-zero constant, and its maximum value is (ϕ, F). From the point of view of the scale-invariance property, $\hat{J}(\hat{p})$ parallels the Rayleigh quotient which arises in connection with homogeneous equations, as we have seen.

8.3.2 Second kind integral equations

By making particular choices of A and F in $A\phi = F$ we can deduce extremum principles for the equation $\phi = f + \lambda K\phi$ from the maximum principle derived in the last section.

Theorem 8.1

Let $K \in B(H)$ be self-adjoint and let $I - \lambda K$ be a positive operator, where λ is a real parameter. Then the functional $J: H \to \mathbb{R}$ defined by

$$J(p) = (f, p) + (p, f) - (p, p) + \lambda(Kp, p)$$

is maximised at $p = \phi$ if and only if ϕ satisfies $\phi = f + \lambda K\phi$; the maximum value of $J(p)$ is (ϕ, f).

Proof

This follows at once from the maximum principle for $A\phi = F$ on setting $A = I - \lambda K$ and $F = f$. □

Suppose that the largest positive eigenvalue of K is μ_1^+ and that its largest negative eigenvalue, in magnitude, is μ_1^-. Then, by using the extremal properties of (Kp, p) (see §5.2) we have

$$((I - \lambda K)p, p) \geqslant \begin{cases} (1 - \lambda \mu_1^+) \|p\|^2 & (\lambda \geqslant 0), \\ (1 - \lambda \mu_1^-) \|p\|^2 & (\lambda \leqslant 0). \end{cases}$$

It follows that, if neither μ_1^+ nor μ_1^- is zero, $I - \lambda K$ is a positive operator, as required in Theorem 8.1, for values of λ such that

$$(\mu_1^+)^{-1} > \lambda > (\mu_1^-)^{-1}. \tag{8.9}$$

This restriction means that λ^{-1} is not an eigenvalue of K. If K is a positive operator, then so is $I - \lambda K$ for $\lambda < (\mu_1^+)^{-1}$ and if K is a negative operator, $I - \lambda K$ is a positive operator for $\lambda > (\mu_1^-)^{-1}$.

Theorem 8.2

Let $K \in B(H)$ be self-adjoint and let λK be a negative operator, where λ is a real parameter. Then the functional $L: H \to \mathbb{R}$ defined by

$$L(p) = \|f + \lambda Kp\|^2 - \lambda(Kp, p)$$

is minimised at $p = \phi$ if and only if ϕ satisfies $\phi = f + \lambda K\phi$; the minimum value of $L(p)$ is (ϕ, f).

Proof

Since λK is a negative operator then the solution of $\lambda K(\lambda K\phi - \phi) = -\lambda Kf$ is also the solution of $\phi - \lambda K\phi = f$. Using the maximum principle derived in §8.3.1 with $A = \lambda^2 K^2 - \lambda K$, which is certainly positive, and $F = -\lambda Kf$, gives

$$(-\lambda Kf, \phi) \geqslant (-\lambda Kf, p) + (p, -\lambda Kf) - (\lambda^2 K^2 p - \lambda Kp, p)$$

where equality is attained at $p = \phi$ if and only if $(\lambda^2 K^2 - \lambda K)\phi = -\lambda Kf$, that is, if and only if $\phi = f + \lambda K\phi$. Since $(\lambda Kf, \phi) = (f, \lambda K\phi) = (f, \phi) - (f, f)$ the inequality can be rearranged as

$$(f, \phi) \leqslant (f, \lambda Kp) + (\lambda Kp, f) + \|\lambda Kp\|^2 + \|f\|^2 - \lambda(Kp, p)$$

so that $(f, \phi) \leqslant \|f + \lambda Kp\|^2 - \lambda(Kp, p)$.

Variational calculus can also be used to establish the result.   □

The minimum principle applies under more restrictive conditions than the maximum principle, but the two co-exist if λK is a negative operator and together give the following dual extremum principles.

Theorem 8.3

Let $K \in B(H)$ be self-adjoint and let λK be a negative operator, where λ is a real parameter. If $\phi \in H$ satisfies $\phi = f + \lambda K\phi$ then for all $p_1 \in H$ and all

$p_2 \in H$, $J(p_1) \leqslant (\phi, f) \leqslant L(p_2)$ where

$$J(p) = (f, p) + (p, f) - (p, p) + \lambda(Kp, p)$$

and

$$L(p) = \| f + \lambda Kp \|^2 - \lambda(Kp, p) = J(p) + \| p - f - \lambda Kp \|^2.$$

The maximum value of J is achieved only at $p_1 = \phi$ and the minimum value of L is achieved only at $p_2 = \phi$.

Proof

The result is a direct consequence of Theorems 8.1 and 8.2 and the fact that $L(p) - J(p) = \| p - f - \lambda Kp \|^2$ which is easily verified. □

Remark

As their separate derivations indicate, the maximum and minimum principles embodied in the foregoing theorem can be used independently to approximate (ϕ, f) and ϕ. Suppose that we seek to approximate (ϕ, f) by finding the maximum value of $J(p_1)$ with $p_1 \in E_N$, a subspace of H of dimension N. It is not necessary to use this same subspace to approximate the minimum value of $L(p_2)$ and even if the same space E_N is used the maximiser p_1 of J will usually differ from the minimiser p_2 of L. Of course we can calculate an approximation $p_1 \in E_N$ for ϕ and choose $p_2 = p_1$ to provide the upper bound $L(p_1)$ for (ϕ, f). In this case it follows that the upper and lower estimates of (ϕ, f) obtained will differ by $\| p_1 - f - \lambda Kp_1 \|^2 = \| (p_1 - \lambda Kp_1) - (\phi - \lambda K\phi) \|^2$, the square of the norm of the residual error incurred in replacing ϕ by p_1. (In approximating the solution ϕ of the equation $A\phi = f$ we shall refer to $Ap - f$ as the *residual error* in the approximation of the solution by p.)

Example 8.1

Let the operator K be defined on $L_2(0, 1)$ by

$$(K\phi)(x) = \int_0^1 \{\min(x, t) - xt\}\phi(t)\, dt \quad (0 \leqslant x \leqslant 1)$$

and let ϕ satisfy $\phi = f + \lambda K\phi$ where $\lambda < 0$ and $f(x) = 1$ $(0 \leqslant x \leqslant 1)$. We wish to approximate $(\phi, f) = \int_0^1 \phi(x)\, dx$.

Note that we can consider the given equation in the Hilbert space of real-valued functions in $L_2(0, 1)$, that the operator K is positive (see Example 6.3) and therefore that the operator λK is negative. The dual extremum principles of Theorem 8.3 therefore apply, and we seek the maximum value of $J(p_1)$ and the minimum value of $L(p_2)$, with p_1 and p_2 restricted to suitable test spaces.

Maximum principle

Note from the given equation that $\phi(0) = f(0) = 1$ and we therefore choose the test space composed of constant functions. Let $p_1 = c_1$, a (real) constant. Then $(p_1, f) = (f, p_1) = c_1$, $\|p_1\|^2 = c_1^2$, $(Kp_1)(x) = c_1 x(1-x)/2$ $(0 \leqslant x \leqslant 1)$ and $(Kp_1, p_1) = c_1^2/12$. Therefore $J(p_1) = 2c_1 - c_1^2(1 - \lambda/12)$ which is maximised at $c_1 = 12/(12 - \lambda)$ and the maximum value of J is also $12/(12 - \lambda)$. Note that the maximisation duplicates that used to derive (8.8), which we could have used directly with $\hat{p} = 1$, $f = 1$ and $A = I - \lambda K$.

Minimum principle

Using the same test space again, let $p_2 = c_2$ a (real) constant. We can make use of the calculations given for the maximum principle, needing in addition $\|f\| = 1$, $(Kp_2, f) = c_2/12$ and $\|Kp_2\|^2 = c_2^2/120$. We find that $L(p_2) = 1 + \lambda c_2/6 - \lambda(1 - \lambda/10)c_2^2/12$, which is minimised at $c_2 = 10/(10 - \lambda)$ and the minimum value of L is $1 + 5\lambda/(60 - 6\lambda)$. By Theorem 8.3 therefore,

$$12/(12 - \lambda) \leqslant (\phi, f) \leqslant 1 + 5\lambda/(60 - 6\lambda).$$

In this example the exact value of (ϕ, f) can be calculated via the associated boundary value problem and $(\phi, f) = 2 \tanh(\tfrac{1}{2}(-\lambda)^{\frac{1}{2}})/(-\lambda)^{\frac{1}{2}}$. With $\lambda = -1$ the extremum principles give $0.92308 \leqslant (\phi, f) \leqslant 0.92424$ and $2 \tanh(\tfrac{1}{2}) = 0.92423$ (to five decimal places).   □

Example 8.2

Let ϕ satisfy the integral equation

$$\phi(x) = \log x + \lambda \int_0^1 \{\min(x, t) - xt\} \phi(t)\, dt \quad (0 < x \leqslant 1)$$

where $\lambda < 0$ and seek approximations to $\int_0^1 \phi(x) \log x\, dx$.

This problem differs from that in Example 8.1 only in the new free term $f(x) = \log x$ $(0 < x \leqslant 1)$. Proceeding as in Example 8.1, let $p_1 = c_1$ and $p_2 = c_2$, where c_1 and c_2 are real constants. It is found that $J(p_1) = -2c_1 - c_1^2(1 - \lambda/12)$, which has the maximum value at $c_1 = -12/(12 - \lambda)$ of $12/(12 - \lambda)$, and that $L(p_2) = 2 - 5\lambda c_2/36 - \lambda(1 - \lambda/10)c_2^2/12$, which has the minimum value of $2 + 125\lambda/(216(10 - \lambda))$. With $\lambda = -1$, the dual extremum principles give $0.9231 \leqslant (\phi, f) \leqslant 1.9474$ (to four decimal places).

It is not difficult to see why these bounds are so much less satisfactory than those obtained in Example 8.1. The choice of constant trial functions has ignored the fact that ϕ is known to be the sum of two terms, one of which is the free term f and the other a bounded function. Clearly, any test space should include f as a member, at least if reasonable approximations are to be obtained. On this basis, a more suitable one-term trial function for use in the maximum principle is $p_1 = c_1 f$ where c_1 is a real constant. For this p_1 we find that $\| p_1^2 \| = 2c_1^2$, $(p_1, f) = 2c_1$ and $(Kp_1, p_1) = 29c_1^2/432$. The maximum value of the corresponding $J(p_1)$ is $1728/(864 - 29\lambda)$, giving $1.9351 \leqslant (\phi, f)$ with $\lambda = -1$.

We can revise the application of the minimum principle as well, although it is now clear that it provided a reasonable upper bound for (ϕ, f) using only constant test functions. Because the minimum principle is derived using the equation $\phi = f + \lambda K\phi$ in the form $-\lambda K(\phi - \lambda K\phi) = -\lambda K f$ (cf. Theorem 8.2), so that Kf effectively plays the part of the free term, it is less sensitive to a free term such as $f(x) = \log x$ $(0 < x \leqslant 1)$.

There is another way of improving the original lower bound. As we have remarked, the deficiency in the simple trial function $p_1 = c_1 = -12/(12 - \lambda)$ is that it does not have the structure of the actual solution. The iterated trial function $\hat{p}_1 = f + \lambda K p_1$ does have the correct form however. Now $\hat{p}_1(x) = \log x - 6\lambda x(1 - x)/(12 - \lambda)$ $(0 < x \leqslant 1)$ and a straightforward, but tedious, calculation gives $J(\hat{p}_1) = 1.9377 \leqslant (\phi, f)$ with $\lambda = -1$. This lower bound exceeds the two previously found.

Notice that this iteration method converts the one-term trial function p_1 into the two-term trial function $\hat{p}_1 = f + \lambda K p_1$, but it does not necessarily maximise J in the test space spanned by f and Kp_1. □

Example 8.3

Let $\phi = f + \lambda K\phi$ in $[0, 1]$ where $\lambda < 0$, $f(x) = 1$ $(0 \leqslant x \leqslant 1)$ and

$$(K\phi)(x) = \int_0^1 |x - t|^{-\frac{1}{2}} \phi(t)\, dt \quad (0 \leqslant x \leqslant 1).$$

Estimate the value of $\int_0^1 \phi(x)\, dx$.

Following the discussions in the previous two examples we take $p_1 = c_1$ and $p_2 = c_2$, where c_1 and c_2 are real constants. Clearly $\| p_1 \|^2 = c_1^2$ and $(f, p_1) = c_1$ and it is not difficult to show that $(Kp_1)(x) = 2c_1\{x^{\frac{1}{2}} + (1 - x)^{\frac{1}{2}}\}$, $(Kp_1, p_1) = 8c_1^2/3$, $(Kp_1, f) = 8c_1/3$ and $\| Kp_1 \|^2 = (4 + \pi)c_1^2$. We find that

$$J(p_1) = 2c_1 - (1 - 8\lambda/3)c_1^2,$$
$$L(p_2) = 1 + 16\lambda c_2/3 - \lambda\{8/3 - (4 + \pi)\lambda\}c_2^2,$$

leading to

$$3/(3-8\lambda) \leqslant \int_0^1 \phi(x)\,dx \leqslant 1+(8/3)^2\lambda/(8/3-(4+\pi)\lambda).$$

With $\lambda = -1$ therefore, the average value of ϕ is bounded below by 0.2727 and above by 0.2750 (to four decimal places).   □

8.3.3 *Further extremum principles and bounds*

We return to the maximum principle based on the inequality

$$(\phi, F) \geqslant J(p) = (F, p) + (p, F) - (Ap, p)   (p \in H)   (8.10)$$

for the solution of $A\phi = F$ in H, where A is a positive operator. In the case where $F = f$ and $A = I - \lambda K$, we have shown that (8.10) leads to both upper and lower bounds for (ϕ, f), provided that λK is a negative operator. We can do better, however. By discarding less than we did in deducing the inequality (8.10) we can improve the lower bound for (ϕ, f) already obtained and we can derive an upper bound for (ϕ, f) which is more widely applicable than our existing one.

Suppose that A is a bounded, positive operator, so that

$$0 < (Ap, p) \leqslant \alpha \| p \|^2   (8.11)$$

for some $\alpha > 0$ and for all non-zero $p \in H$. This condition is satisfied for many of the operators we are interested in. Suppose, for example, that $A = I - \lambda K$ where K is a compact self-adjoint operator on H whose largest eigenvalue is μ_1^+ and whose largest negative eigenvalue, in magnitude, is μ_1^-. If neither μ_1^+ nor μ_1^- is zero and if $(\mu_1^-)^{-1} < \lambda < (\mu_1^+)^{-1}$ then (8.11) is satisfied with

$$\alpha = \begin{cases} (1 - \lambda\mu_1^+) & \text{(if } \lambda \leqslant 0\text{)}, \\ (1 - \lambda\mu_1^-) & \text{(if } \lambda \geqslant 0\text{)}. \end{cases}$$

Our derivation of (8.10), given in §8.3.1, made use of the identity

$$(A\phi - Ap, \phi - p) = (\phi, F) - (F, p) - (p, F) + (Ap, p)   (8.12)$$

together with $(A\phi - Ap, \phi - p) \geqslant 0$. If we use (8.11) in (8.12) we find that $(\phi, F) \leqslant (F, p) + (p, F) - (Ap, p) + \alpha \| p - \phi \|^2$, but this is not a helpful inequality as a rule because its right hand side involves the unknown function ϕ. However, since A is a positive operator by hypothesis, and assuming that it is of the form $aI + bK$, where a and b are constants and K is a compact operator, then the operator $A^{\frac{1}{2}}$ is uniquely defined on H (see Problem 6.2). It follows from (8.11) that $\|Ap\|^2 = (AA^{\frac{1}{2}}p, A^{\frac{1}{2}}p) \leqslant \alpha \|A^{\frac{1}{2}}p\|^2 = \alpha(Ap, p)$. In particular, replacing p by $\phi - p$ and using $A\phi = F$, we have $(A\phi - Ap, \phi - p) \geqslant \alpha^{-1}\|Ap - F\|^2$, and we deduce from (8.12) that

$$(\phi, F) \geqslant (F, p) + (p, F) - (Ap, p) + \alpha^{-1}\|Ap - F\|^2 \quad (p \in H). \qquad (8.13)$$

We can regard this inequality as providing a lower bound for (ϕ, F) which sharpens the underestimate given by the maximum principle (8.10), as the following example shows.

Example 8.4

Let $\phi = f + \lambda K\phi$ where $f(x) = 1$ $(0 \leqslant x \leqslant 1)$, $\lambda < 0$ and

$$(K\phi)(x) = \int_0^1 \{\min(x, t) - xt\}\phi(t)\,dt \quad (0 \leqslant x \leqslant 1).$$

This illustration builds on Example 8.1, where we used the test function $p_1 = c_1$ and showed that $J(p_1)$ is maximised with $c_1 = 12/(12 - \lambda)$. The maximum value of $J(p_1)$ in this case is $12/(12 - \lambda)$, which is a lower bound for $\int_0^1 \phi(x)\,dx$. Now the operator K is positive and its largest eigenvalue is $\mu_1^+ = \pi^{-2}$. Therefore $((I - \lambda K)p, p) \leqslant \alpha \|p\|^2$ with $\alpha = 1 - \lambda\pi^{-2}$, since $\lambda < 0$. From our calculation in Example 8.1 we have $p_1(x) - \lambda(Kp_1)(x) - f(x) = c_1\{1 - \lambda x(1 - x)/2\} - 1$ and we find that $\|p_1 - \lambda Kp_1 - f\|^2 = \lambda^2/(5(12 - \lambda)^2)$. Using (8.13) with $A = I - \lambda K$, $F = f$ and $p_1 = c_1$ therefore gives

$$12/(12 - \lambda) + \pi^2\lambda^2/(5(\pi^2 - \lambda)(12 - \lambda)^2) \leqslant \int_0^1 \phi(x)\,dx.$$

The lower bound at $\lambda = -1$ is 0.92415 (to five decimal places), which significantly tightens the dual bounds obtained in Example 8.1. □

Having used (8.13) purely to supply a lower bound, we now examine whether it also provides us with a maximum principle. This task is made easier by noticing that (8.13) can be arranged in the form

$$(\phi, (\alpha I - A)F) \geqslant ((\alpha I - A)F, p) + (p, (\alpha I - A)F) - ((\alpha I - A)Ap, p), \qquad (8.14)$$

which has the same structure as the inequality $(\phi, F) \geqslant (F, p) + (p, F) - (Ap, p)$ we used in §8.3.1. The relationship between the two inequalities allows us to deduce at once that the functional on the right hand side of (8.14) is stationary at $p = \phi$, where $(\alpha I - A)A\phi = (\alpha I - A)F$. Thus it is stationary at $p = \phi$ where $A\phi = F + \psi$, ψ being an eigenvector of A belonging to its largest eigenvalue α. Expressed another way, the functional on the right of (8.14) is stationary at $p=\phi+\psi$, where $A\phi=F$, and $A\psi=\alpha\psi$. A straightforward calculation shows that equality is obtained in (8.13) with $p=\phi+\psi$.

The problem of solving $A\phi=F$ and that of maximising the functional on the right hand side of (8.13) are therefore not equivalent, in the sense that they do not have the same solution sets. Consequently, (8.13) is not useful from the point of view of deriving approximations to the solution of $A\phi=F$, although its use as a lower bound is not invalidated by the lack of equivalence.

Even though (8.13) cannot be interpreted in terms of a maximum principle for the solution of $A\phi=F$, we can easily perturb it so as to give such a principle. If the real number v is such that $(Ap,p)<v\|p\|^2$ for all non-zero $p\in H$ then the functional $J_1:H\rightarrow\mathbb{R}$ defined by

$$J_1(p)=(F,p)+(p,F)-(Ap,p)+v^{-1}\|Ap-F\|^2 \quad (p\in H) \qquad (8.15)$$

is maximised at, and only at, $p=\phi$ where $A\phi=F$. This follows from our discussion of (8.13) and the fact that the operator $vI-A$ is positive. The functional $J_1(p)$ does form a basis for approximating ϕ, but it is obviously more cumbersome to work with than our original $J(p)$.

In order to obtain an upper bound for (ϕ,F) corresponding to the lower bound (8.13) we have to assume that the operator A is positive and bounded below. That is, we suppose that there exists a real number $\beta>0$ such that

$$\beta\|p\|^2\leqslant(Ap,p) \qquad (8.16)$$

for all $p\in H$. We have in fact already made use of this condition, for the minimum principle of Theorem 8.2 required that λK be a negative operator, which is equivalent to requiring that the operator $A=I-\lambda K$ satisfy $(Ap,p)\geqslant\|p\|^2$. More generally, if K is an indefinite, self-adjoint compact operator, if $\mu_1^{\pm}$ have their usual meanings and if $(\mu_1^-)^{-1}<\lambda<(\mu_1^+)^{-1}$, then $A=I-\lambda K$ satisfies (8.16) with

$$\beta=\begin{cases}1-\lambda\mu_1^+ & (\lambda\geqslant0),\\ 1-\lambda\mu_1^- & (\lambda\leqslant0).\end{cases}$$

Restricting attention to the case in which A is the sum of the identity operator and a self-adjoint compact operator (for which we have proved that $A^{\frac{1}{2}}$ is defined, by Problem 6.2), note that (8.16) implies $(Ap,p)=\|A^{\frac{1}{2}}p\|^2\leqslant\beta^{-1}\|Ap\|^2$ for all $p\in H$. In particular, $(A\phi-Ap,\phi-p)\leqslant\beta^{-1}\|Ap-F\|^2$, where $A\phi=F$, and from (8.12) we deduce that

$$(\phi,F)\leqslant(F,p)+(p,F)-(Ap,p)+\beta^{-1}\|Ap-F\|^2, \qquad (8.17)$$

for any $p\in H$, where $A\phi=F$.

An argument similar to that which we used in relation to (8.13) shows that minimising the functional on the right hand side of (8.17) is not equivalent to solving $A\phi = F$. Nevertheless (8.17) does provide an upper bound for (ϕ, F) and we can deduce from it the minimum principle, that the functional $L_1 : H \to \mathbb{R}$ defined by

$$L_1(p) = (F, p) + (p, F) - (Ap, p) + \omega^{-1} \| Ap - F \|^2 \quad (p \in H),$$

where $\omega \| p \|^2 < (Ap, p)$ for all non-zero $p \in H$ and $\omega > 0$, is minimised at, and only at, the solution of $A\phi = F$.

Whereas (8.13) merely refines the lower bound (8.10), (8.17) provides an upper bound for (ϕ, F) in circumstances in which we previously did not have one. Similarly, the minimum principle generated by $L_1(p)$ is more general than that given in Theorem 8.2, as it applies if λK is an indefinite operator, as long as $I - \lambda K$ is positive and bounded below.

We have now derived dual extremum principles for the solution of $A\phi = F$, in the case where $A = aI + bK$ and K is compact, and upper and lower bounds for (ϕ, F), and it is convenient to gather our results together at this point.

Theorem 8.4

Let $K \in B(H)$ be a compact self-adjoint operator, let $A\phi = F$ have a solution in H, where $A = aI + bK$, a and b being constants and let $J(p) = (F, p) + (p, F) - (Ap, p)$ $(p \in H)$.

(i)   If $\beta \| p \|^2 \leqslant (Ap, p) \leqslant \alpha \| p \|^2$ for all $p \in H$ and some α and $\beta > 0$, then $J(p) + \alpha^{-1} \| Ap - F \|^2 \leqslant (\phi, F)$ for all $p \in H$, the equality holding for $p = \phi + \psi$, where $A\psi = \alpha\psi$; and $(\phi, F) \leqslant J(p) + \beta^{-1} \| Ap - F \|^2$ for all $p \in H$, the equality holding for $p = \phi + \chi$, where $A\chi = \beta\chi$.

(ii)   If $\omega \| p \|^2 < (Ap, p) < \nu \| p \|^2$ for all non-zero $p \in H$ and some ν and $\omega > 0$, then $J(p) + \nu^{-1} \| Ap - F \|^2$ is maximised in H at, and only at, $p = \phi$ and its maximum value is (ϕ, F); and $J(p) + \omega^{-1} \| Ap - F \|^2$ is minimised in H at, and only at, $p = \phi$ and its minimum value is (ϕ, F).   □

Remark

The lower bound and maximum principle apply if A is a positive, bounded operator. The upper bound and minimum principle require A to be additionally bounded below away from zero. Although it will not concern us here, the requirement that A be of the form of compact plus a multiple of the identity can also be relaxed.

Example 8.5

Let $\phi = f + \lambda K \phi$ where $f = 1$, $0 < \lambda < \pi^2$ and K is defined on $L_2(0, 1)$ by

$$(K\phi)(x) = \int_0^1 \{\min(x, t) - xt\}\phi(t)\, dt \quad (0 \leq x \leq 1).$$

Obtain upper and lower bounds for $\int_0^1 \phi(x)\, dx$.

The operator K is compact and self-adjoint and we have noted previously that it is positive with largest eigenvalue $\mu_1^+ = \pi^{-2}$. Therefore $0 < (Kp, p) \leq \pi^{-2} \|p\|^2$ for all $p \in L_2(0, 1)$ whence (since $0 < \lambda < \pi^2$) $(1 - \lambda\pi^{-2})\|p\|^2 \leq ((I - \lambda K)p, p) \leq \|p\|^2$ for all $p \in L_2(0, 1)$ and the bounds given in Theorem 8.4 apply with $A = I - \lambda K$, $F = f$, $\alpha = 1$ and $\beta = 1 - \lambda\pi^{-2}$.

We can hope to obtain fairly close bounds for $(\phi, f) = \int_0^1 \phi(x)\, dx$ if we take p to be an approximation to ϕ. A suitable p can be derived by using the maximum principle of Theorem 8.1, maximising $J(p)$ over some selected subspace of $L_2(0, 1)$. In particular, by way of a simple illustration, we can choose the subspace composed of constant functions and take advantage of the calculations carried out in Examples 8.1 and 8.4; the fact that λ belongs to a different interval here is not material to those calculations. Thus with $p = c$, a real constant, $J(p)$ is maximised at $c = 12/(12 - \lambda)$, the maximum value of $J(p)$ is also $12/(12 - \lambda)$ and $\|Ap - F\|^2 = \lambda^2/(5(12 - \lambda)^2)$.

It now follows that

$$\frac{12}{12 - \lambda} + \frac{\lambda^2}{5(12 - \lambda)^2} \leq \int_0^1 \phi(x)\, dx \leq \frac{12}{12 - \lambda} + \frac{\pi^2\lambda^2}{5(\pi^2 - \lambda)(12 - \lambda)^2}$$

for $0 < \lambda < \pi^2$. With $\lambda = 1$, for example, we have $1.09256 \leq \int_0^1 \phi(x)\, dx \leq 1.09275$ (to five decimal places). (The exact value of $\int_0^1 \phi(x)\, dx$ at $\lambda = 1$ is $2\tan(\frac{1}{2})$.)

Notice that we can estimate the accuracy with which $p = c = 12/(12 - \lambda)$ approximates ϕ, in norm, since $\|Ap - F\| = \|A(p - \phi)\| = \lambda/(5^{\frac{1}{2}}(12 - \lambda))$, whence $\beta \|p - \phi\| \leq \lambda/(5^{\frac{1}{2}}(12 - \lambda))$ and $\|p - \phi\| \leq \lambda\pi^2/(5^{\frac{1}{2}}(\pi^2 - \lambda)(12 - \lambda))$. With $\lambda = 1$ this gives $\|p - \phi\| \leq 0.04524$ (to five decimal places). $\square$

Remark

We have obtained good bounds in this example with very little trouble, using one simple approximation to ϕ. The machinery is available to produce tighter bounds of course, provided that extra effort is invested. For example, the minimum principle in Theorem 8.4 (ii) used with constant test functions and with $\omega = 0.89$ ($< 1 - \pi^{-2}$) gives $\int_0^1 \phi(x)\, dx \leq 1.09267$ (to five decimal places). The minimum principle of Theorem 8.2 does not apply of course.

8.3.4 First kind integral equations

Let K be a positive operator on H and suppose that $K\phi = f$ has a solution in H. Then from §8.3.1 we see that $J(p) = (f, p) + (p, f) - (Kp, p)$ is maximised only at $p = \phi$ and that the maximum value is (ϕ, f). The lower bound and maximum principle of Theorem 8.4 also apply with $A = K$ and $F = f$.

However, the minimum principle and upper bound we have derived for $A\phi = F$ require A to be positive and bounded below. In general a positive integral operator is not bounded below away from zero and we need a new approach to derive a minimum principle for $K\phi = f$.

Let $p \in H$, $u \in H$ and note that if $(p - u, u) = 0$ then $\|p\|^2 = \|u + p - u\|^2 = \|u\|^2 + \|p - u\|^2 \geqslant \|u\|^2$. Now let $K\phi = f$ where $K \in B(H)$ is a positive operator and let T be such that $K = T^*T$. Putting $u = T\phi$ in the foregoing inequality, we find that $\|p\|^2 \geqslant \|T\phi\|^2 = (\phi, K\phi) = (\phi, f)$ provided that $(p - T\phi, T\phi) = (T^*p - f, \phi) = 0$. This leads us to the following result.

Theorem 8.5

Let $K \in B(H)$ be a positive operator and let $K\phi = f$ have a solution $\phi \in H$. Then the minimum value of $\|p\|^2$ taken over all $p \in H$ satisfying $T^*p = f$ is attained at, and only at, $p = T\phi$, where $T^*T = K$. The minimum value of $\|p\|^2$ is (ϕ, f). $\qquad\square$

This result is quite different in character from other extremum principles we have met in this chapter. The minimisation has to be carried out subject to a constraint, and it is obviously necessary to determine an operator T explicitly, having the property that $T^*T = K$. It is not difficult to show that if $T^*p = f$ has a unique solution then the corresponding value of $\|p\|^2$ is equal to (ϕ, f). In this case no minimisation arises. It may happen however that $T^*p = f$ can be solved for p and that its solution is not unique, in which case the minimisation process does apply. What is at issue here is the theoretical proliferation of T^*T factorisations of a given positive operator K, which we discussed in Chapter 6. For a particular T such that $T^*T = K$, it may not be possible to solve $T^*p = f$ for p, or such a solution may be accessible and may or may not be unique.

8.3.5 Bounds on other quantities

In many practical problems the number (ϕ, F), where $A\phi = F$, is a quantity of prime interest and we have now derived bounds for this number. In some applications it is (ϕ, G) which is required however, where G is not a constant multiple of F and we now widen the scope of our methods and seek bounds for (ϕ, G). In fact, since all our applications from now on will be such that

$F = f$, the free term in a given equation, we shall use f and g in place of F and G.

Let $A\phi = f$ in H, where A is a self-adjoint operator, and suppose that we wish to determine (ϕ, g) where $g \in H$ is given. We assume that $A\phi = f$ has a unique solution, and that $A\psi = g$ also has a unique solution $\psi \in H$. Since $(\phi, g) = (\phi, A\psi) = (A\phi, \psi) = (f, \psi)$ then (ϕ, g) is known if we can determine either ψ or ϕ. The interplay between the given equation $A\phi = f$ and the auxiliary equation $A\psi = g$ is at the heart of the matter and, however the problem of bounding (ϕ, g) is approached, it is inevitable that the second equation for ψ will arise. The identity $(\phi, g) = (f, \psi)$ is frequently referred to as a *reciprocal principle*.

To get matters under way we look for a generalisation of the maximum principle derived in §8.3.1, which gave rise to bounds for (ϕ, F). It is not difficult to deduce that the new functional of two variables $J : H \times H \to \mathbb{C}$ defined by

$$J(p, q) = (f, q) + (p, g) - (Ap, q) \quad (p, q \in H) \tag{8.18}$$

is suitable for our present purpose. To see this, note that

$$J(\phi + \delta\phi, \psi + \delta\psi) = J(\phi, \psi) + (f - A\phi, \delta\psi) + (\delta\phi, g - A\psi) - (A\delta\phi, \delta\psi),$$

from which we deduce that $J(p, q)$ is stationary if and only if $p = \phi$ and $q = \psi$, where $A\phi = f$ and $A\psi = g$. (We extend the notions of §7.2 in the obvious way.) The stationary value is $J(\phi, \psi) = (\phi, g) = (f, \psi)$ and therefore

$$J(p, q) = (\phi, g) - (A(p - \phi), q - \psi) \tag{8.19}$$

for any $p \in H$ and any $q \in H$.

We therefore have the stationary principle that $J(p, q)$ is stationary at, and only at, the point (ϕ, ψ) in $H \times H$, where $A\phi = f$ and $A\psi = g$. It is clear from (8.19) that we do not have an extremum principle, whether or not A is a definite operator. Stationary principles governed by a functional which maps a product space into its field of scalars are sometimes called *bivariational principles*. It is worth remarking here that the functionals $J_1(q) = J(\phi, q) \, (q \in H)$ and $J_2(p) = J(p, \psi) \, (p \in H)$ both take only the constant value (ϕ, g); this is consistent with our earlier observation that (ϕ, g) is determined once either ϕ or ψ is known.

Assuming that the exact solution of $A\phi = f$ cannot be found, the stationary principle based on $J(p, q)$ can be used to approximate ϕ and (ϕ, g) by an extension of the Rayleigh–Ritz method. We find the stationary point $(p, q) \in E_N \times E'_{N'}$ of $J(p, q)$ where E_N and $E'_{N'}$ denote (generally different) finite-dimensional subspaces of H. The resulting element $p \in E_N$

approximates ϕ, $q \in E'_N$ approximates ψ (which may be of no interest) and
the stationary value $J(p,q)$ so obtained approximates (ϕ, g) to 'second order'
in the sense that the error term is $O(\|p - \phi\| \|q - \psi\|)$ as (8.19) shows. Because
the sign of the error term is indefinite, it is not known whether the
approximation $J(p,q)$ $(p \in E_N, q \in E'_{N'})$ to the stationary point is an
overestimate or an underestimate of (ϕ, g). Ideally we would like upper and
lower bounds for (ϕ, g) of course and so we try to carry over the ideas of
§8.3.3 to the present circumstances.

To avoid undue complication we make the simplifying assumption at
this stage, and for the rest of this section, that we are working in a real
Hilbert space H. Thus (ϕ, g) is a real number.

We suppose that $A = I - \lambda K$, where K is a compact self-adjoint operator
and λ is a real parameter. We also suppose (as we did to derive bounds for
(ϕ, f)) that

$$\beta \|p\|^2 \leqslant (Ap, p) \leqslant \alpha \|p\|^2 \tag{8.20}$$

for all $p \in H$ and some α and $\beta > 0$ so $A^{\frac{1}{2}}$ is uniquely defined on H. Then
$(Ap, p) = \|A^{\frac{1}{2}}p\|^2 \leqslant \beta^{-1} \|Ap\|^2$ for all $p \in H$. Also, by the Schwarz inequality
$(Ap, q)^2 = (A^{\frac{1}{2}}p, A^{\frac{1}{2}}q)^2 \leqslant \|A^{\frac{1}{2}}p\|^2 \|A^{\frac{1}{2}}q\|^2 \leqslant \beta^{-2} \|Ap\|^2 \|Aq\|^2$ for all $p, q \in H$.
In particular $|(A(p - \phi), q - \psi)| \leqslant \beta^{-1} \|Ap - f\| \|Aq - g\|$, using $A\phi = f$ and
$A\psi = g$ and therefore from (8.19) we obtain the upper and lower bounds

$$J(p,q) - \beta^{-1} \|Ap - f\| \|Aq - g\| \leqslant (\phi, g) \leqslant J(p,q) + \beta^{-1} \|Ap - f\| \|Aq - g\| \tag{8.21}$$

which hold for all $p, q \in H$.

If we set $q = p$ and $g = f$ in (8.21) we obtain

$$J(p) - \beta^{-1} \|Ap - f\|^2 \leqslant (\phi, f) \leqslant J(p) + \beta^{-1} \|Ap - f\|^2$$

where $J(p) = 2(f, p) - (Ap, p)$. Allowing for the facts that we are now working
in a real space H, and that F has been replaced by f, the upper bound here is
the same as that given in Theorem 8.4. The present lower bound is a good
deal weaker than the earlier version however, suggesting that there is some
scope for improvement in (8.21).

Accordingly, we make a fresh approach to bounding (ϕ, g), basing the
analysis on the bounds given in Theorem 8.4(i). In the setting of a real Hilbert
space H these bounds can be expressed as follows. Let $A = I - \lambda K \in B(H)$ be
self-adjoint, where K is compact and self-adjoint, and suppose that
$\beta \|p\|^2 \leqslant (Ap, p) \leqslant \alpha \|p\|^2$ for all $p \in H$ where $\alpha > \beta > 0$. Then if $A\phi = F$ we
have

$$L(p, F) \leqslant (\phi, F) \leqslant U(p, F) \tag{8.22}$$

where

$$L(p, F) = 2(F, p) - (Ap, p) + \alpha^{-1} \| Ap - F \|^2, \\ U(p, F) = 2(F, p) - (Ap, p) + \beta^{-1} \| Ap - F \|^2. \Big\} \tag{8.23}$$

The two bounds hold independently for all $p \in H$.

Now the equations $A\phi = f$ and $A\psi = g$ can be replaced by the pair $A(a\phi + \psi) = af + g$ and $A(a\phi - \psi) = af - g$ where a is a non-zero (real) constant. Applying (8.22) to each of these new equations in turn, we have

$$L(ap + q, af + g) \leqslant (a\phi + \psi, af + g) \leqslant U(ap + q, af + g), \\ L(ap - q, af - g) \leqslant (a\phi - \psi, af - g) \leqslant U(ap - q, af - g),$$

for any $p \in H$ and any $q \in H$. Since $(\phi, g) = (f, \psi)$, subtraction of these inequalities gives

$$L(ap + q, af + g) - U(ap - q, af - g) \leqslant 4a(\phi, g) \\ \leqslant U(ap + q, af + g) - L(ap - q, af - g),$$

which, after substituting for L and U from (8.23) and simplifying, reduces to

$$G(p, q) - R(p, q) \leqslant (\phi, g) \leqslant G(p, q) + R(p, q) \tag{8.24}$$

where

$$G(p, q) = (f, q) + (p, g) - (Ap, q) + \tfrac{1}{2}(\alpha^{-1} + \beta^{-1})(Ap - f, Aq - g)$$

or

$$G(p, q) = J(p, q) + \tfrac{1}{2}(\alpha^{-1} + \beta^{-1})(Ap - f, Aq - g) \tag{8.25}$$

using our earlier notation, and

$$R(p, q) = \tfrac{1}{4}(\beta^{-1} - \alpha^{-1})\{a \| Ap - f \|^2 + a^{-1} \| Aq - g \|^2\}.$$

The bounds (8.24) hold for any $p, q \in H$ and for any $a > 0$. In particular, $R(p, q)$ is a minimum with $a = \| Aq - g \| \| Ap - f \|^{-1}$ and the optimal version of (8.24) is given with

$$R(p, q) = \tfrac{1}{2}(\beta^{-1} - \alpha^{-1}) \| Ap - f \| \| Aq - g \|. \tag{8.26}$$

Note that

$$\tfrac{1}{2}(\alpha^{-1} + \beta^{-1})(Ap - f, Aq - g) - \tfrac{1}{2}(\beta^{-1} - \alpha^{-1}) \| Ap - f \| \| Aq - g \| \\ \geqslant \{-\tfrac{1}{2}(\alpha^{-1} + \beta^{-1}) - \tfrac{1}{2}(\beta^{-1} - \alpha^{-1})\} \| Ap - f \| \| Aq - g \| \\ = -\beta^{-1} \| Ap - f \| \| Aq - g \|$$

and that

$$\frac{1}{2}(\alpha^{-1}+\beta^{-1})(Ap-f, Aq-g)+\frac{1}{2}(\beta^{-1}-\alpha^{-1})\|Ap-f\|\|Aq-g\|$$
$$\leqslant\{\frac{1}{2}(\alpha^{-1}+\beta^{-1})+\frac{1}{2}(\beta^{-1}-\alpha^{-1})\}\|Ap-f\|\|Aq-g\|$$
$$=\beta^{-1}\|Ap-f\|\|Aq-g\|$$

so that the optimal form of (8.24) provides bounds which are certainly no worse than those in (8.21). We have proved the following result.

Theorem 8.6

Let K be a self-adjoint compact operator on the real Hilbert space H, let $A = I - \lambda K$ where λ is a real parameter and let $\beta\|p\|^2 \leqslant (Ap, p) \leqslant \alpha\|p\|^2$ for some α and $\beta > 0$ and all $p \in H$. Then if $A\phi = f$ and $g \in H$,

$$G(p, q) - R(p, q) \leqslant (\phi, g) \leqslant G(p, q) + R(p, q)$$

for all $p, q \in H$, where

$$G(p, q) = (f, q) + (p, g) - (Ap, q) + \frac{1}{2}(\alpha^{-1} + \beta^{-1})(Ap - f, Aq - g)$$

and $R(p, q) = \frac{1}{2}(\beta^{-1} - \alpha^{-1})\|Ap - f\|\|Aq - g\|$. $\square$

In order to obtain tight bounds for (ϕ, g) we must use a p which is a reasonable approximation to the solution of $A\phi = f$ and a q which is a reasonable approximation to the solution of $A\psi = g$. Such approximations can be determined by using the stationary principle based on $J(p, q)$, as we described earlier.

Example 8.6

Let the operator K be defined on the real elements of $L_2(0, 1)$ by

$$(K\phi)(x) = \int_0^1 \{\min(x, t) - xt\}\phi(t)\,dt,$$

and determine bounds for (ϕ, g) where $g(x) = x$ $(0 \leqslant x \leqslant 1)$ and ϕ satisfies $\phi = f + \lambda K\phi$ with $\lambda \leqslant 0$ and $f = 1$.

The criteria of Theorem 8.6 regarding $A = I - \lambda K$ are met and, in particular, $\beta\|p\|^2 \leqslant (Ap, p) \leqslant \alpha\|p\|^2$ with $\beta = 1$ and $\alpha = 1 - \lambda\pi^{-2}$, since K is a positive operator with largest eigenvalue $\mu_1^+ = \pi^{-2}$. The bounds in Theorem 8.6 therefore apply and we must seek approximations $p \approx \phi$ and $q \approx \psi$, where $\psi = g + \lambda K\psi$, to obtain tight bounds.

We choose simple test functions to use in conjunction with $J(p, q)$ and, recalling our observation in Example 8.2 that each test space should include the corresponding free term, we let $p = c_1$ and $q = c_2 x$ where c_1 and c_2 are real constants. Straightforward calculations give $(Kp)(x) = c_1 x(1-x)/2$ $(0 \leqslant x \leqslant 1)$, $(Kp, q) = c_1 c_2/24$, $(p, g) = c_1/2$, $(f, q) = c_2/2$ and $(p, q) = c_1 c_2/2$. Therefore

$$J(p, q) = (f, q) + (p, g) - (p, q) + \lambda(Kp, q)$$
$$= \{c_1 + c_2 - c_1 c_2(1 - \lambda/12)\}/2$$

which is stationary for $c_1 = c_2 = 12/(12 - \lambda)$ and the stationary value of J is $6/(12 - \lambda)$. Therefore $(\phi, g) \approx 6/(12 - \lambda)$, $\phi \approx p = 12/(12 - \lambda)$ and $\psi(x) \approx q(x) = 12x/(12 - \lambda)$.

Now $\| Ap - f \|^2 = \lambda^2/(5(12 - \lambda)^2)$ (from Example 8.4), $(Kq)(x) = c_2 x(1 - x^2)/6$ $(0 \leqslant x \leqslant 1)$ whence $\| Aq - g \|^2 = 11\lambda^2/(105(12 - \lambda)^2)$ and $(Ap - f, Aq - g) = \lambda^2/(10(12 - \lambda)^2)$. Using these values together with $J(p, q) = 6/(12 - \lambda)$ in the bounds of Theorem 8.6 gives $|(\phi, g) - G| \leqslant R$ where

$$G = 6/(12 - \lambda) + \lambda^2 \{1 + \pi^2(\pi^2 - \lambda)^{-1}\}/(20(12 - \lambda)^2)$$

and

$$R = \lambda^2 11^{\frac{1}{2}} \{1 - \pi^2(\pi^2 - \lambda)^{-1}\}/(10(21)^{\frac{1}{2}}(12 - \lambda)^2).$$

With $\lambda = -1$ we find that $0.46206 \leqslant (\phi, g) \leqslant 0.46214$, to five decimal places. (For $\lambda = -1$ the weaker bounds (8.21) give $0.46068 \leqslant (\phi, g) \leqslant 0.46240$, substantiating the observation made earlier that the lower bound in (8.21) is not particularly good.) □

8.3.6 Pointwise bounds

So far in this discussion of variational principles we have concentrated much of our effort on bounding quantities of the form (ϕ, g). As we have seen, these bounds are derived by making use of an approximation to ϕ, derived by applying the Rayleigh–Ritz method to a variational principle. We have used these approximations to ϕ rather as a means to an end, and have devoted little direct attention to them. The reason for this is that only at this stage are we in a position to consider fully the issue of approximating ϕ. We can now add to the Rayleigh–Ritz approximations pointwise bounds for ϕ, obtained by applying the result in Theorem 8.6.

Consider the equation $\phi = f + \lambda K \phi$ in $L_2(0, 1)$, where K is the self-adjoint integral operator generated by the L_2-kernel k and is therefore compact. It follows that $k(x, y)$, regarded as a function of x, is in $L_2(0, 1)$ for almost all $y \in [0, 1]$. Let $y \in [0, 1]$ be fixed and such that $g \in L_2(0, 1)$, where $g(x) =$

$k(y, x) = k(x, y)$ $(0 \leqslant x \leqslant 1)$. For this y and for $\lambda \neq 0$ we therefore have

$$(\phi, g) = \int_0^1 \phi(x) k(y, x) \, dx = \frac{1}{\lambda} \left(\phi(y) - f(y) \right).$$

Hence, by estimating (ϕ, g) for this choice of g, we can estimate $\phi(y)$ for almost all $y \in [0, 1]$. In particular, if all the functions involved are real-valued and we can consider $\phi = f + \lambda K \phi$ in the Hilbert space of real-valued elements of $L_2(0, 1)$ then Theorem 8.6 can be used to provide upper and lower bounds for $\phi(y)$, for almost all $y \in [0, 1]$. The following example illustrates this situation.

Example 8.7

Suppose we wish to estimate $\phi(\frac{1}{2})$ where ϕ satisfies the integral equation

$$\phi(x) = 1 - \int_0^1 \{\min(x, t) - xt\} \phi(t) \, dt \quad (0 \leqslant x \leqslant 1).$$

Let $g(x) = \min(x, \frac{1}{2}) - \frac{1}{2}x$ $(0 \leqslant x \leqslant 1)$, so that, working in the Hilbert space of real-valued functions in $L_2(0, 1)$, we have $(\phi, g) = 1 - \phi(\frac{1}{2})$. In order to obtain reasonably close bounds for $\phi(\frac{1}{2})$ using Theorem 8.6, we need to determine approximations $p \approx \phi$ and $q \approx \psi$ where $\phi = f - K\phi$, with $f = 1$, and $\psi = g - K\psi$, K being defined by

$$(K\phi)(x) = \int_0^1 \{\min(x, t) - xt\} \phi(t) \, dt \quad (0 \leqslant x \leqslant 1).$$

Our choice of simple trial functions is governed by the respective free terms and we take $p(x) = c_1$ and $q(x) = c_2 x$ $(0 \leqslant x \leqslant 1)$ where c_1 and c_2 are real constants. The calculation now follows that given in Example 8.6 fairly closely, and is entirely straightforward if rather tedious. The functional $J(p, q)$ is stationary for $c_1 = 12/13$ and $c_2 = 3/13$ and the stationary value is $3/26$. Using the fact that $\|p\|^2 \leqslant (Ap, p) \leqslant (1 + \pi^{-2}) \|p\|^2$, where $A = I + K$, we find that the bounds on (ϕ, g) give $0.88652 \leqslant \phi(\frac{1}{2}) \leqslant 0.88701$ to five decimal places.

Note that we have determined different types of approximation to ϕ here. From the variational principle we have $\phi \approx p = c_1$. For this p we find that $\|Ap - f\| = 1/13(5)^{\frac{1}{2}}$ and since $\|Ap - f\| = \|A(p - \phi)\| \geqslant \|p - \phi\|$, the approximation $\phi \approx p = c_1$ is such that $\|p - \phi\| \leqslant 1/13(5)^{\frac{1}{2}}$; this of course gives no information about the accuracy of the approximation $\phi(\frac{1}{2}) \approx 12/13$. The stationary value of $J(p, q)$ approximates $(\phi, g) = 1 - \phi(\frac{1}{2})$, giving $\phi(\frac{1}{2}) \approx 23/26$, but with no immediate error estimate available. Finally we have the bounds on $\phi(\tfrac{1}{2})$, and we can similarly bound $\phi(y)$ for any $y \in [0, 1]$.   □

8.3.7 Non-self-adjoint operator equations

This is a convenient point at which to mention briefly the equation $A\phi = f$ where $A \in B(H)$ is not self-adjoint. We assume that the equation has one and only one solution and suppose that we wish to approximate (ϕ, g), where $g \in H$ is given.

The natural reaction is perhaps to consider the equation $A^*A\phi = A^*f$, supposing that $A^*\psi = 0$ has only the trivial solution, and make use of our existing theory for self-adjoint operators. In particular, the stationary principle based on the functional $J(p, q) = (f, q) + (p, g) - (Ap, q)$ can be used, suitably adapted to the modified equation. Replacing f by A^*f and A by A^*A in $J(p, q)$ gives the new functional $J_1 : H \times H \rightarrow \mathbb{C}$ defined by $J_1(p, q) = (A^*f, q) + (p, g) - (A^*Ap, q) = (f, Aq) + (p, g) - (Ap, Aq) = J(p, Aq)$. By setting $\hat{q} = Aq$, the original form of the functional is restored, showing that we might as well work with $J(p, q)$ as it stands.

Therefore let $J : H \times H \rightarrow \mathbb{C}$ be defined by

$$J(p, q) = (f, q) + (p, g) - (Ap, q) \quad (p, q \in H). \tag{8.27}$$

It is easy to show that $J(p, q)$ is stationary at the point $(\phi, \psi) \in H \times H$ where $A\phi = f$ and $A^*\psi = g$, that the stationary value is $J(\phi, \psi) = (\phi, g) = (f, \psi)$ and that

$$J(p, q) = (\phi, g) - (A(p - \phi), q - \psi) \tag{8.28}$$

for all $p, q \in H$. The only difference between this position and the corresponding one for a self-adjoint operator A is the appearance of the adjoint of A in the auxiliary equation $A^*\psi = g$. The remarks we made in §8.3.5 about how the stationary principle may be used to approximate ϕ and (ϕ, g) need not be repeated here. Since our stated aim is to estimate (ϕ, g) and since we have only a stationary principle, upper and lower bounds provide valuable additional information.

To show how bounds may be calculated, suppose now that H is a real Hilbert space. We can again look for guidance to our earlier work where A is self-adjoint and consider the equation $A^*A\phi = A^*f$. This suggests that we should require A^*A to be positive and bounded below, which is equivalent to supposing that there is a positive number γ such that $\|Ap\| \geqslant \gamma \|p\|$ for all $p \in H$. It then follows that

$$|(Ap, q)| = |(p, A^*q)| \leqslant \|p\| \|A^*q\| \leqslant \gamma^{-1} \|Ap\| \|A^*q\|$$

for all $p, q \in H$, and in particular that

$$|(A(p-\phi), q-\psi)| \leqslant \gamma^{-1} \|Ap-f\| \|A^*q-g\|$$

for all $p, q \in H$, where $A\phi = f$ and $A^*\psi = g$. We deduce from (8.28) that

$$|J(p,q)-(\phi,g)| \leqslant \gamma^{-1} \|Ap-f\| \|A^*q-g\| \qquad (8.29)$$

for all $p, q \in H$.

If we put $A^* = A$ at this stage, we recover the bounds (8.21) from (8.29). To see this, note that the condition $\|Ap\| \geqslant \gamma \|p\|$ $(p \in H)$ implies, when A is positive, that $(Ap, p) \geqslant \gamma \|p\|^2$ for all $p \in H$ and therefore we obtain the bounds (8.21) with $\beta = \gamma$.

The way in which the bounds (8.29) can be allied to the stationary principle to produce practically useful estimates of (ϕ, g) follows exactly as in the self-adjoint case.

8.3.8 The Rayleigh—Ritz method

We have referred to the Rayleigh–Ritz method a number of times in relation to variational principles and the basic idea of the procedure has already been mentioned in this chapter. We have actually used the Rayleigh–Ritz method in our examples, but only in the most elementary form, using test functions belonging to one-dimensional spaces, to avoid excessive complication.

At this point we consider the general application of the Rayleigh–Ritz method, in the context of seeking to approximate ϕ and (ϕ, g) given that $A\phi = f$, where $A \in B(H)$ is not self-adjoint, and $g \in H$ is assigned. Other cases, in which A is self-adjoint for example, require straightforward specialisations of the following account.

We recall from §8.3.7 that the functional $J: H \times H \to \mathbb{C}$ defined by

$$J(p,q) = (f,q) + (p,g) - (Ap,q) \quad (p, q \in H) \qquad (8.27)$$

is stationary at the point $(\phi, \psi) \in H \times H$ where $A\phi = f$ and $A^*\psi = g$, and the stationary value is (ϕ, g). Suppose that we seek the stationary point of $J(p,q)$ with the point (p, q) restricted to $E_N^{(1)} \times E_N^{(2)}$, $E_N^{(1)}$ and $E_N^{(2)}$ being two subspaces of H, both of dimension N. This stationary point will yield approximations to ϕ and ψ, and the corresponding stationary value will give an estimate of (ϕ, g). We have shown that the accuracy of the approximation to ϕ obtained in this way is given in the form of a norm error (see Example 8.5 for instance) and that upper and lower bounds for (ϕ, g) follow if these are desired.

Let $E_N^{(i)}$ be spanned by the given functions $\chi_n^{(i)}$ $(n = 1, \ldots, N)$ for $i = 1, 2,$

and let

$$p(x) = \sum_{n=1}^{N} \alpha_n \chi_n^{(1)}(x), \quad q(x) = \sum_{n=1}^{N} \beta_n \chi_n^{(2)}(x). \tag{8.30}$$

Our objective is to choose $\alpha_1, \ldots, \alpha_N, \beta_1, \ldots, \beta_N$ so that the functional $J(p, q)$ resulting from this choice of p and q is stationary. Substituting (8.30) into (8.27) gives

$$
\left.
\begin{aligned}
J &= \sum_{n=1}^{N} \{\bar{\beta}_n(f, \chi_n^{(2)}) + \alpha_n(\chi_n^{(1)}, g)\} - \sum_{n=1}^{N} \sum_{m=1}^{N} \alpha_n \bar{\beta}_m(A\chi_n^{(1)}, \chi_m^{(2)}) \\
&= \sum_{n=1}^{N} \bar{\beta}_n(f, \chi_n^{(2)}) + \sum_{m=1}^{N} \alpha_m \left\{ (\chi_m^{(1)}, g) - \sum_{n=1}^{N} \bar{\beta}_n(A\chi_m^{(1)}, \chi_n^{(2)}) \right\} \\
&= \sum_{m=1}^{N} \bar{\beta}_m \{ (f, \chi_m^{(2)}) - \sum_{n=1}^{N} \alpha_n(A\chi_n^{(1)}, \chi_m^{(2)}) \} + \sum_{n=1}^{N} \alpha_n(\chi_n^{(1)}, g)
\end{aligned}
\right\} \tag{8.31}
$$

where J is now to be regarded as a function of the $2N$ variables $\alpha_1, \ldots, \alpha_N$, $\beta_1, \ldots, \beta_N$. It is easy to show that for J to be stationary we require

$$
\left.
\begin{aligned}
\sum_{n=1}^{N} \bar{\beta}_n(A\chi_m^{(1)}, \chi_n^{(2)}) &= (\chi_m^{(1)}, g) \quad \text{(a)} \\
\sum_{n=1}^{N} \alpha_n(A\chi_n^{(1)}, \chi_m^{(2)}) &= (f, \chi_m^{(2)}) \quad \text{(b)}
\end{aligned}
\right\} \quad (m = 1, \ldots, N). \tag{8.32}
$$

The corresponding stationary value, J_s say, of J is given by (8.31) with the coefficients α_n and $\bar{\beta}_n$ satisfying (8.32), and obviously

$$J_s = \sum_{n=1}^{N} \bar{\beta}_n(f, \chi_n^{(2)}) = \sum_{n=1}^{N} \alpha_n(\chi_n^{(1)}, g). \tag{8.33}$$

We recall that the exact solutions of $A\phi = f$ and $A^*\psi = g$ are such that the reciprocal principle $(f, \psi) = (\phi, g)$ is satisfied. Since (8.33) can be written as $J_s = (f, q) = (p, g)$, we see that the approximations $p \approx \phi$ and $q \approx \psi$ satisfy the same reciprocal principle. These approximations are obtained by solving the two systems (8.32) and substituting the resulting values of α_n and β_n ($n = 1, \ldots, N$) into (8.30). From the point of view of the given problem, only the approximation $p \approx \phi$ is required, of course, and a knowledge of the coefficients α_n ($n = 1, \ldots, N$) is sufficient also to determine J_s, as (8.33) shows.

If only the approximation $J_s \approx (\phi, g)$ is required, then there is a way of proceeding which avoids the need to solve either of the systems (8.32). Using the notation

$$a_{nm} = (A\chi_n^{(1)}, \chi_m^{(2)}), \quad f_m = (f, \chi_m^{(2)}), \quad g_m = (\chi_m^{(1)}, g)$$

we deduce from (8.32) that

$$
\left.
\begin{aligned}
J_s \sum_{n=1}^{N} \bar{\beta}_n a_{mn} = J_s g_m = g_m \sum_{n=1}^{N} \bar{\beta}_n f_n \\
J_s \sum_{n=1}^{N} \alpha_n a_{nm} = J_s f_m = f_m \sum_{n=1}^{N} \alpha_n g_n
\end{aligned}
\right\} \quad (m=1,\ldots,N).
$$

Therefore

$$
\left.
\begin{aligned}
\sum_{n=1}^{N} \bar{\beta}_n (J_s a_{mn} - g_m f_n) = 0 \\
\sum_{n=1}^{N} \alpha_n (J_s a_{nm} - g_n f_m) = 0
\end{aligned}
\right\} \quad (m=1,\ldots,N),
$$

both of these systems of equations having non-trivial solutions if

$$ \det(g_m f_n - J_s a_{mn}) = 0. $$

By using row manipulations, this determinant can be solved for J_s to give

$$
J_s = g_1 \begin{vmatrix} f_1 f_2 & \ldots & f_N \\ \mathbf{h}_1 \mathbf{h}_2 & \ldots & \mathbf{h}_N \end{vmatrix} \begin{vmatrix} a_{11} a_{12} & \ldots & a_{1N} \\ \mathbf{h}_1 & \mathbf{h}_2 & \ldots & \mathbf{h}_N \end{vmatrix}^{-1},
$$

where the vector $\mathbf{h}_n$ is defined by

$$ \mathbf{h}_n = (g_2 a_{1n} - g_1 a_{2n},\, g_3 a_{1n} - g_1 a_{3n},\,\ldots,\, g_N a_{1n} - g_1 a_{Nn})^T \quad (n=1,\ldots,N). $$

8.4 Galerkin's method and related topics

The methods we have used so far to calculate approximate solutions to integral equations have concentrated on how best to approximate the desired quantity by restricting attention to some finite-dimensional subspace of the space concerned, the subspace being assumed given. In this section we shall focus on the choice of the subspace and, in particular, on choosing a subspace to ensure that the approximation eventually found has the desired accuracy.

Suppose that we seek approximations to the solution of the equation $A\phi = f$ where A is some bounded linear map on the Hilbert space H and $f \in H$ is given. We choose a subspace E_n of H of finite dimension and seek $p_n \in E_n$ which in some sense is the 'best' approximation in E_n to the true solution ϕ. To be of any significance this technique requires that we can find an effective way of calculating p_n, and this is likely to be influenced by the sense in which we define the 'best' approximation. We may now expect (or hope) that by choosing successively larger subspaces E_n ($n = 1, 2, 3, \ldots$) the approximation p_n will tend to the true solution ϕ.

Within this broad framework there are several issues to be tackled. We require the subspaces to be chosen so that $p_n \to \phi$ as $n \to \infty$, which ensures that we obtain any desired degree of accuracy if n is chosen sufficiently large; a judicious choice of subspaces may allow us to obtain more accurate approximations for a given value of n than an injudicious one. There remains, too, the matter of the sense in which $\phi_n \to \phi$ as $n \to \infty$ and the extent to which we can estimate the difference between ϕ_n and ϕ in advance.

Projection methods

Suppose that A is a bounded linear map on the Hilbert space H and that we wish to solve the equation $A\phi = f$. Let E_n be a subspace of H of dimension n from which we wish to choose an approximation p_n to ϕ. One obvious choice of p_n is to let it be the element of E_n which minimises $\| Ap_n - f \|$, that is, we choose p_n so that $\| Ap_n - f \| = \min\{\| Ap - f \| : p \in E_n\}$. Then Ap_n will be the point of the subspace $\{Ap : p \in E_n\}$ which is closest to f, that is, $Q_n f$ where Q_n denotes the orthogonal projection of H onto the subspace $\{Ap : p \in E_n\}$. This approach to the approximation of $A\phi = f$, which is often called the least squares method, gives rise to a finite set of equations by noticing that we wish to choose the element p_n of E_n for which $Ap_n - f$ is orthogonal to $\{Ap : p \in E_n\}$. If the vectors $\chi_1, \ldots, \chi_n$ form a basis for E_n, then $p_n = \sum_{i=1}^{n} \alpha_i \chi_i$ where the scalars $\alpha_1, \ldots, \alpha_n$ are chosen so that

$$\sum_{i=1}^{n} \alpha_i (A\chi_i, A\chi_j) = (f, A\chi_j) \quad (j = 1, 2, \ldots, n).$$

Comparing this with (8.32), we see that if we choose $\chi_i^{(1)} = \chi_i$ and $\chi_i^{(2)} = A\chi_i$ ($i = 1, 2, \ldots, N$) in the Rayleigh–Ritz method of §8.3.8, we obtain exactly this equation. This connection is not surprising, for minimising $\| Ap - f \|$ is equivalent to maximising $(A^*f, p) + (p, A^*f) - (A^*Ap, p)$, which is the functional J of (8.7) adapted to the equation $A^*A\phi = A^*f$.

We are now in a position to see how the approximation p_n varies as n varies (noticing that our remarks also apply to the particular version of the Rayleigh–Ritz method referred to above). We suppose here that each of the subspaces E_n contains its predecessor, so that $E_{n-1} \subseteq E_n$ ($n = 2, 3, 4, \ldots$) and therefore that E_n is spanned by the first n terms of the sequence (χ_n). Let P_n be the orthogonal projection onto E_n. We require that, for each $\phi \in H$, $P_n\phi \to \phi$ as $n \to \infty$, that is, that the closest approximation in E_n to ϕ should tend to ϕ as $n \to \infty$. This is equivalent to the statement that every vector is a limit of a sequence of (finite) linear combinations of the vectors (χ_n) or, equivalently, that $(\phi, \chi_n) = 0$ $(n = 1, 2, 3, \ldots)$ implies $\phi = 0$. If (χ_n) is ortho-normal the last mentioned condition is exactly that it be a complete orthonormal sequence.

If, in addition, A is an invertible operator then it follows that $Ap_n \to f$ as $n \to \infty$. Firstly, there is a vector $\phi = A^{-1}f$ for which $A\phi = f$, and since $P_n\phi \to \phi$ as $n \to \infty$, $AP_n\phi \to f$ as $n \to \infty$. But $\|Ap_n - f\| \leqslant \|AP_n\phi - f\|$ by the choice of p_n, so that $Ap_n \to f$ as $n \to \infty$. The invertibility now shows us that $p_n \to A^{-1}f$ as $n \to \infty$.

Theorem 8.7

Suppose that A is an invertible, bounded, linear map on the Hilbert space H and that $f \in H$. If the sequence (χ_n) is linearly independent and every element of H is the limit of a sequence of linear combinations of finitely many of the χ_n, then the procedure of choosing p_n to be that linear combination of $\chi_1, \ldots, \chi_n$ which minimises $\|Ap_n - f\|$ has the property that $p_n \to \phi$ as $n \to \infty$ where $A\phi = f$. Equivalently, p_n may be chosen to be that linear combination of $\chi_1, \ldots, \chi_n$ such that $(Ap_n - f, A\chi_j) = 0$ $(j = 1, 2, \ldots, n)$.

$\square$

We can find a slightly neater expression for the approximation p_n to the solution ϕ of $A\phi = f$ in the calculation above. Let Q_n be the orthogonal projection onto $\{A\phi : \phi \in E_n\}$. Then $Ap_n = Q_nf$, since Q_nf is the point of $\{A\phi : \phi \in E_n\}$ nearest to f. Hence $p_n = A^{-1}Q_nf$ $(n \in \mathbb{N})$ and $p_n - \phi = A^{-1}(Q_nf - f)$ and so

$$\|p_n - \phi\| \leqslant \|A^{-1}\| \|Q_nf - f\|. \tag{8.34}$$

Once we know more about the value of $\|Q_nf - f\|$ as a function of n we can estimate $\|p_n - \phi\|$.

The method of approximation just described, of finding that member p_n of E_n which minimises the norm of the residual $Ap_n - f$, may not be convenient to carry out. Before we pass to examples, therefore, we describe an alternative process. In the method just described we chose $p_n \in E_n$ satisfying $(Ap_n - f, A\chi_j) = 0$ $(j = 1, 2, \ldots, n)$, that is, $Ap_n - f$ is to be ortho-gonal to all vectors of the form Ap for $p \in E_n$. If, instead, we demand that $Ap_n - f$ be orthogonal to E_n itself we obtain *Galerkin's method*. The equations satisfied are $(Ap_n - f, \chi_j) = 0$ $(j = 1, 2, \ldots, n)$, so that if P_n denotes the orthogonal projection of H onto E_n, $P_nAp_n = P_nf$. (It is worth noting that we have not yet shown that such a vector p_n exists, nor, even if it exists, that it is unique.) If we now recall that the invertible operators with which we have dealt in the context of integral equations are all of the form $I - \lambda K$

where K is an integral operator (usually compact) then we may reformulate the equation above. Suppose that $A = I - K$ is invertible. Then the Galerkin approximation p_n to the solution ϕ of $\phi - K\phi = f$ satisfies $p_n - P_n K p_n = P_n f$ (since $P_n p_n = p_n$) and so $p_n = (I - P_n K)^{-1} P_n f$, provided $I - P_n K$ is invertible. In this case, writing $\phi = (I - K)^{-1} f$, we have

$$
\begin{aligned}
\phi - p_n &= ((I - K)^{-1} - (I - P_n K)^{-1} P_n) f \\
&= (I - P_n K)^{-1} ((I - P_n K) - P_n (I - K))(I - K)^{-1} f \\
&= (I - P_n K)^{-1} (I - P_n)(I - K)^{-1} f.
\end{aligned}
$$

Given that we chose (χ_n) so that, for each $\psi \in H$, $P_n \psi \to \psi$ as $n \to \infty$ (the assumption of Theorem 8.7) we see that $(I - P_n)(I - K)^{-1} f \to 0$ as $n \to \infty$ and we are left to consider $(I - P_n K)^{-1}$. If K is compact, Lemma 7.2 now shows us that $\| P_n K - K \| \to 0$ as $n \to \infty$ whence $I - P_n K = (I - K) + (K - P_n K)$ is invertible if n is sufficiently large that $\| K - P_n K \| < 1/\|(I - K)^{-1}\|$. Also, if $\| K - P_n K \| < 1/(2\|(I - K)^{-1}\|)$ we have $\|(I - P_n K)^{-1}\| \leqslant 2\|(I - K)^{-1}\|$. Therefore, for sufficiently large n,

$$
\| \phi - p_n \| \leqslant C \|(I - P_n)(I - K)^{-1} f \| \tag{8.35}
$$

where $C = 2\|(I - K)^{-1}\|$, a constant. Summing up gives us the following result.

Theorem 8.8

Suppose that K is a compact operator on the Hilbert space H, that $I - K$ is invertible and that $f \in H$. Suppose also that the sequence (χ_n) is linearly independent and every element of H is the limit of a sequence of linear combinations of finitely many of the χ_n. If p_n is chosen to be a linear combination of $\chi_1, \ldots, \chi_n$ for which $p_n - K p_n - f$ is orthogonal to $\chi_1, \ldots, \chi_n$, then $\| p_n - \phi \| \to 0$ as $n \to 0$, where ϕ is the solution of $(I - K)\phi = f$.   ☐

Remark

Notice that the system of equations defining the Galerkin approximation to ϕ, namely $(A p_n - f, \chi_j) = \sum_{i=1}^{n} \alpha_i (A \chi_i, \chi_j) - (f, \chi_j) = 0$ $(j = 1, 2, \ldots, n)$, coincides with the system (8.32b) defining the Rayleigh–Ritz approximation to ϕ when $\chi_j^{(1)} = \chi_j^{(2)} = \chi_j$. Therefore the convergence properties of the Galerkin approximation apply equally to this version of the Rayleigh–Ritz approximation. The auxiliary equation $A^* \psi = g$ and the resulting system (8.32a) do not arise in Galerkin's method which seeks to approximate only the solution of $A\phi = f$ and not also the value of (ϕ, g).

Example 8.8

Let us try Galerkin's method and the minimisation of the residual method for the problem

$$\phi(x) - \int_0^1 \{\min(x, t) - xt\} \phi(t) \, dt = 1 \quad (0 \leqslant x \leqslant 1) \tag{8.36}$$

or $\phi - K\phi = 1$ where $(K\phi)(x) = \int_0^1 \{\min(x, t) - xt\} \phi(t) \, dt$. We have used this integral operator before and have observed that its eigenvalues are $1/(n^2\pi^2)$ $(n = 1, 2, 3, \ldots)$, so we know a unique solution of (8.36) exists. It is not hard to convert the equation to a differential equation and check that $\phi(x) = (\cos(x - \frac{1}{2}))/\cos(\frac{1}{2})$.

Let us approximate ϕ with a function which is constant on the three intervals $(0, \frac{1}{3})$, $(\frac{1}{3}, \frac{2}{3})$ and $(\frac{2}{3}, 1)$, that is, a linear combination of χ_1, χ_2 and χ_3 where

$$\chi_i(x) = \begin{cases} 1 & \left(\dfrac{i-1}{3} < x < \dfrac{i}{3}\right) \\ 0 & \text{otherwise.} \end{cases}$$

Then $(\chi_i, \chi_j) = \frac{1}{3}$ if $i = j$ and $(\chi_i, \chi_j) = 0$ if $i \neq j$. Moreover $(K\chi_1, \chi_1) = 1/108$, $(K\chi_1, \chi_2) = 1/108$, $(K\chi_1, \chi_3) = 1/324$, $(K\chi_2, \chi_1) = 1/108$, $(K\chi_2, \chi_2) = 7/324$, $(K\chi_2, \chi_3) = 1/108$, $(K\chi_3, \chi_1) = 1/324$, $(K\chi_3, \chi_2) = 1/108$ and $(K\chi_3, \chi_3) = 1/108$. For $f = 1$, we have $(f, \chi_i) = \frac{1}{3}$ $(i = 1, 2, 3)$.

Galerkin's method now sets $p = \alpha_1\chi_1 + \alpha_2\chi_2 + \alpha_3\chi_3$ where the scalars α_1, α_2 and α_3 are chosen so that

$$\sum_{i=1}^3 \alpha_i\{(\chi_i, \chi_j) - (K\chi_i, \chi_j)\} = \tfrac{1}{3} \quad (j = 1, 2, 3).$$

The solutions, correct to four decimal places, are $\alpha_1 = \alpha_3 = 1.0711$ and $\alpha_2 = 1.1329$.

To carry out the minimisation of the residual $\|(I - K)p - f\|$ we need to know, in addition to the above, that $(K\chi_1, K\chi_1) = 0.00060585$, $(K\chi_1, K\chi_2) = (K\chi_2, K\chi_1) = 0.0010288$, $(K\chi_2, K\chi_2) = 0.0020919$, $(K\chi_1, K\chi_3) = (K\chi_3, K\chi_1) = 0.00045725$, $(K\chi_2, K\chi_3) = (K\chi_3, K\chi_2) = 0.0010288$ and $(K\chi_3, K\chi_3) = 0.00060585$. Then $p = \alpha_1\chi_1 + \alpha_2\chi_2 + \alpha_3\chi_3$ where

$$\sum_{i=1}^3 \alpha_i((\chi_i, K\chi_j) - (K\chi_i, K\chi_j)) = (f, K\chi_j),$$

where $(f, K\chi_1) = (f, K\chi_3) = 0.021605$ and $(f, K\chi_2) = 0.040123$ noting $f = 1$.

In this case the solutions are $\alpha_1 = \alpha_3 = 1.0787$ and $\alpha_2 = 1.1462$ (to four decimal places).

The true solution $\phi(x) = \cos(x - \frac{1}{2})/\cos(\frac{1}{2})$ varies between 1 and $\sec(\frac{1}{2}) = 1.1395$, its value at $x = \frac{1}{3}$ and $x = \frac{2}{3}$ being 1.1237. □

The calculation of Example 8.8 is, of course, not especially startling in the accuracy it produces in that the approximating function p is required to be constant on three rather 'large' intervals. We could, of course, choose p to be constant on smaller intervals, at the expense of increasing the labour of the calculation. Now we have error estimates for the two approximation methods, these being contained in (8.34) and (8.35), so it is worth noticing a few features of these. For definiteness we shall refer to the Galerkin approximation (8.35) where $\|\phi - p_n\| \leqslant C\|(I - P_n)\phi\|$ for sufficiently large n, ϕ being the solution of $(I - K)\phi = f$ and p_n being the approximation to it, with $p_n \in E_n$. Since P_n is a projection, so is $I - P_n$ and therefore $\|I - P_n\| = 1$ for all n. Although this implies that there is a vector ψ of norm 1 for which $\|(I - P_n)\psi\| = 1$ we do, in fact, apply the operator $I - P_n$ to the fixed (but unknown) vector ϕ and our assumptions about the subspaces E_n ensure that $(I - P_n)\phi \to 0$ as $n \to \infty$. Notice, however, that because $\|(I - P_n)\| \nrightarrow 0$, we need additional information to be able to make any statement about the rate at which $\|(I - P_n)\phi\| \to 0$. Fortunately, this sort of information is usually available.

In Example 8.8 the function we seek, ϕ, satisfies

$$\phi(x) = 1 + \int_0^x t(1 - x)\phi(t)\,\mathrm{d}t + \int_x^1 x(1 - t)\phi(t)\,\mathrm{d}t \quad (0 \leqslant x \leqslant 1) \quad (8.37)$$

so that it is easily seen that ϕ is continuous. Utilising this information shows that the right hand side of (8.37) is differentiable, whence ϕ is and we may now deduce that ϕ is twice differentiable and, indeed, differentiable to all orders. This additional information allows us to make a suitable choice of the subspace E_n, for if we choose E_n to be the subspace of functions which have constant values on the n subintervals $((j-1)/n, j/n)$ $(j = 1, 2, \ldots, n)$ of $[0, 1]$ then it is not difficult to prove that if P_n is the projection of $L_2(0, 1)$ onto E_n, then $\|\psi - P_n\psi\| \leqslant \|\psi'\|/(n\sqrt{6})$ for every differentiable function ψ whose derivative belongs to L_2. Moreover, if ψ is known to be twice differentiable, with $\psi'' \in L_2(0, 1)$ and we choose our subspace to consist of functions which are linear on each of these n subintervals and continuous on $[0, 1]$, giving a subspace of dimension $n + 1$ (spanned by the functions which take the value 1 at $x = j/n$ and zero at all other multiples of $1/n$), then, calling the subspace E_{n+1}, we have $\|\psi - P_{n+1}\psi\| \leqslant \|\psi''\|/(\pi^2 n^2)$. (These assertions are the substance of Problems 8.15 and 8.16.) Therefore, although we do not know ϕ explicitly, we do know enough to be able to construct subspaces for which $\|(I - P_n)\phi\|$ can be estimated and from it an estimate of $\|p_n - \phi\|$ can be made.

The examples we have given, in which the subspaces used to construct the approximation p to the true solution ϕ of the equation $(I - K)\phi = f$ are chosen so that the functions in them are piecewise constant or continuous and piecewise linear, have the merit that the functions belonging to E_n are of a particularly simple kind and one might expect this to cause the set of simultaneous equations corresponding to the finite-dimensional problem of finding the appropriate coefficients to be correspondingly tractable. Nevertheless, setting aside the matter of the simplicity or otherwise of the finite-dimensional problem, these methods have the disadvantage that we may be able to see at a glance that the approximate solution lacks some of the analytical properties of the true solution. In Example 8.8 we know (by equation (8.37)) that the true solution ϕ is continuous; a piecewise constant approximation to it will yield an approximate solution which is certainly not continuous (unless it happens to be constant over $[0, 1]$). The piecewise linear approximation passes this test, but it fails at the next hurdle: it is not differentiable, whereas ϕ is. To see what we can do to ameliorate this, we recall that there is one situation which we have already encountered where we reduced the problem to one in a finite-dimensional space: that of solving an equation with degenerate kernel. A second glance at what we did in this case may clarify matters.

Suppose that the operator K has finite-dimensional image, spanned by $\chi_1, \ldots, \chi_n$ say. Then, proceeding as in §3.2, we do not immediately express the solution ϕ of $(I - K)\phi = f$ in the form of a linear combination of $\chi_1, \ldots, \chi_n$, but notice that, since $\phi = f + K\phi$, $\phi = f + \psi$ where $\psi = K\phi$ so ψ is a linear combination of $\chi_1, \ldots, \chi_n$. We may reduce the problem to one in the finite-dimensional space of linear combinations of $\chi_1, \ldots, \chi_n$ by noticing that ψ satisfies the equation $\psi = Kf + K\psi$, where Kf belongs to the subspace spanned by $\chi_1, \ldots, \chi_n$. Then, in §3.2, we solved this problem in the finite-dimensional subspace (where possible, that is, where $I - K$ is invertible) and found ψ. The solution of the original problem is $\phi = f + \psi$.

Is there anything to be gained by following the procedure of §3.2 for operators which arise from non-degenerate kernels? Suppose we wish to solve $(I - K)\phi = f$ where $I - K$ is invertible. Then the solution ϕ can be expressed as $\phi = f + K\phi = f + \psi$ where ψ satisfies $\psi = Kf + K\psi$. If we apply Galerkin's method to this equation, by finding the solution $q_n \in E_n$ for which $q_n - Kq_n - Kf \in E_n^{\perp}$ then, writing P_n for the orthogonal projection onto E_n, we have $q_n = (I - P_n K)^{-1} P_n K f$ and our approximation to ϕ is $f + q_n$. Now q_n, being the projection of a vector onto E_n, is no more continuous than the approximation p_n we found before and at first sight we have gained nothing. However, noting that $\phi = f + K\phi$ where $K\phi$ satisfies $\psi = Kf + K\psi$ so that $K\phi = K(I - K)^{-1} f$ we see that

$$
\begin{aligned}
\| (f + q_n) - \phi \| &= \| (I - P_n K)^{-1} P_n K f - K (I - K)^{-1} f \| \\
&= \| (I - P_n K)^{-1} (P_n K (I - K) - (I - P_n K) K)(I - K)^{-1} f \| \\
&= \| (I - P_n K)^{-1} (P_n K - K)(I - K)^{-1} f \| \\
&\leqslant \| (I - P_n K)^{-1} \| \, \| P_n K - K \| \, \| (I - K)^{-1} f \|,
\end{aligned}
\tag{8.38}
$$

where we have presumed here (and above) that $(I - P_n K)^{-1}$ exists. (As in the derivation of Galerkin's method, this is true for sufficiently large n by Lemma 7.2.) There is a significant difference here in that, by Lemma 7.2, $\| P_n K - K \| \to 0$ as $n \to \infty$ and therefore an upper bound for (8.38) can be found in terms of $\| f \|$ which is valid for all f, namely

$$
\| (f + q_n) - \phi \| \leqslant \| (I - P_n K)^{-1} \| \, \| P_n K - K \| \, \| (I - K)^{-1} \| \, \| f \|.
$$

In this case we should choose the subspaces E_n so that $\| P_n K - K \|$ is small; we do not need to pay attention to the effect of P_n on the particular vector $(I - K)^{-1} f$. The method of solution just outlined is called *Kantorovich's method*. Notice that the approximation, $\tilde{p}_n$, to the true solution of $(I - K)\phi = f$ which is obtained by Kantorovich's method is $\tilde{p}_n = f + (I - P_n K)^{-1} P_n K f = (I - P_n K)^{-1} f$ whereas that obtained by Galerkin's method is $p_n = (I - P_n K)^{-1} P_n f$. Looking at the difference between these and $\phi = (I - K)^{-1} f$ we see that

$$
p_n - \phi = (I - P_n K)^{-1} (P_n - I)(I - K)^{-1} f = S_n f
$$

and

$$
\tilde{p}_n - \phi = (I - P_n K)^{-1} (P_n K - K)(I - K)^{-1} f = \tilde{S}_n f.
$$

The principal point here is the difference in the nature of the convergence of the solution operators; with S_n and $\tilde{S}_n$ defined as above, then for all f, $S_n f \to 0$ as $n \to \infty$ whereas $\| \tilde{S}_n \| \to 0$ as $n \to \infty$, a stronger condition (which implies, of course, that for each f, $\tilde{S}_n f \to 0$ as $n \to \infty$).

Theorem 8.9 (Kantorovich's method)

Suppose that K is a compact operator on the Hilbert space H, that $I - K$ is invertible, that $f \in H$ and that $\phi = (I - K)^{-1} f$. Suppose also that (χ_n) is linearly independent and that every element of H is the limit of a sequence of finite linear combinations of (χ_n). Let E_n be the subspace spanned by

$\chi_1, \ldots, \chi_n$. Then if $q_n \in E_n$ is chosen such that $q_n - Kq_n - Kf$ is orthogonal to E_n and $\tilde{p}_n = f + q_n$ then for large n there is $\tilde{S}_n \in B(H)$ for which $\tilde{p}_n - \phi = \tilde{S}_n f$, where

$$\|\tilde{S}_n\| \leqslant \|(I - P_n K)^{-1}\| \| P_n K - K \| \|(I - K)^{-1}\| \to 0$$

as $n \to \infty$, where P_n denotes the projection onto E_n. □

The error estimate (8.38) for the Kantorovich method contains the term $\|P_n K - K\|$, where P_n is the orthogonal projection onto the subspace E_n of dimension n. $P_n K$ is therefore an operator of rank at most n so, by Example 5.1, if K is self-adjoint as well as compact, with eigenvalue sequence (μ_n) ordered in the usual way, we have $\|P_n K - K\| \geqslant |\mu_{n+1}|$. For a given compact self-adjoint operator K this shows us that there is a lower bound on $\|P_n K - K\|$ no matter how the subspace E_n is chosen, the subspace yielding the lowest value of $\|P_n K - K\|$ being that spanned by the eigenvectors $\phi_1, \ldots, \phi_n$ corresponding to $\mu_1, \ldots, \mu_n$ respectively. (In practice, of course, we may not know what these eigenvectors are.) This may be viewed in a different light; if the kernel $k(x, t)$ generating K is sufficiently smooth that the kernel obtained by replacing it with one which is constant on squares of side $1/n$ (i.e. a piecewise constant approximation to $k(x, t)$ on the square $[0, 1] \times [0, 1]$) is close to k, then $\|P_n K - K\|$ will be small, where P_n is the projection onto the space of functions which have constant values on each of the intervals $((j-1)/n, j/n)$ $(j = 1, 2, \ldots, n)$. This in turn can be pursued to show that if the kernel is sufficiently smooth, then $\mu_n \to 0$ correspondingly rapidly. A precise statement of this result (which was proved by Weyl in 1912) is contained in Problem 8.20.

The Kantorovich method does not eliminate the uncomfortable feature of our finite-dimensional approximations that the approximate solution, p, which is found may be discontinuous when we know the true solution ϕ to be continuous, or may fail in similar ways to belong to the class of functions known to contain the true solution. We know that if $\phi - K\phi = f$ then ϕ must be of the form $f + \psi$ where ψ belongs to $\mathscr{R}(K)$ (and thus ψ has such analytic properties as we may deduce from the properties of the kernel $k(x, t)$ generating K). One way of ensuring that the approximation found is of this form is by iteration. We used this device in Example 8.2, to improve a simple, but rather crude, approximation. In the *iterated Galerkin* method one finds the Galerkin approximation $p_n \in E_n$ to the solution ϕ of $\phi - K\phi = f$ and then considers $\hat{p}_n = f + Kp_n$. The approximation $\hat{p}_n$ need not belong to E_n, and, indeed, $P_n \hat{p}_n = P_n f + P_n Kp_n = p_n$, but $\hat{p}_n$ is certainly the sum of f and an element of $\mathscr{R}(K)$. In this case, even if E_n consists of piecewise constant functions we see that if the kernel k which generates K is continuous then $\hat{p}_n - f$ is continuous. In fact, we can show more. By construction $\hat{p}_n = f + K(I - P_n K)^{-1} P_n f$ so $\hat{p}_n = (I - KP_n)^{-1} f$ and

$$
\begin{aligned}
\hat{p}_n - \phi &= ((I - KP_n)^{-1} - (I - K)^{-1}) f \\
&= (I - KP_n)^{-1} (KP_n - K)(I - K)^{-1} f \\
&= (I - KP_n)^{-1} K(P_n - I)\phi \\
&= (I - KP_n)^{-1} K(P_n - I)(\phi - p_n)
\end{aligned}
$$

(since $(P_n - I)p_n = 0$). As Lemma 7.2 shows that $\|KP_n - K\| \to 0$ as $n \to \infty$ and therefore that $(I - KP_n)^{-1}$ exists for sufficiently large n, we obtain

$$
\|\hat{p}_n - \phi\| \leqslant \|(I - KP_n)^{-1}\| \|KP_n - K\| \|\phi - p_n\|. \tag{8.39}
$$

Noticing that there is an N and a constant C such that for all $n \geqslant N$ $\|(I - KP_n)^{-1}\| \leqslant C$ this gives

$$
\|\hat{p}_n - \phi\| \leqslant C \|KP_n - K\| \|p_n - \phi\|
$$

showing that $\hat{p}_n$ is eventually closer to ϕ than p_n is, the improved accuracy being greater for large n because $\|KP_n - K\| \to 0$ as $n \to \infty$.

Theorem 8.10 (Iterated Galerkin method)

Suppose that K is a compact operator on the Hilbert space H, that $I - K$ is invertible, that $f \in H$ and that $\phi = (I - K)^{-1} f$. Suppose also that (χ_n) is linearly independent and that every element of H is the limit of a sequence of finite linear combinations of (χ_n). Let E_n be the subspace spanned by $\chi_1, \ldots, \chi_n$ and P_n the projection onto E_n. If p_n is the Galerkin approximation to ϕ (that is, $p_n \in E_n$ and $p_n - Kp_n - f \in E_n^\perp$) and we set $\hat{p}_n = f + Kp_n$, then, for sufficiently large n, $\hat{p}_n = (I - KP_n)^{-1} f$ and $\hat{p}_n - \phi = \hat{S}_n f$ where $\hat{S}_n \in B(H)$ and $\|\hat{S}_n\| \to 0$ as $n \to \infty$.

Moreover, there is a constant C such that for n sufficiently large $\|\hat{p}_n - \phi\| \leqslant C \|KP_n - K\| \|p_n - \phi\|$, so that if n is large enough $\hat{p}_n$ is a closer approximation to ϕ than p_n. $\qquad\square$

Example 8.9

In Example 8.8 we found the Galerkin approximation to the true solution ϕ of

$$
\phi(x) - \int_0^1 (\min(x, t) - xt)\phi(t)\, dt = 1 \quad (0 \leqslant x \leqslant 1)
$$

where the subspace E_n had dimension 3 and the basis functions were χ_1, χ_2, χ_3 where χ_i had the value 1 on $((i-1)/3, i/3)$ and zero elsewhere. This approximation was $p = \alpha_1 \chi_1 + \alpha_2 \chi_2 + \alpha_3 \chi_3$ where $\alpha_1 = \alpha_2 = 1.0711$ and $\alpha_2 = 1.1329$. We may now calculate the iterated Galerkin approximation $\hat{p} = 1 + Kp$, giving

$$\hat{p}(x) = \begin{cases} 1 + 0.5459x - 0.5356x^2 & (0 \leqslant x \leqslant \tfrac{1}{3}), \\ 0.9966 + 0.5665x - 0.5665x^2 & (\tfrac{1}{3} \leqslant x \leqslant \tfrac{2}{3}), \\ 1.0103 + 0.5253x - 0.5356x^2 & (\tfrac{2}{3} \leqslant x \leqslant 1). \end{cases}$$

Comparing the accuracy of the Galerkin and iterated Galerkin methods for $x = 0, \tfrac{1}{4}$ and $\tfrac{1}{2}$ we have

	Exact solution	Galerkin	Iterated Galerkin
$x = 0$	1	1.0711	1
$x = \tfrac{1}{4}$	1.1041	1.0711	1.1030
$x = \tfrac{1}{2}$	1.1395	1.1329	1.1382

$\square$

The iterated Galerkin method does not extend, that is, we cannot necessarily improve the approximation by calculating $f + K\hat{p}$. However, we can derive some results in this direction by considering more closely how to choose the subspace E_n. So far we have only demanded of the subspaces chosen that the associated projection P_n has the property that for each $\psi \in H$ (where H is the Hilbert space concerned) $P_n \psi \to \psi$ as $n \to \infty$. In fact, we have rather assumed that E_n (and thus P_n) is chosen for ease of computation, for example by choosing E_n to be the subspace of functions which take constant values on n subintervals of equal length. In cases like this we can, of course, obtain simple estimates for $\| P_n \psi - \psi \|$, given additional information about ψ such as smoothness. This sort of consideration, as well as habit, tends to make us notice the continuity, differentiability etc. of the solution and its approximations. In what follows we shall choose a subspace related directly to the problem.

Suppose that we wish to solve the equation $\phi - K\phi = f$ in the Hilbert space H, where K is a compact operator on H. Choose E_n to be the subspace spanned by $f, Kf, \ldots, K^{n-1}f$. Given that $I - K$ is invertible, we wish to show that the procedures described above (minimisation of the residual, Galerkin, Kantorovich and iterated Galerkin) all give approximating sequences which converge, eventually rapidly, to the true solution. There are some immediate issues to be considered here, for we have not presupposed conditions under which a typical member of H can be expressed as a limit of elements of E_n. The question is therefore whether

$(I-K)^{-1}f$ can be expressed as the limit of a sequence of elements of the form $\sum_{j=0}^{n}\alpha_{j}K^{j}f$, which would be true if $(I-K)^{-1}$ were the limit of a sequence of polynomials (h_n) in K (so that $\|h_n(K)-(I-K)^{-1}\| \to 0$ as $n \to \infty$). This is easily seen to be so if $\rho(K) < 1$ for then the required result is an immediate consequence of the Neumann series $(I-K)^{-1} = \sum_{j=0}^{\infty} K^j$. If $\rho(K) \geqslant 1$ it is not so clear that $(I-K)^{-1}$ is the limit of a sequence of polynomials; a general sufficient condition for this to hold for a bounded operator K is that 1 belongs to the unbounded component of the complement of $\sigma(K)$, although we shall not require to prove this to establish the result we wish for compact operators. What we shall prove is that $P_n\phi \to \phi$ as $n \to \infty$, and we shall restrict attention to the case where K is self-adjoint, which avoids a substantial amount of complication.

Theorem 8.11

Let K be a self-adjoint compact operator on the Hilbert space H, $I-K$ be invertible, $f \in H$ and $\phi = (I-K)^{-1}f$. Suppose also that E_n is the subspace spanned by $f, Kf, \ldots, K^{n-1}f$ and P_n is the orthogonal projection onto E_n. Then $\|P_n\phi - \phi\|^{1/n} \to 0$ as $n \to \infty$.

Proof

Choose $\rho > 0$, and let E be the (finite-dimensional) subspace spanned by all eigenvectors of K corresponding to eigenvalues of modulus greater than ρ; let the dimension of E be N. Let $F = E^{\perp}$. Both E and F are invariant subspaces of K (in the sense that $\psi \in E \Rightarrow K\psi \in E$ and $\psi \in F \Rightarrow K\psi \in F$) and so the restriction of K to E or F yields a self-adjoint bounded linear map on that smaller space. By choice of F, all the eigenvalues of the restriction of K to F have modulus not exceeding ρ, so $\|K|_F\| \leqslant \rho$ where $K|_F$ denotes the restriction of K to F. Let $f = f_E + f_F$ be the decomposition of f into its components in E and F respectively. Since E is a finite-dimensional space and $K: E \to E$ there is a polynomial p (of degree at most N) for which $(I-K)p(K)\psi = \psi$ for all $\psi \in E$. Let $\phi_n = (I + K + \cdots + K^{n-1})f + K^n p(K)f \in E_{n+N+1}$ (the subspace spanned by $f, Kf, \ldots, K^{n+N}f$). Then

$$
\begin{aligned}
(I-K)\phi_n - f &= -K^n f + K^n(I-K)p(K)f \\
&= K^n((I-K)p(K) - I)f_F \quad \text{(since } (I-K)p(K)f_E = f_E) \\
&= K^n g_F \qquad\qquad\qquad\qquad\qquad\qquad (8.40)
\end{aligned}
$$

where $g_F = ((I-K)p(K) - I)f_F \in F$. Since $\|K|_F\| \leqslant \rho$, it follows that $\|K^n g_F\| \leqslant \rho^n \|g_F\|$ for all n. Then

$$
\begin{aligned}
\|\phi_n - \phi\| &= \|(I-K)^{-1}((I-K)\phi_n - f)\| \\
&\leqslant \rho^n \|g_F\| \|(I-K)^{-1}\|.
\end{aligned}
$$

Since $\phi_n \in E_{n+N+1}$, it follows that $\|P_{n+N+1}\phi - \phi\| \leqslant \rho^n \|g_F\| \|(I-K)^{-1}\|$ for all n and so, letting $d_n = \|P_n\phi - \phi\|$, there is a constant C such that for all $n \geqslant N+1$ $d_n \leqslant C\rho^{n-N-1}$. Hence we see that $\lim \sup_{n \to \infty} d_n^{1/n} \leqslant \rho$ and, since ρ was arbitrary, $\lim_{n \to \infty} d_n^{1/n} = 0$. $\qquad\qquad\square$

It follows that, at least in the limit as $n \to \infty$, if we choose our subspace E_n to be that spanned by $f, Kf, \ldots, K^{n-1}f$, then $\|P_n\phi - \phi\| \to 0$ very rapidly, in that $\|P_n\phi - \phi\|^{1/n} \to 0$ as $n \to \infty$. To see the effect of this, we recall that for the method of minimisation of the residual $\|(I-K)p_n - f\| \leqslant \|(I-K)\psi - f\|$ for all $\psi \in E_n$ and hence, in particular, $\|(I-K)p_n - f\| \leqslant \|(I-K)P_n\phi - f\| = \|(I-K)(P_n\phi - \phi)\|$. In this case, $\|p_n - \phi\| \leqslant \|(I-K)^{-1}\| \|(I-K)\| \|P_n\phi - \phi\|$. In the Galerkin approximation (8.35) shows us that $\|\phi - p_n\| \leqslant C\|\phi - P_n\phi\|$ so we can again see that $\|\phi - p_n\|^{1/n} \to 0$ as $n \to \infty$. A slight modification is required to obtain the corresponding result for the Kantorovich method since (8.38) shows that if $\tilde{p}_n$ is the estimate obtained for ϕ (with $\tilde{p}_n = f + q_n$) then

$$\|\tilde{p}_n - \phi\| \leqslant \|(I-P_nK)^{-1}\| \|(P_n - I)K\phi\|.$$

However (8.40) shows that $\phi_n - \phi = (I-K)^{-1}K^n g_F$ whence $\|K\phi_n - K\phi\| \leqslant \|(I-K)^{-1}\| \|K^{n+1}g_F\|$ and, pursuing the same argument as with $\|P_n\phi - \phi\|$, we deduce that $\|P_nK\phi - K\phi\|^{1/n} \to 0$ as $n \to \infty$. That the iterated Galerkin method has the property that $\|\hat{p}_n - \phi\|^{1/n} \to 0$ is immediate from (8.39) because we already know $\|p_n - \phi\|^{1/n} \to 0$.

Example 8.10

To illustrate the effect of the choice of subspace for the Galerkin method discussed in the paragraphs above, consider the equation

$$\phi(x) - 4\int_0^1 \min(x,t)\phi(t)\,dt = 1 \quad (0 \leqslant x \leqslant 1). \qquad (8.41)$$

Letting K be the integral operator defined by $K\phi(x) = 4\int_0^1 \min(x,t)\phi(t)\,dt$ $(0 \leqslant x \leqslant 1)$, and $f = 1$, this equation is of the form $\phi - K\phi = f$, where the eigenvalues of K are $\{16/((2n-1)^2\pi^2): n = 1, 2, 3, \ldots\}$. By differentiating both sides of (8.41) and solving the boundary value problem obtained, we can see that the exact solution of (8.41) is $\phi(x) = (\cos 2(1-x))/\cos 2$, $(0 \leqslant x \leqslant 1)$.

In this case $f = 1$, $(Kf)(x) = 4x - 2x^2$ and

$$(K^2f)(x) = (16/3)x - (8/3)x^3 + (2/3)x^4$$

so if we choose E to be the subspace spanned by f, Kf and K^2f and apply Galerkin's method on this subspace we need to calculate (K^jf, f) for

$j = 0, 1, \ldots, 5$. Solving the equations obtained gives the Galerkin solution $p = \alpha_1 f + \alpha_2 K f + \alpha_3 K^2 f$ where $\alpha_1 = 0.9877$, $\alpha_2 = 1.6328$ and $\alpha_3 = -1.9956$ (to four decimal places).

If, instead, we choose the three-dimensional subspace E to be spanned by the functions which take the value 1 on one of the intervals $[0, \frac{1}{3}]$, $(\frac{1}{3}, \frac{2}{3})$ and $[\frac{2}{3}, 1]$ and are zero elsewhere, and apply Galerkin's method to the resulting problem we obtain the approximation p to the solution, where $p = \alpha_i$ on the interval $((i-1)/3, i/3)$ $(i = 1, 2, 3)$, where in this case $\alpha_1 = 0.1702$, $\alpha_2 = -1.4220$ and $\alpha_3 = -2.4257$, again to four decimal places. Because the true solution ϕ varies relatively rapidly (its derivative is about -2.4 at one point) there is no prospect of a piecewise constant solution such as this being everywhere near to the true solution, but we may expect more by using the iterated Galerkin method and forming $\hat{p} = f + Kp$ with p as just calculated, and this does indeed give a better approximation. (We could, of course, calculate the iterated Galerkin approximation for the subspace used in the last paragraph, giving $\hat{p} = f + \alpha_1 K f + \alpha_2 K^2 f + \alpha_3 K^3 f$ for α_1, α_2 and α_3 as found there. This, in practice, does not involve new calculation since we have already determined the required quantities.)

We quote below the values of the true solution ϕ, the Galerkin solution p_1 on the subspace spanned by f, Kf and $K^2 f$ and the iterated Galerkin solution p_2 on the subspace of functions constant on the intervals $[0, \frac{1}{3}]$, $(\frac{1}{3}, \frac{2}{3})$ and $[\frac{2}{3}, 1]$.

x	0	0.2	0.4	0.6	0.8	1
$\phi(x)$	1	0.0702	-0.8707	-1.6742	-2.2133	-2.4030
$p_1(x)$	0.9877	0.0768	-0.8526	-1.6091	-2.1031	-2.3987
$p_2(x)$	1	0.0006	-1.0014	-2.4065	-2.4113	-2.6054

$\square$

Problems

8.1   Prove Theorem 8.3 by using variational calculus.

8.2   Let ϕ satisfy $\phi + K\phi = f$ where

$$(K\phi)(x) = \int_0^1 \{1 + e^{i\pi(x-t)}\}\phi(t) \, dt \quad (0 \leqslant x \leqslant 1)$$

and $f(x) = 1$ $(0 \leqslant x \leqslant 1)$. Use constant test functions in the dual extremum principles of Theorem 8.3 to obtain upper and lower bounds for $\int_0^1 \phi(x) \, dx$. Compare the bounds with the exact value of $\int_0^1 \phi(x) \, dx$.

8.3   Use constant test functions in conjunction with the dual extremum principles of Theorem 8.3 to obtain upper and lower bounds for $\int_0^1 \phi(x)\,dx$ where $\lambda < 0$ and

$$\phi(x) = 1 + \lambda \int_0^1 \min(x,t)\phi(t)\,dt \quad (0 \leqslant x \leqslant 1).$$

Use the result in Theorem 8.4(i) to improve the lower bound.

(The exact value of $\int_0^1 \phi(x)\,dx$ can be determined by converting the integral equation to an equivalent boundary value problem.)

8.4 Obtain upper and lower bounds for (ϕ, f) where $\phi + K\phi = f$, $(K\phi)(x) = \int_0^1 \min(x,t)\phi(t)\,dt$ $(0 \leqslant x \leqslant 1)$ and $f(x) = x^{-\frac{1}{4}}$ $(0 < x \leqslant 1)$ by using a constant trial function in the corresponding minimum principle and a constant multiple of f in the corresponding maximum principle.

8.5 Use constant trial functions in the dual extremum principles of Theorem 8.3 for

$$\phi(x) + \int_0^1 |x-t|^{-\frac{1}{4}}\phi(t)\,dt = 1 \quad (0 \leqslant x \leqslant 1)$$

to show that $0.3962 \leqslant \int_0^1 \phi(x)\,dx \leqslant 0.3969$.

(Note that $B(1.75, 1.75) = 0.25416$, to five decimal places, where B denotes the beta function.)

8.6 Let $f(x) = x$ $(0 \leqslant x \leqslant 1)$ and $(K\phi)(x) = \frac{1}{4}\int_0^1 \{1 - 2|x-t|\}\phi(t)\,dt$ $(0 \leqslant x \leqslant 1)$. Use the maximum principle of Theorem 8.1, with the test function f, and Theorem 8.4(i) to bound (ϕ, f) where $\phi = f + K\phi$.

8.7 Let K be the operator on $L_2(0,1)$ defined by

$$(K\phi)(x) = \int_0^1 \log\left|\frac{x^{\frac{1}{2}} + t^{\frac{1}{2}}}{x^{\frac{1}{2}} - t^{\frac{1}{2}}}\right| \phi(t)\,dt.$$

Note from Example 7.8 that $\mu_1^+(K) \leqslant \pi$ and from Example 6.7 that $K = TT^*$ where

$$(T\phi)(x) = \int_0^x \frac{\phi(t)\,dt}{(x-t)^{\frac{1}{2}}}.$$

Let $\psi(x) = (1-x)^{\frac{1}{2}}$ $(0 \leqslant x \leqslant 1)$ and show that $(T^*\psi)(x) = \pi(1-x)/2$ $(0 \leqslant x \leqslant 1)$, that $(K\psi, \psi) = \|T^*\psi\|^2 = \pi^2/12$ and that $(K\psi)(x) = \pi x^{\frac{1}{2}}(3 - 2x)/3$ $(0 \leqslant x \leqslant 1)$.

Now suppose that ϕ satisfies $(I - \frac{1}{4}K)\phi = f$ where $f(x) = 1 - x$ $(0 \leqslant x \leqslant 1)$. Use the maximum principle of Theorem 8.1, with the trial function $c\psi$, c being a constant, together with the bounds of Theorem 8.4(i) to show that

$$0.550 \leqslant (\phi, f) \leqslant 0.575.$$

8.8 Obtain bounds for (ϕ, f) where $(I - K)\phi = f$ in $L_2(0,1)$, $f(x) = x$ and

$$(K\phi)(x) = \int_0^1 \{xt - \max(x,t)\}\phi(t)\,dt$$

as follows.

(i) Use the basic maximum principle with the trial function $p=cf$, c being a constant, to show that $(\phi, f) \geqslant 0.263$.

(ii) Deduce that $(\phi, f) \leqslant 0.291$ by using the appropriate part of Theorem 8.4(i). To obtain an estimate of the quantity β required there, write $K = K_1 + K_2$ where

$$(K_1\phi)(x) = -\int_0^1 \max(x, t)\phi(t)\, dt, \quad (K_2\phi)(x) = \int_0^1 xt\phi(t)\, dt.$$

Refer to Example 7.2 to obtain bounds for $\mu_1^+(K_1)$, evaluate the non-zero eigenvalue of K_2 and use Theorem 5.4 to give an upper bound for $\mu_1^+(K)$.

(An improved lower bound for (ϕ, f) can also be found using the calculations of Example 7.7 and Theorem 5.4.)

8.9   Prove Theorem 8.4(ii) by using variational calculus.

8.10   Let ϕ satisfy $\phi = f + \lambda K\phi$ where $f(x) = 1$ $(0 \leqslant x \leqslant 1)$, $(K\phi)(x) = \int_0^1 \max(x, t)$ $\phi(t)\, dt$ $(0 \leqslant x \leqslant 1)$ and $(\mu_1^-)^{-1} < \lambda < (\mu_1^+)^{-1}$, $\mu_1^\pm$ denoting the dominant eigenvalues of K. Obtain estimates for $\Phi = \int_0^1 \phi(x)\, dx$ as follows.

(i) Use the maximum principle of Theorem 8.1 to determine the best approximation to ϕ in the space of constant functions. Refer to Examples 7.2 and 7.7 to obtain bounds for μ_1^+ and μ_1^-. Use Theorem 8.4(i) to deduce bounds for Φ, showing in particular that, for $\lambda = -1$,

$$0.6068 \leqslant \Phi \leqslant 0.6093$$

and, for $\lambda = 0.5$,

$$1.5116 \leqslant \Phi \leqslant 1.5192.$$

(ii) Use the dual extremum principles of Theorem 8.4(ii) in conjunction with constant test functions to show that, if $\lambda = -1$,

$$0.6089 \leqslant \Phi \leqslant 0.6092$$

and, if $\lambda = 0.5$,

$$1.5119 \leqslant \Phi \leqslant 1.5121.$$

(This problem combines the approximation methods of the last chapter and this one. The advantage of having good bounds for the dominant eigenvalues of an operator is very much in evidence here.)

8.11   Let ϕ satisfy

$$\phi(x) = 1 - \int_0^1 \{1 - \max(x, t)\}\phi(t) \quad (0 \leqslant x \leqslant 1)$$

and obtain upper and lower bounds for $\phi(0)$ by the method of §8.3.6 as follows. Construct the appropriate auxiliary equation $\psi = g - K\psi$ and use the functional $J(p,q)=(f,q)+(p,g)-(p,q)-(Kp,q)$ to approximate ϕ and ψ in the space of constant functions. Deduce the required bounds by using Theorem 8.6.

8.12 Let K be a self-adjoint compact operator on the real Hilbert space H and let $A\phi=f$ and $A\psi=g$ have solutions in H, where $A=I-\lambda K$, λ being a real parameter. Suppose that $\omega\|p\|^2<(Ap,p)<v\|p\|^2$ for some v and $\omega>0$ and all $p\neq0$, and let $G_1:H\times H\to\mathbb{R}$ and $R_1:H\times H\to\mathbb{R}$ be defined by $G_1(p,q)=(f,q)+(p,g)-(Ap,q)+\frac{1}{2}(\omega^{-1}+v^{-1})(Ap-f,Aq-g)$ and $R_1(p,q)=\frac{1}{2}(\omega^{-1}-v^{-1})\|Ap-f\|\|Aq-g\|$.

Prove that G_1-R_1 is maximised if, and only if, $p=\phi$ or $q=\psi$ and that its maximum value is (ϕ,g); and that G_1+R_1 is minimised if, and only if, $p=\phi$ or $q=\psi$ and that its minimum value is (ϕ,g).

(These results correspond to the dual bounds of Theorem 8.6. A proof of the principles by variational calculus is not as daunting as it may at first appear to be.)

8.13 Let A be an operator on the real Hilbert space H satisfying $\|Ap\|>\eta\|p\|$ for some $\eta>0$ and all non-zero $p\in H$. Suppose that the equations $A\phi=f$, $A^*\psi=g$ have solutions in H and define the functionals $J,S:H\times H\to\mathbb{R}$ by $J(p,q)=(f,q)+(p,g)-(Ap,q)$, $S(p,q)=\eta^{-1}\|Ap-f\|\|A^*q-g\|$. Prove that $J-S$ is maximised if, and only if, $p=\phi$ or $q=\psi$ and that $J+S$ is minimised if, and only if, $p=\phi$ or $q=\psi$. Prove further that the maximum value of $J-S$ and the minimum value of $J+S$ are both (ϕ,g).

8.14 (i) Let $A\phi=f$ in a Hilbert space H, where A is self-adjoint, and let the functions χ_n $(n=1,2,\ldots,N)$ span the subspace E_N of H. Show that the approximation $p_N\in E_N$ to ϕ obtained by using the Rayleigh–Ritz method in conjunction with the maximum principle of Theorem 8.1 coincides with the Galerkin approximation to ϕ in E_N.

Show further that

$$(p_N,f)=f_1\begin{vmatrix}f_1 f_2\ldots f_N\\ \mathbf{h}_1\ \mathbf{h}_2\ldots\mathbf{h}_N\end{vmatrix}\begin{vmatrix}a_{11}a_{12}\ldots a_{1N}\\ \mathbf{h}_1\ \ \mathbf{h}_2\ \ldots\mathbf{h}_N\end{vmatrix}^{-1},$$

where $a_{nm}=(A\chi_n,\chi_m)$, $f_n=(f,\chi_n)$ and

$$\mathbf{h}_n=(f_2 a_{1n}-f_1 a_{2n},\ f_3 a_{1n}-f_1 a_{3n},\ldots,\ f_N a_{1n}-f_1 a_{Nn})^T.$$

(ii) Let ϕ satisfy

$$\phi(x)=1-\int_0^1\frac{\phi(t)\,dt}{|x-t|^{\frac{1}{2}}}\quad(0\leqslant x\leqslant1)$$

and seek the approximation $p_3=\Sigma_{i=1}^3\,\alpha_i\chi_i$ to ϕ, where

$$\chi_i(x)=\begin{cases}1&(i-1)/3<x<i/3,\\0&\text{otherwise}\end{cases}$$

By calculating (p_3,f) show that $\int_0^1\phi(x)\,dx\geqslant0.273$ (to three decimal places). Show also that the Galerkin approximation to ϕ is given with $\alpha_1 = \alpha_3 = 0.284$ and $\alpha_2 = 0.251$.

(The calculation in (ii) is not as arduous as it may appear, for, in the notation of (i), it can be shown in advance of evaluating the elements required that $a_{11} = a_{22} = a_{33}$, $a_{12} = a_{23}$ and $f_1 = f_2 = f_3$.)

8.15 (i)   Let K be the operator on $L_2(a, b)$ defined by

$$(K\phi)(x) = \int_a^x \frac{t-a}{b-a} \phi(t) \, dt + \int_x^b \frac{t-b}{b-a} \phi(t) \, dt \quad (a \leqslant x \leqslant b).$$

Show that, for $\phi \in L_2(a, b)$, $\int_a^b (K\phi)(x) \, dx = 0$ and $(K\phi)'(x) = \phi(x)$ (almost everywhere). Show also that $\|K\| \leqslant (b-a)/6^{\frac{1}{2}}$.

Let E be the subspace of $L_2(a, b)$ consisting of constant functions and P be the projection of $L_2(a, b)$ onto E. Show that if ψ is differentiable and $\psi' \in L_2(a, b)$, then $\psi - P\psi = K(\psi')$, and thus $\|\psi - P\psi\| \leqslant (b-a)\|\psi'\|/6^{\frac{1}{2}}$.

(ii)   Let $H = L_2(0, 1)$ and E_n be the subspace consisting of those functions constant on each of the intervals $((j-1)/n, j/n)$ $(j = 1, 2, \ldots, n)$. Show that if P_n is the projection of H onto E_n then, for every differentiable function ψ whose derivative belongs to $L_2(0, 1)$, $\|\psi - P_n\psi\| \leqslant \|\psi'\|/(n(6)^{\frac{1}{2}})$.

8.16 (i)   Let K be the operator on $L_2(a, b)$ given by

$$(K\phi)(x) = \int_a^x \frac{(x-b)(t-a)}{b-a} \phi(t) \, dt + \int_x^b \frac{(x-a)(t-b)}{b-a} \phi(t) \, dt \quad (a \leqslant x \leqslant b).$$

Show that $(K\phi)(a) = (K\phi)(b) = 0$ and that $(K\phi)''(x) = \phi(x)$ (almost everywhere). By finding the eigenvalues of K, show that $\|K\| \leqslant (b-a)^2/\pi^2$.

(ii)   Let $H = L_2(0, 1)$ and E_{n+1} be the subspace of functions which are continuous on $[0, 1]$ and which are linear on each of the subintervals $[(j-1)/n, j/n]$ $(j = 1, 2, \ldots, n)$. Show that if ψ is twice differentiable and $\psi'' \in L_2(0, 1)$ then $\|\psi - P_{n+1}\psi\| \leqslant \|\psi''\|/(n^2\pi^2)$, where P_{n+1} is the projection of H onto E_{n+1}.

8.17   Let ϕ satisfy $\phi = f + K\phi$ where $f(x) = 1$ $(0 \leqslant x \leqslant 1)$ and $(K\phi)(x) = \int_0^1 \max(x, t) \phi(t) \, dt$ $(0 \leqslant x \leqslant 1)$.

(i)   Show that $\phi(x) = e \cosh x$ $(0 \leqslant x \leqslant 1)$.

(ii)   Use Galerkin's method to obtain an approximation to ϕ in the spaces spanned by

(a) χ_1, χ_2 and χ_3 where

$$\chi_i(x) = \begin{cases} 1 & (i-1)/3 < x < i/3 \\ 0 & \text{otherwise} \end{cases} \quad i = 1, 2, 3,$$

(b) f and Kf.

Determine the iterated Galerkin approximation in each case, showing in particular that in case (b) this gives $\phi \approx \hat{p}$ where

$$288\hat{p}(x) = 35x^4 + 390x^2 + 783 \quad (0 \leqslant x \leqslant 1).$$

8.18 Let $\phi = 1 + K\phi$ in $L_2(0, 1)$ where K is the operator generated by the kernel

$$k(x, t) = \begin{cases} 1 & (x+t \leqslant 1) \\ -1 & (x+t > 1) \end{cases} \quad (0 \leqslant x, t \leqslant 1).$$

Show that $(Kf, f) = 0$, $(K^2f, f) = \frac{1}{3}$, $(K^3f, f) = 0$, $(K^4f, f) = 2/15$, $(K^5f, f) = 0$, where $f(x) = 1$ $(0 \leqslant x \leqslant 1)$.

Show that the Galerkin method used with the trial function $p = \alpha_1 f + \alpha_2 Kf + \alpha_3 K^2 f$ gives the approximation $\phi \approx p$ where $p(x) = 2(4 - 5x^2)/3$, and that the iterated Galerkin approximation $\phi \approx \hat{p}$ is given by $9\hat{p}(x) = 20x^3 - 60x^2 + 12x + 23$.

Show also that the minimisation of the residual method with the trial function $q = \beta_1 f + \beta_2 Kf + \beta_3 K^2 f$ is such that its iterate $\hat{q} = 1 + Kq$ is equal to the Galerkin approximation p.

(The equation $\phi = 1 + K\phi$ is related to a mixed boundary value problem for the differential equation $\phi'' + 4\phi = 0$ and it is not difficult to show that $\phi(x) = \sec(1)\{\cos(2x - 1) - \sin(2x - 1)\}$. The largest eigenvalue of K, in magnitude, is estimated in Problem 7.8. Notice that it is easy to adapt the calculations to consider $\phi = 1 + \lambda K\phi$ for any $\lambda \neq 0$ and in particular for values of λ for which the Neumann series does not converge. It is also possible to examine the second iterate of the Galerkin approximation, that is, $f + K\hat{p}$, without much difficulty in this example.)

8.19 (i) The functions $k_i(x, t)$ $(i = 1, 2)$ are continuous for $0 \leqslant x, t \leqslant 1$ and $\alpha_i = \sup\{|k_i(x, t)| : 0 \leqslant x, t \leqslant 1\}$. Show that if f is continuous and $|\lambda| < \min(\alpha_1^{-1}, \alpha_2^{-1})$ then the solutions ϕ_1 and ϕ_2 of the two integral equations

$$\phi_i(x) - \lambda \int_0^1 k_i(x, t)\phi_i(t)\, dt = f(x) \quad (0 \leqslant x \leqslant 1, \ i = 1, 2)$$

are such that

$$|\phi_1(x) - \phi_2(x)| \leqslant (1 - |\lambda|\alpha_1)^{-1}(1 - |\lambda|\alpha_2)^{-1}|\lambda|\,\|f\|\beta$$

for all $x \in [0, 1]$, where $\|f\| = \sup\{|f(x)| : 0 \leqslant x \leqslant 1\}$ and

$$\beta = \sup\{|k_1(x, t) - k_2(x, t)| : 0 \leqslant x, t \leqslant 1\}.$$

(ii) Let ϕ satisfy the integral equation

$$\phi(x) = x + \int_0^1 \sin(xt)\phi(t)\, dt \quad (0 \leqslant x \leqslant 1).$$

Find the approximation to ϕ which results from replacing $\sin(xt)$ by the first two terms of its Maclaurin series and determine the maximum pointwise error in this approximation.

8.20   Suppose that k is a continuous kernel on $[0, 1] \times [0, 1]$ whose first order partial derivatives are also continuous. Suppose also that $|\partial k/\partial x| \leqslant C$ and $|\partial k/\partial t| \leqslant C$ at all points. Let k_0 be the L_2-kernel obtained from k by dividing $[0, 1] \times [0, 1]$ into n^2 squares of side $1/n$ and, on each square giving k_0 the constant value which k takes at the centre of the square. Show that if K and K_0 are the integral operators generated by k and k_0 respectively then $\|K - K_0\| < C'/n$ where $C' = \sqrt{(7/24)}C$. By observing that $K_0\phi$ belongs to the subspace E_n of functions constant on each of the intervals $((j-1)/n, j/n)$, show that $\|P_n K - K\| < C'/n$ where P_n is the projection onto E_n.

8.21   Let K be the operator on $L_2(0, a)$ defined by

$$(K\phi)(x) = \sum_{n=1}^{\infty} |\gamma_n|^2 \phi_n(x) \int_0^a \overline{\phi_n(t)} \phi(t) \, dt \quad (0 \leqslant x \leqslant a)$$

and let $K\phi = \hat{\phi}_0$, where $\hat{\phi}_0$ is the restriction to $(0, a)$ of $\phi_0 \in L_2(0, b)$, $(\phi_n : n = 0, 1, 2, \ldots)$ being a complete orthonormal sequence in $L_2(0, b)$, where $b > a$, and (γ_n) being a sequence of complex numbers such that $\gamma_n \to 0$ as $n \to \infty$. Obtain upper and lower bounds for $\alpha = \int_0^a \phi(t) \overline{\phi_0(t)} \, dt$ as follows:

(i)   by referring to Problem 6.20, determine an operator $T : L_2(0, a) \to L_2(0, b)$, such that $K = T^*T$.

(ii)   deduce from an appropriate maximum principle that

$$\alpha \geqslant |(p, \phi_0)|^2 \left\{ \sum_{n=1}^{\infty} |\gamma_n|^2 |(p, \phi_n)|^2 \right\}^{-1}$$

for any p of the form

$$p(x) = \begin{cases} p_1(x) & (0 \leqslant x \leqslant a) \\ 0 & (a < x \leqslant b), \end{cases}$$

where $p_1 \in L_2(0, a)$; $(\, , \,)$ denotes the inner product on $L_2(0, b)$.

(iii)   Show that a solution $\psi \in L_2(0, b)$ of $T^*\psi = \hat{\phi}_0$ is given by

$$\psi = (\psi, \phi_0)\phi_0 + \sum_{n=1}^{\infty} \bar{\gamma}_n^{-1}(q, \phi_n)\phi_n$$

for any q of the form

$$q(x) = \begin{cases} 0 & (0 \leqslant x < a), \\ q_1(x) & (a \leqslant x \leqslant b), \end{cases}$$

where $q_1 \in L_2(a, b)$ is such that $\sum_{n=1}^{\infty} |\gamma_n|^{-2} |(q, \phi_n)|^2 < \infty$ and that

$(q, \phi_0) = -1$. Deduce from Theorem 8.5 that

$$\alpha \leqslant |(q, \phi_0)|^{-2} \sum_{n=1}^{\infty} |\gamma_n|^{-2} |(q, \phi_n)|^2$$

for any $q \in L_2(a, b)$ such that the sum converges (and $(q, \phi_0) \neq 0$).

(Integral equations of the type considered here often arise following separation of variables in a mixed boundary value problem. The following problem offers an example.)

8.22 Let the continuous function $u(x, y)$ be defined for $x \geqslant 0$ and $0 \leqslant y \leqslant b$ by the boundary value problem

$$
\begin{aligned}
&u_{xx} + u_{yy} + k^2 u = 0 && (x > 0,\ 0 < y < b), \\
&u_y(x, 0) = 0 && (x \geqslant 0), \\
&u_y(x, b) = 0 && (x > 0), \\
&u(0, y) = 1 && (0 \leqslant y \leqslant a), \\
&u_x(0, y) = 0 && (a < y \leqslant b), \\
&u \sim \tau e^{ikx} && (x \to \infty).
\end{aligned}
$$

Here k is a given real number satisfying $0 < kb < \pi$ and it is required to obtain $\tau \in \mathbb{C}$.

Use separation of variables to show that u is of the form

$$u(x, y) = \tau e^{ikx} + \sum_{n=1}^{\infty} a_n e^{-\beta_n x} \cos(n\pi y/b)$$

where $\beta_n = \{(n\pi/b)^2 - k^2\}^{\frac{1}{2}}$ $(n \in \mathbb{N})$. Deduce that the function $\phi(y) = -u_x(0, y)/((1 - \tau)b^{\frac{1}{2}})$ $(0 \leqslant y \leqslant a)$ satisfies the integral equation

$$\phi_0(y) = \sum_{n=1}^{\infty} \beta_n^{-1} \phi_n(y) \int_0^a \phi_n(t)\phi(t)\, dt \quad (0 \leqslant y \leqslant a) \tag{8.41}$$

where $\phi_0(y) = b^{-\frac{1}{2}}$ and $\phi_n(y) = (2/b)^{\frac{1}{2}} \cos(n\pi y/b)$ $(n \in \mathbb{N})$. Show also that $\alpha = -ik\tau(1 - \tau)^{-1}$ is given by

$$\alpha = \int_0^a \phi_0(t)\phi(t)\, dt.$$

By using (8.41) to show that the imaginary part of ϕ is zero, show that α is real. Suppose that upper and lower bounds for α are determined in the form $\alpha_1 \geqslant \alpha \geqslant \alpha_0 \geqslant 0$. Show that

$$\alpha_0^2(\alpha_0^2 + k^2)^{-1} \leqslant |\tau|^2 \leqslant \alpha_1^2(\alpha_1^2 + k^2)^{-1}$$

and that $\theta = \arg(\tau)$ is such that $\cos\theta = |\tau|$.

Consider the particular case in which $b = 2a = \pi$ and $k = \frac{1}{2}$. Use the upper and lower bounds given in the previous problem with the test functions

$p_1(x) = 1 \ (0 \leqslant x \leqslant \frac{1}{2}\pi)$, $q_1(x) = x - \frac{1}{2}\pi \ (\frac{1}{2}\pi \leqslant x \leqslant \pi)$ and show that $0.89 \leqslant |\tau| \leqslant 0.97$. (The sums appearing in the bounds can be evaluated to two decimal places using a calculator and an estimate of the error in truncating the series.)

(The boundary value problem represents the scattering of a plane sound wave in a duct which is partially closed by a plane barrier. The quantity τ is the (complex) amplitude of the plane wave transmitted beyond the obstructing barrier; $|\tau|$ is the real amplitude of the transmitted wave and $\arg(\tau)$ its phase. The physical problem can be used to suggest more sophisticated test functions but these inevitably lead to bounds which are a good deal more difficult to evaluate.)

Some singular integral equations

9.1 Introduction

This final chapter contains a mixture of old and new ideas. We continue to work with bounded operators on the space $L_2(a, b)$ but broaden the scope of our methods in two ways. We deal with operators which are not compact and with integral equations posed on infinite intervals.

As in the rest of the book, the theory developed is motivated by problems which occur in application areas. One aim of this chapter is to devise a solution method for integral equations with strongly (that is, Cauchy) singular kernels. These are important equations arising, for example, in fluid mechanics, particularly aerodynamics, and are a source of non-compact operators.

In order to establish the operator manipulations used to solve Cauchy singular equations, we begin by considering weakly singular Volterra equations. This allows us to make use of some of the TT^* factorisations of positive operators derived in Chapter 6 to solve a number of practically useful Fredholm equations of the first kind. It also means that we start on familiar ground for, as we have already established, integral equations with weakly singular kernels give rise to compact operators.

The Hilbert operator, generated by the basic Cauchy singular kernel $(t-x)^{-1}$ is introduced via the Fourier transform, which provides an example of an operator on $L_2(-\infty, \infty)$. We conclude by considering a particular class of integral equations which can be solved by means of the Fourier transform.

9.2 Volterra operators with weakly singular kernels

In §1.3.4 we demonstrated that Abel's problem leads to the first kind Volterra integral equation

$$\int_0^x \frac{\phi(t)\,\mathrm{d}t}{(x-t)^{\frac{1}{2}}} = f(x) \quad (0 \leqslant x \leqslant 1). \tag{9.1}$$

In the same section we devised a solution method for this equation, showing that ϕ also satisfies the integral equation

$$\pi \int_0^x \phi(t)\, dt = \int_0^x \frac{f(t)\, dt}{(x-t)^{\frac{1}{2}}} \quad (0 \leqslant x \leqslant 1)$$

and concluding that

$$\phi(x) = \frac{1}{\pi} \frac{d}{dx} \int_0^x \frac{f(t)\, dt}{(x-t)^{\frac{1}{2}}} \quad (0 \leqslant x \leqslant 1).$$

Let us now examine (9.1) and the solution process in the setting which we have developed since Chapter 1. According to Lemma 6.7, the operator defined on $L_2(0, 1)$ by

$$(K\phi)(x) = \int_0^x \frac{\phi(t)\, dt}{(x-t)^{\frac{1}{2}}} \tag{9.2}$$

is bounded and we consider (9.1) via the equation $K\phi = f$ in $L_2(0, 1)$. It is easy to check that $(K^2\phi)(x) = \pi \int_0^x \phi(t)\, dt$.

Suppose that $K\phi = f$ has a solution ϕ in $L_2(0, 1)$. Then ϕ satisfies $K^2\phi = Kf$ and hence

$$\pi \int_0^x \phi(t)\, dt = (Kf)(x) \quad (0 \leqslant x \leqslant 1).$$

Since $\phi \in L_2(0, 1)$, $(Kf)(x)$ must be differentiable almost everywhere in $[0, 1]$ by Theorem B6 and therefore

$$\phi(x) = \frac{1}{\pi} \frac{d}{dx} \int_0^x \frac{f(t)\, dt}{(x-t)^{\frac{1}{2}}} \tag{9.3}$$

for almost all x in $[0, 1]$. Clearly $K\phi = 0$ has only the trivial solution in $L_2(0, 1)$ and so $K\phi = f$ has at most one solution in that space which, if it exists, is given by (9.3). It follows from our arguments that $K\phi = f$ has a unique solution in $L_2(0, 1)$ if and only if f is such that Kf is equal to the integral of a function in $L_2(0, 1)$.

This is not an unfamiliar situation. We have seen in other circumstances that for an equation of the form $K\phi = f$ to have a solution in a given Hilbert space it is not sufficient that f belong to that space. It may happen of course that for a particular $f \in L_2(0, 1)$ the formula (9.3) defines a ϕ which lies outside $L_2(0, 1)$, but which does satisfy (9.1). Only direct verification will establish this fact.

We have made use of the operator K defined in (9.2) previously (see Example 6.7 for instance) and indeed other Volterra operators with weakly singular kernels have proved to be valuable in T^*T type factorisations of Fredholm operators. We therefore consider the generalisation K_α of the operator K, where

$$(K_\alpha \phi)(x) = \int_0^x \frac{\phi(t)\,dt}{(x-t)^\alpha} \quad (0 \leqslant x \leqslant 1, \ 0 < \alpha < 1). \tag{9.4}$$

K_α is a bounded operator on $L_2(0, 1)$.

There is a simple, direct way of reducing K_α to an operator with a continuous kernel. Let $\beta \in (0, 1)$ and let $\hat{k}$ denote the kernel of $K_\beta K_\alpha$. Then

$$\hat{k}(x, t) = \int_t^x \frac{ds}{(x-s)^\beta (s-t)^\alpha} \quad (0 \leqslant t \leqslant x \leqslant 1)$$

and setting $s = t + (x-t)u$ easily gives

$$\hat{k}(x, t) = (x-t)^{1-\alpha-\beta} \int_0^1 (1-u)^{-\beta} u^{-\alpha}\,du$$

$$= B(1-\beta, 1-\alpha)(x-t)^{1-\alpha-\beta},$$

where B is the beta function defined in Appendix C. (We shall use the beta function frequently in this chapter.) In particular the kernel of $K_{1-\alpha}K_\alpha$ is a constant, equal to $B(\alpha, 1-\alpha) = \pi \operatorname{cosec}(\pi \alpha)$.

Now consider the integral equation

$$\int_0^x \frac{\phi(t)\,dt}{(x-t)^\alpha} = f(x) \quad (0 \leqslant x \leqslant 1)$$

where $0 < \alpha < 1$, which is a generalisation of (9.1). The corresponding equation $K_\alpha \phi = f$ in $L_2(0, 1)$ can be modified to $K_{1-\alpha}K_\alpha \phi = K_{1-\alpha}f$ and, as we have shown, $(K_{1-\alpha}K_\alpha \phi)(x) = \pi \operatorname{cosec}(\pi \alpha) \int_0^x \phi(t)\,dt$. By following the argument we used in relation to $K\phi = f$ (that is, $K_{\frac{1}{2}}\phi = f$ in the present notation) we see that $K_\alpha \phi = f$ has a solution in $L_2(0, 1)$ if and only if $K_{1-\alpha}f$ is the integral of a function in $L_2(0, 1)$. This solution, if it exists, is unique and given almost everywhere in $[0, 1]$ by

$$\phi(x) = \frac{\sin(\alpha \pi)}{\pi} \frac{d}{dx} \int_0^x \frac{f(t)\,dt}{(x-t)^{1-\alpha}}.$$

A further generalisation of the classical Abel equation (9.1) is possible. Let $h: [0, 1] \rightarrow \mathbb{R}$ have a continuous derivative which is everywhere positive and let ϕ satisfy the integral equation

$$\int_0^x \frac{\phi(t)\,dt}{(h(x)-h(t))^\alpha} = f(x) \quad (0 \leqslant x \leqslant 1) \tag{9.5}$$

for some $\alpha \in (0, 1)$. The associated operator L_α, where

$$(L_\alpha \phi)(x) = \int_0^x \frac{\phi(t)\, dt}{(h(x) - h(t))^\alpha} \quad (0 \leqslant x \leqslant 1),$$

is bounded on $L_2(0, 1)$ and so is the operator M_α defined by

$$(M_\alpha \phi)(x) = \int_0^x \frac{h'(t)\phi(t)}{(h(x) - h(t))^\alpha}\, dt \quad (0 \leqslant x \leqslant 1).$$

The kernel of $M_{1-\alpha} L_\alpha$ is

$$\int_t^x \frac{h'(s)\, ds}{(h(x) - h(s))^{1-\alpha}(h(s) - h(t))^\alpha} = \int_0^1 (1-u)^{\alpha-1} u^{-\alpha}\, du = \pi \operatorname{cosec}(\pi\alpha)$$

for $0 \leqslant t \leqslant x \leqslant 1$, the substitution $h(s) = h(t) + [h(x) - h(t)]u$ having been made to simplify the integral. It follows that if $M_{1-\alpha} f$ is the integral of a function belonging to $L_2(0, 1)$, then there is one and only one solution of $L_\alpha \phi = f$ in $L_2(0, 1)$ and it is given by

$$\phi(x) = \frac{\sin(\alpha\pi)}{\pi} \frac{d}{dx} \int_0^x \frac{h'(t) f(t)\, dt}{(h(x) - h(t))^{1-\alpha}} \tag{9.6}$$

almost everywhere in $[0, 1]$.

This method of solving first kind Volterra equations is not restricted to those in which the kernel is purely algebraic; other examples are given in Problem 9.1. The method relies on the identification of what may be called a 'simplifying operator', the application of which transforms the given equation into one of a more manageable type. Thus, for example, $K_{1-\alpha}$ is a simplifying operator for K_α. This notion of simplifying operators recurs throughout the chapter. In particular, now that we have a method for solving Volterra equations such as $K_\alpha \phi = f$ we can seek operators which reduce other more complicated equations to this form.

In the process of implementing this idea we shall also encounter equations like $L_\alpha^* \phi = f$, and these are dealt with by the same simplifying operator approach. Thus, for suitable functions f, $L_\alpha^* \phi = f$ has a unique solution in $L_2(0, 1)$ which is given for almost all x in $[0, 1]$ by

$$\phi(x) = -\frac{\sin(\alpha\pi)}{\pi} \frac{d}{dx} \int_x^1 \frac{h'(t) f(t)\, dt}{(h(t) - h(x))^{1-\alpha}}. \tag{9.7}$$

There is another way of dealing with certain equations involving adjoints of operators, which will prove to be useful in later sections. The operator K_α defined on $L_2(0, 1)$ by (9.4) is such that $UK_\alpha U = K_\alpha^*$, where U is the unitary operator on $L_2(0, 1)$ given by $(U\phi)(x) = \phi(1 - x)$. Therefore

$K_\alpha^* \phi = f$ implies that $UK_\alpha U\phi = f$ and hence that $K_\alpha(U\phi) = Uf$. Since we can solve $K_\alpha \phi = f$, the solution $U\phi$ of $K_\alpha(U\phi) = Uf$ follows and ϕ is easily recovered using $U^2\phi = \phi$. The operators L_α and L_α^* can be related in a similar way by using a different unitary operator.

9.3 First kind Fredholm equations

The Volterra equations considered in the last section are of some interest in their own right but our main purpose in solving them is because of the way their solutions may be used to construct solutions of other equations.

In Chapter 6 we were able to express a number of positive operators in the form TT^*, T being a Volterra operator with a weakly singular kernel. Our main purpose in the earlier chapter was to demonstrate the positivity of certain operators, but the factorisations may also be used to obtain explicit solutions of some first kind Fredholm equations.

Suppose that ϕ satisfies $K\phi = f$, where K is a positive operator. If we can find an operator T such that $K = TT^*$ then the given equation can be expressed as the coupled pair $T\psi = f$ and $T^*\phi = \psi$. A solution ϕ follows if these two component equations can be solved.

Example 9.1

Solve the integral equation

$$\int_0^1 \log\left|\frac{x^{\frac{1}{2}} + t^{\frac{1}{2}}}{x^{\frac{1}{2}} - t^{\frac{1}{2}}}\right| \phi(t)\,dt = f(x) \quad (0 \leqslant x \leqslant 1).$$

Let T be the bounded operator on $L_2(0, 1)$ defined by

$$(T\phi)(x) = \int_0^x \frac{\phi(t)\,dt}{(x-t)^{\frac{1}{2}}}.$$

We have shown in Example 6.7 that

$$(TT^*\phi)(x) = \int_0^1 \log\left|\frac{x^{\frac{1}{2}} + t^{\frac{1}{2}}}{x^{\frac{1}{2}} - t^{\frac{1}{2}}}\right| \phi(t)\,dt \quad (0 \leqslant x \leqslant 1),$$

and therefore the equation to be solved can be written as $T\psi = f$ with $\psi = T^*\phi$, for this T. Now suppose f to be such that these two equations have solutions ψ and ϕ in $L_2(0, 1)$. Then, by the method used in the last section, we have

$$\psi(x) = \frac{1}{\pi}\frac{d}{dx}\int_0^x \frac{f(t)\,dt}{(x-t)^{\frac{1}{2}}}, \quad \phi(x) = -\frac{1}{\pi}\frac{d}{dx}\int_x^1 \frac{\psi(t)\,dt}{(t-x)^{\frac{1}{2}}}$$

almost everywhere in $[0, 1]$. The solution of the given equation is therefore

$$\phi(x) = -\frac{1}{\pi^2} \frac{d}{dx} \int_x^1 \frac{ds}{(s-x)^{\frac{1}{2}}} \frac{d}{ds} \int_0^s \frac{f(t)\, dt}{(s-t)^{\frac{1}{2}}}$$

for almost all $x \in [0, 1]$, when a solution exists. The function ϕ given by the formula will indeed be a solution provided the two derivatives exist at all but finitely many points and the resulting function $\phi \in L_2(0, 1)$. To identify conditions which guarantee that ϕ as given by the formula is a solution, is a little complicated. Let $\Psi(s) = \int_0^s (s-t)^{-\frac{1}{2}} f(t)\, dt$ $(0 \leqslant s \leqslant 1)$. If Ψ is differentiable at all but finitely many points of $[0, 1]$, Ψ is continuous in $[0, 1]$ with $\Psi(0) = 0$ and $\psi = \Psi' \in L_2(0, 1)$ then $T\psi = f$, with similar conditions guaranteeing $T^*\phi = \psi$. In other words, we require that the two derivatives in the formula for ϕ exist at all but finitely many points, that the functions to be differentiated be continuous, particularly at the end points $x = 1$ and $s = 0$, and that the differentiated functions belong to $L_2(0, 1)$. (See Theorem B6(ii).) If the given equation has a solution in $L_2(0, 1)$ it is unique (since the operator involved is positive) and it is given by the derived formula. $\qquad \square$

Example 9.2

Solve the integral equation

$$\int_0^1 \frac{\phi(t)\, dt}{|x-t|^\nu} = 1 \quad (0 \leqslant x \leqslant 1)$$

where $0 < \nu < 1$.

We first recall from Example 6.8 that the bounded operator on $L_2(0, 1)$ defined by

$$(T\phi)(x) = \int_0^x \left(\frac{x}{t}\right)^{\frac{1}{2}(1-\nu)} \frac{\phi(t)\, dt}{(x-t)^{\frac{1}{2}(1+\nu)}} \quad (0 \leqslant x \leqslant 1)$$

is such that

$$(TT^*\phi)(x) = B(\nu, \tfrac{1}{2}(1-\nu)) \int_0^1 \frac{\phi(t)\, dt}{|x-t|^\nu}.$$

We therefore consider the equation $TT^*\phi = f$ in $L_2(0, 1)$ where $f(x) = B(\nu, \tfrac{1}{2}(1-\nu))$ $(0 \leqslant x \leqslant 1)$. The equivalent coupled pair is $T\psi = f$ and $T^*\phi = \psi$ and, solving the first of these by the technique used in §9.2, we formally have

$$\psi(x) = B(\nu, \tfrac{1}{2}(1-\nu)) \cos(\tfrac{1}{2}\pi\nu) \frac{x^{\frac{1}{2}(1-\nu)}}{\pi} \frac{d}{dx} \int_0^x \frac{t^{\frac{1}{2}(\nu-1)}\, dt}{(x-t)^{\frac{1}{2}(1-\nu)}}.$$

Setting $t = sx$,

$$\int_0^x \frac{t^{\frac{1}{2}(v-1)}\,dt}{(x-t)^{\frac{1}{2}(1-v)}} = x^v \int_0^1 (1-s)^{\frac{1}{2}(v-1)}s^{\frac{1}{2}(v-1)}\,ds = B(\tfrac{1}{2}+\tfrac{1}{2}v, \tfrac{1}{2}+\tfrac{1}{2}v)x^v$$

and using the properties of beta and gamma functions it is not difficult to show that $B(\tfrac{1}{2}+\tfrac{1}{2}v, \tfrac{1}{2}+\tfrac{1}{2}v)B(v, \tfrac{1}{2}-\tfrac{1}{2}v)\cos(\tfrac{1}{2}\pi v) = \pi v^{-1}$. Therefore $\psi(x) = x^{\frac{1}{2}(v-1)}$. Since the function ψ is differentiable at all points except 0, is continuous, $\psi' \in L_1(0,1)$ and $\psi \in L_2(0,1)$, Theorem B6 confirms that we have obtained the unique solution of $T\psi = f$ in that space.

Now $T^*\phi = \psi$ gives

$$\phi(x) = -\cos(\tfrac{1}{2}\pi v)\,\frac{x^{\frac{1}{2}(v-1)}}{\pi}\,\frac{d}{dx}\int_x^1 \frac{t^{\frac{1}{2}(1-v)}\psi(t)}{(t-x)^{\frac{1}{2}(1-v)}}\,dt$$

$$= -\cos(\tfrac{1}{2}\pi v)\,\frac{x^{\frac{1}{2}(v-1)}}{\pi}\,\frac{d}{dx}\int_x^1 \frac{dt}{(t-x)^{\frac{1}{2}(1-v)}}$$

$$= \pi^{-1}\cos(\tfrac{1}{2}\pi v)\,\{x(1-x)\}^{\frac{1}{2}(v-1)}.$$

This ϕ belongs to $L_2(0,1)$, and is therefore the unique solution of $T^*\phi = \psi$ and of $TT^*\phi = f$ in that space.

Notice that ϕ is continuous in $(0,1)$ and has integrable singularities at the end points of $[0,1]$. This feature is typical of the solution of a first kind Fredholm equation with a weakly singular kernel. $\qquad\square$

Example 9.3

Solve the integral equation

$$\int_0^1 \left\{\frac{1}{|x-t|^v} - \frac{1}{(x+t)^v}\right\}\phi(t)\,dt = f(x) \quad (0<x<1)$$

where $0 < v < 1$.

Reference to §1.3.3 shows that this equation arises in connection with a boundary value problem for the generalised Tricomi equation, which can be related to a problem in aerodynamics. Apart from some minor notational changes the equation to be solved is (1.59).

Problem 6.9 provides the necessary factorisation; the bounded operator on $L_2(0,1)$ defined by

$$(T\phi)(x) = 2\int_0^x \left(\frac{t}{x^2-t^2}\right)^{\frac{1}{2}(v+1)}\phi(t)\,dt$$

is such that

$$(TT^*\phi)(x)=B(\tfrac{1}{2}v,\tfrac{1}{2}-\tfrac{1}{2}v)\int_0^1\{|x-t|^{-v}-(x+t)^{-v}\}\phi(t)\,dt.$$

The given equation may therefore be written as two Volterra equations, namely

$$2\int_0^x\left(\frac{t}{x^2-t^2}\right)^{\tfrac{1}{2}(v+1)}\psi(t)\,dt=g(x),\quad 2\int_x^1\left(\frac{x}{t^2-x^2}\right)^{\tfrac{1}{2}(v+1)}\phi(t)\,dt=\psi(x),$$

where $g(x)=B(\tfrac{1}{2}v,\tfrac{1}{2}-\tfrac{1}{2}v)f(x)$ $(0\leqslant x\leqslant 1)$. These equations are examples of (9.5) and its adjoint, the corresponding solution formulae being (9.6) and (9.7), subject to the restrictions stated in their derivation. We find that

$$\psi(x)=\cos(\tfrac{1}{2}\pi v)\frac{x^{-\tfrac{1}{2}(1+v)}}{\pi}\frac{d}{dx}\int_0^x\frac{tg(t)\,dt}{(x^2-t^2)^{\tfrac{1}{2}(1-v)}}$$

and

$$\phi(x)=-\frac{\cos(\tfrac{1}{2}\pi v)}{\pi}\frac{d}{dx}\int_x^1\frac{t^{\tfrac{1}{2}(1-v)}\psi(t)\,dt}{(t^2-x^2)^{\tfrac{1}{2}(1-v)}},$$

and therefore that if the given equation has a solution $\phi\in L_2(0,1)$ then

$$\phi(x)=-\frac{B(\tfrac{1}{2}v,\tfrac{1}{2}-\tfrac{1}{2}v)}{\pi^2\sec^2(\tfrac{1}{2}\pi v)}\frac{d}{dx}\int_x^1\frac{s^{-v}\,ds}{(s^2-x^2)^{\tfrac{1}{2}(1-v)}}\frac{d}{ds}\int_0^s\frac{tf(t)\,dt}{(s^2-t^2)^{\tfrac{1}{2}(1-v)}},\quad(9.8)$$

almost everywhere in $[0,1]$.   □

Example 9.4

Solve the integral equation

$$\int_0^1|x-t|^{\tfrac{1}{2}}\phi(t)\,dt=1\quad(0\leqslant x\leqslant 1).$$

This equation is different from those arising in the previous three examples in several respects. The operator L defined on $L_2(0,1)$ by $(L\phi)(x)=\int_0^1|x-t|^{\tfrac{1}{2}}\phi(t)\,dt$ is not positive (its trace Λ_1 is zero) and so it cannot be represented in the form TT^*. We can convert the equation $L\phi=f$, where $f(x)=1$ $(0\leqslant x\leqslant 1)$, into one involving a positive operator, however. This is achieved by recalling from Problem 7.17 that $(TKT^*\phi)(x)=2\pi\int_0^1\{t^{\tfrac{1}{2}}+x^{\tfrac{1}{2}}-|x-t|^{\tfrac{1}{2}}\}\phi(t)\,dt$ $(0\leqslant x\leqslant 1)$,

where

$$(T\phi)(x)=\int_0^x (x-t)^{-\frac{1}{4}}\phi(t)\,\mathrm{d}t, \quad (K\phi)(x)=\int_0^1 |x-t|^{-\frac{1}{4}}\phi(t)\,\mathrm{d}t \quad (0\leqslant x\leqslant 1).$$

Since $L\phi=f$ implies that $\int_0^1 \{t^{\frac{1}{4}}-|x-t|^{\frac{1}{4}}\}\phi(t)\,\mathrm{d}t=0$ $(0\leqslant x\leqslant 1)$ (because $f=1$ here) we deduce that ϕ also satisfies $(TKT^*\phi)(x)=2\pi Cx^{\frac{1}{4}}$ $(0\leqslant x\leqslant 1)$, where $C=\int_0^1 \phi(t)\,\mathrm{d}t$. Now K is a positive operator and, indeed, from Example 6.8 we have $SS^*=B(\frac{1}{2},\frac{1}{4})K$ where

$$(S\phi)(x)=\int_0^x \frac{x^{\frac{1}{4}}\phi(t)\,\mathrm{d}t}{t^{\frac{1}{4}}(x-t)^{\frac{3}{4}}} \quad (0\leqslant x\leqslant 1).$$

Therefore $B(\frac{1}{2},\frac{1}{4})TKT^*=TSS^*T^*$ and we can solve $TKT^*\phi=g$ for any g by solving four Volterra-type equations in succession.

Before carrying out this procedure, note that the operator L is such that $ULU=L^*=L$, where $(U\phi)(x)=\phi(1-x)$, and because $Uf=f$ the given equation $L\phi=f$ implies that $LU\phi=f$. Therefore ϕ satisfies $U\phi=\phi$.

Regarding C as known at this stage, $(TKT^*\phi)(x)=2\pi Cx^{\frac{1}{4}}$ $(0\leqslant x\leqslant 1)$ implies that

$$(KT^*\phi)(x)=2C\,\frac{\mathrm{d}}{\mathrm{d}x}\int_0^x \frac{t^{\frac{1}{4}}\,\mathrm{d}t}{(x-t)^{\frac{3}{4}}}=C\pi,$$

on putting $t=xu$ to simplify the integral. Thus $(SS^*T^*\phi)(x)=C\pi B(\frac{1}{2},\frac{1}{4})$ $(0\leqslant x\leqslant 1)$, giving

$$(S^*T^*\phi)(x)=\frac{CB(\frac{1}{2},\frac{1}{4})}{2^{\frac{1}{2}}}x^{\frac{1}{4}}\frac{\mathrm{d}}{\mathrm{d}x}\int_0^x \frac{t^{-\frac{1}{4}}\,\mathrm{d}t}{(x-t)^{\frac{1}{4}}}$$

$$=CB(\tfrac{1}{2},\tfrac{1}{4})B(\tfrac{3}{4},\tfrac{3}{4})x^{-\frac{1}{4}}/2(2^{\frac{1}{2}}),$$

using the substitution $t=xu$ again to reduce the integral to the beta function. Now $B(\frac{1}{2},\frac{1}{4})B(\frac{3}{4},\frac{3}{4})=2\pi(2^{\frac{1}{2}})$ so we now have $(S^*T^*\phi)(x)=C\pi x^{-\frac{1}{4}}$ $(0<x\leqslant 1)$, which in turn gives

$$(T^*\phi)(x)=-\frac{C}{2^{\frac{1}{2}}}x^{-\frac{1}{4}}\frac{\mathrm{d}}{\mathrm{d}x}\int_x^1 \frac{\mathrm{d}t}{(t-x)^{\frac{1}{4}}}=\frac{C}{2^{\frac{1}{2}}}x^{-\frac{1}{4}}(1-x)^{-\frac{1}{4}} \tag{9.9}$$

for $0<x<1$.

At each of the three stages in the solution process so far we have obtained the unique solution of the respective equation in $L_2(0,1)$. Equation (9.9) is a convenient point to determine C, for it implies that

$$\int_x^1 \mathrm{d}s\int_s^1 (t-s)^{-\frac{1}{4}}\phi(t)\,\mathrm{d}t=\frac{C}{2^{\frac{1}{2}}}\int_x^1 t^{-\frac{1}{4}}(1-t)^{-\frac{1}{4}}\,\mathrm{d}t \quad (0\leqslant x\leqslant 1)$$

which reduces to

$$\int_x^1 (t-x)^{\frac{1}{4}}\phi(t)\,dt = \frac{C}{2(2)^{\frac{1}{4}}}\int_x^1 t^{-\frac{1}{4}}(1-t)^{-\frac{1}{4}}\,dt \quad (0\leqslant x\leqslant 1).$$

Using the fact that $U\phi = \phi$, it follows that

$$\int_0^x (x-t)^{\frac{1}{4}}\phi(t)\,dt = \frac{C}{2(2)^{\frac{1}{4}}}\int_0^x t^{-\frac{1}{4}}(1-t)^{-\frac{1}{4}}\,dt \quad (0\leqslant x\leqslant 1)$$

and hence that $(L\phi)(x) = CB(\frac{3}{4},\frac{3}{4})/(2(2)^{\frac{1}{4}})\,(0\leqslant x\leqslant 1)$. We must therefore choose $C = 2(2^{\frac{1}{4}})/B(\frac{3}{4},\frac{3}{4})$.

The function ϕ is most easily deduced from (9.9) by using the formula

$$\int_x^1 \frac{t^{-\frac{3}{4}}(1-t)^{-\frac{3}{4}}}{(t-x)^{\frac{1}{2}}}\,dt = B(\frac{1}{4},\frac{1}{2})x^{-\frac{1}{4}}(1-x)^{-\frac{1}{4}} \quad (0<x<1),$$

which can be verified by making the substitution $t = x(x+s-sx)^{-1}$. We see that $\phi(x) = C\{2^{\frac{1}{4}}B(\frac{1}{4},\frac{1}{2})\}^{-1}x^{-\frac{3}{4}}(1-x)^{-\frac{3}{4}} = x^{-\frac{3}{4}}(1-x)^{-\frac{3}{4}}/(\pi(2^{\frac{1}{4}}))\ (0<x<1)$, when the calculated value of C is used. This ϕ is not in $L_2(0,1)$, although the above calculations and Theorem B6 can be used to show that it does satisfy the given equation. We infer that $L\phi = f$ has no solution in $L_2(0,1)$, but that the given equation is satisfied by the function with values $\phi(x) = x^{-\frac{3}{4}}(1-x)^{-\frac{3}{4}}/(\pi(2^{\frac{1}{4}}))\ (0<x<1)$. $\square$

In the last section we used the term 'simplifying operator' to describe an operator which reduces an equation to a more amenable form. The procedure followed in the last four examples is closely related to the simplifying operator idea. For, if a positive operator K can be factorised in the form $K = TT^*$, where $T\psi = f$ and $T^*\phi = \psi$ can be solved, then $K\phi = f$ can be converted into the form $T^*\phi = \psi$, where ψ is known. There is an implied simplifying operator which reduces $K\phi = f$ to $T^*\phi = \psi$. Thus the factorisation of an operator K as TT^* (or, more generally, as ST^*) is equivalent to finding a simplifying operator for K, at least if $T\psi = f$ and $T^*\phi = \psi$ can be solved.

Our main application of the simplifying operator method is to integral equations having Cauchy singular kernels and in this case no factorisation of the associated operator is available. As we have not previously dealt with operators generated by Cauchy singular kernels, some preliminary analysis is needed and we tackle this next.

9.4 Fourier transforms and the Hilbert transform

In this section we shall extend the classical results about Fourier transforms to yield results about transforms of functions belonging to $L_2(-\infty, \infty)$. The immediate use of these results is on the Hilbert transform, but application can also be made to certain other types of singular integral equations, a topic we shall consider in §9.6.

The principal piece of analysis we shall require is one which allows us to deduce that various results which can be relatively easily established for a restricted class of ('sufficiently well-behaved') functions can be extended to the whole of a larger class. The technique is to establish that an operator is a bounded operator on a dense subspace (i.e. one whose closure is the whole space) of a Hilbert space and extend the operator to the whole space.

Lemma 9.1

Let H be a Hilbert space and E be a dense subspace of H, and suppose that $T: E \to H$ is a bounded linear map. Then T extends to a bounded linear map from H to itself and this extension is unique.

Proof

Suppose that for all $\phi \in E \parallel T\phi \parallel \leqslant M \parallel \phi \parallel$; we are given that such a constant exists. Now let $\phi \in H$. Since E is dense in H there is a sequence (ϕ_n) of elements of E for which $\parallel \phi_n - \phi \parallel \to 0$ as $n \to \infty$. Then (ϕ_n) is a Cauchy sequence so, because $\parallel T\phi_m - T\phi_n \parallel \leqslant M \parallel \phi_m - \phi_n \parallel$ $(m, n \in \mathbb{N})$, $(T\phi_n)$ is also a Cauchy sequence. H is complete, whence $T\phi_n$ converges to some limit in H; call it $S\phi$. To show that this process defines a function $S: H \to H$, we only need to show that $S\phi$ is independent of the choice of the particular sequence (ϕ_n). Let $\psi_n \to \phi$ as $n \to \infty$ (where $\psi_n \in E$ for each $n \in \mathbb{N}$). Then $\psi_n - \phi_n \to 0$ as $n \to \infty$, whence $\parallel T\psi_n - T\phi_n \parallel \leqslant M \parallel \psi_n - \phi_n \parallel \to 0$ showing that $(T\psi_n)$ tends to the same limit as $(T\phi_n)$.

The function S is a bounded linear map. To see this, choose $\phi, \psi \in H$ and λ a scalar. Then if $\phi_n \to \phi$ and $\psi_n \to \psi$ as $n \to \infty$ (with $\phi_n, \psi_n \in E$ $(n \in \mathbb{N})$), $\phi_n + \lambda \psi_n \to \phi + \lambda \psi$ as $n \to \infty$, so that $S(\phi + \lambda \psi) = \lim_{n \to \infty} T(\phi_n + \lambda \psi_n) = \lim_{n \to \infty} (T\phi_n + \lambda T\psi_n) = S\phi + \lambda S\psi$. That S is bounded follows since $\phi_n \to \phi \neq 0$ implies that for sufficiently large n, $\parallel \phi_n \parallel \leqslant 2 \parallel \phi \parallel$ whence $\parallel T\phi_n \parallel \leqslant 2M \parallel \phi \parallel$ and $\parallel S\phi \parallel \leqslant 2M \parallel \phi \parallel$, the last inequality being trivially true if $\phi = 0$.

Finally suppose that U is another extension of T. So S and U are bounded linear maps on H and $S\phi = U\phi$ ($\phi \in E$). Then for $\phi \in H$, choose $\phi_n \in E$ $(n \in \mathbb{N})$ with $\parallel \phi_n - \phi \parallel \to 0$. Then $S\phi_n \to S\phi$, $U\phi_n \to U\phi$ and $S\phi_n = U\phi_n$ whence $S\phi = U\phi$, establishing that $S = U$. ☐

The usefulness of this lemma depends, of course, on the identification of a subspace E satisfying the hypotheses. This turns out to be a simple matter in several useful cases.

Definition 9.1

Suppose that $\phi: \mathbb{R} \to \mathbb{C}$ is integrable and that $\int_{-\infty}^{\infty} |\phi(t)| \, dt < \infty$. We define the *Fourier transform* of ϕ, $\hat{\phi}$, by

$$\hat{\phi}(s) = \frac{1}{(2\pi)^{\frac{1}{2}}} \int_{-\infty}^{\infty} \phi(t) \, e^{ist} \, dt \quad (s \in \mathbb{R}). \qquad \square$$

The assumed properties of ϕ ensure the existence of $\hat{\phi}$ while the classical theory of Fourier transforms shows us that if ϕ and ϕ' are continuous and are zero outside some interval of finite length $\int_{-\infty}^{\infty} |\hat{\phi}(s)|^2 \, ds = \int_{-\infty}^{\infty} |\phi(t)|^2 \, dt$. If we now let E denote the subspace of $L_2(-\infty, \infty)$ consisting of functions with a continuous derivative which are zero outside some interval (depending on the function) of finite length, and set $F\phi = \hat{\phi}$ then $F: E \to L_2(-\infty, \infty)$ is a bounded linear map and can therefore be extended to a bounded linear map on $L_2(-\infty, \infty)$, since E is dense, by Lemma 9.1. It follows, in fact, that $\|F\phi\| = \|\phi\|$ ($\phi \in L_2(-\infty, \infty)$).

Using the boundedness of the operator F, and the functions ψ_n defined by $\psi_n(t) = \psi(t)$ ($|t| \leqslant n$), $\psi_n(t) = 0$ ($|t| > n$) we see that $\|\hat{\psi}_n - \hat{\psi}\| = \|\psi_n - \psi\| \to 0$ as $n \to \infty$ if $\psi \in L_2(-\infty, \infty)$ and therefore that

$$\int_{-n}^{n} \psi(t) \, e^{ist} \, dt \to (F\psi)(s) \qquad (9.10)$$

in the sense of L_2-convergence of the functions (of s), as $n \to \infty$.

Let E be, as before, the set of functions which have a continuous derivative and are zero outside some interval (depending on the function) of finite length. Then for $\phi \in E$ we may set

$$(F^*\phi)(s) = \frac{1}{(2\pi)^{\frac{1}{2}}} \int_{-\infty}^{\infty} \phi(t) \, e^{-ist} \, dt \quad (s \in \mathbb{R})$$

and trivial modifications to the above reasoning show that this operator can be extended to an operator $F^* \in B(L_2(-\infty, \infty))$ for which $\|F^*\phi\| = \|\phi\|$ ($\phi \in L_2(-\infty, \infty)$). Moreover, for all $\phi, \psi \in E$

$$(F\phi, \psi) = \frac{1}{(2\pi)^{\frac{1}{2}}} \int_{-\infty}^{\infty} \left\{ \int_{-\infty}^{\infty} \phi(t) \, e^{ist} \, dt \right\} \overline{\psi(s)} \, ds$$

$$= \frac{1}{(2\pi)^{\frac{1}{2}}} \int_{-\infty}^{\infty} \phi(t) \overline{\left\{ \int_{-\infty}^{\infty} \psi(s) \, e^{-ist} \, ds \right\}} \, dt = (\phi, F^*\psi).$$

From this we deduce the corresponding identity for $\phi, \psi \in L_2(-\infty, \infty)$: let $\phi, \psi \in L_2(-\infty, \infty)$ and choose $\phi_n, \psi_n \in E$ with $\|\phi_n - \phi\| \to 0$ and $\|\psi_n - \psi\| \to 0$ as $n \to \infty$. Then (since F and F^* are bounded)

$$(F\phi, \psi) = \lim_{n \to \infty} (F\phi_n, \psi_n) = \lim_{n \to \infty} (\phi_n, F^*\psi_n) = (\phi, F^*\psi).$$

This justifies the notation F^*, because it proves that F^* is indeed the adjoint of F.

The classical theory also shows that if ϕ has a continuous derivative and is zero outside some interval of finite length then

$$\lim_{n \to \infty} \frac{1}{2\pi} \int_{-n}^{n} e^{-ixs} \left\{ \int_{-\infty}^{\infty} \phi(t) e^{ist} \, dt \right\} ds = \phi(x) \quad (x \in \mathbb{R}).$$

Setting $\psi = F\phi$, and $\psi_n(s) = \psi(s)$ $(|s| < n)$, $\psi_n(s) = 0$ $(|s| > n)$ we see that $\psi \in L_2(-\infty, \infty)$ whence, by the remarks above and (9.10), $F^*\psi_n \to F^*\psi$ (in the L_2 norm) as $n \to \infty$ so that $F^*\psi = \phi$, and $F^*F\phi = \phi$ (as elements of $L_2(-\infty, \infty)$). Similar reasoning proves that $FF^*\phi = \phi$ $(\phi \in L_2(-\infty, \infty))$ whence we see that $FF^* = F^*F = I$ so that F is a unitary operator on $L_2(-\infty, \infty)$. In particular, notice that the operator F is not compact.

Lemma 9.2

There is a bounded operator F on $L_2(-\infty, \infty)$, called the Fourier transform, which is defined by

$$(F\phi)(s) = \frac{1}{(2\pi)^{\frac{1}{2}}} \int_{-\infty}^{\infty} e^{ist} \phi(t) \, dt \quad (s \in \mathbb{R})$$

for all $\phi \in L_2(-\infty, \infty)$ for which the integral exists (and which extends to all of $L_2(-\infty, \infty)$ using Lemma 9.1). The adjoint operator F^* is defined by

$$(F^*\phi)(s) = \frac{1}{(2\pi)^{\frac{1}{2}}} \int_{-\infty}^{\infty} e^{-ist} \phi(t) \, dt \quad (s \in \mathbb{R})$$

(when the integral exists) and $FF^* = F^*F = I$. In particular, $\|F\phi\| = \|\phi\|$ $(\phi \in L_2(-\infty, \infty))$. $\qquad \square$

The purpose for which we shall first use the Fourier transform is to discuss the integral operator which maps ϕ into $\int_0^1 \phi(t)(t-x)^{-1} \, dt$. This operator, as well as others related to it, is significant in application areas but we shall need care in bringing it within the scope of the theory we have established and, indeed, we need care in showing that the integral has a meaning at all.

Suppose that $\phi: [a, b] \to \mathbb{C}$ is continuous, that ϕ' exists and is continuous on (a, b) and that $a < x < b$. Then the integral $\int_a^b \phi(t)(t-x)^{-1} \, dt$ exists in the sense of a *principal value*, that is, $\lim_{\epsilon \to 0}\{\int_a^{x-\epsilon}\phi(t)(t-x)^{-1}\,dt+\int_{x+\epsilon}^b \phi(t)$ $(t-x)^{-1}\,dt\}$ exists; it is easily checked that the value of the limit and thus of the principal value integral is

$$\int_a^b \frac{\phi(t)\,dt}{t-x}=\phi(x)\log\left(\frac{b-x}{x-a}\right)+\int_a^b \frac{\phi(t)-\phi(x)}{t-x}\,dt \quad (a<x<b),$$

where the second integral is well-defined because its integrand is a continuous function of t if it is allocated the value $\phi'(x)$ at $t=x$.

Retaining our assumption that ϕ' is continuous, we may set

$$\psi_\epsilon(x)=\int_a^{x-\epsilon}\frac{\phi(t)}{t-x}\,dt+\int_{x+\epsilon}^b \frac{\phi(t)}{t-x}\,dt \quad (a<x<b)$$

(with the obvious modification if $x-\epsilon<a$ or $x+\epsilon>b$), and set $\psi(x)=\int_a^b \phi(t)(t-x)^{-1}\,dt$, giving

$$\psi(x)-\psi_\epsilon(x)=\begin{cases}\displaystyle\int_{x-\epsilon}^{x+\epsilon}\frac{\phi(t)-\phi(x)}{t-x}\,dt & (a+\epsilon\leqslant x\leqslant b-\epsilon),\\[2mm]\displaystyle\phi(x)\log\left(\frac{\epsilon}{x-a}\right)+\int_a^{x+\epsilon}\frac{\phi(t)-\phi(x)}{t-x}\,dt & (a<x<a+\epsilon).\end{cases}$$

Observing that $\int_a^{a+\epsilon}\log(\epsilon/(x-a))\,dx=\epsilon$, we now see that for $a<\xi<b$,

$$\int_a^\xi \psi_\epsilon(x)\,dx\to\int_a^\xi \psi(x)\,dx \quad \text{as } \epsilon\to 0+. \tag{9.11}$$

Now for $0<\epsilon<\min(\xi-a,b-\xi)$ reversal of the order of integration below is legitimate so

$$\int_a^\xi \psi_\epsilon(x)\,dx=\int_a^\xi dx\left\{\int_a^{x-\epsilon}\frac{\phi(t)}{t-x}\,dt+\int_{x+\epsilon}^b \frac{\phi(t)}{t-x}\,dt\right\}$$

$$=\int_a^b \phi(t)\log\left|\frac{t-a}{\xi-t}\right|\,dt-\int_{\xi-\epsilon}^{\xi+\epsilon}\phi(t)\log\left|\frac{\epsilon}{\xi-t}\right|\,dt$$

$$-\int_a^{a+\epsilon}\phi(t)\log\frac{t-a}{\epsilon}\,dt$$

$$\to\int_a^b \phi(t)\log\left|\frac{t-a}{\xi-t}\right|\,dt \quad \text{as } \epsilon\to 0+.$$

Therefore, for $a<\xi<b$,

$$\int_a^\xi dx\int_a^b \frac{\phi(t)}{t-x}\,dt=-\int_a^b \phi(t)\log\left|1-\frac{\xi-a}{t-a}\right|\,dt, \tag{9.12}$$

provided ϕ has a continuous derivative.

Consider now the operator from $L_2(-\infty, \infty)$ to itself given by F^*MF where F and F^* denote, respectively, the Fourier transform and its inverse and M is defined by $(M\phi)(x) = -\operatorname{isgn}(x)\phi(x)$. Since each of the operators F^*, M and F has norm at most one, the same is true of F^*MF. Let $\phi: \mathbb{R} \to \mathbb{R}$ be a twice differentiable function, with ϕ'' continuous, which is zero outside some interval of finite length, so that $\hat{\phi} = F\phi$ is continuous and $\int_{-\infty}^{\infty}|\hat{\phi}| < \infty$. (To see this last point, notice that, by integrating by parts, we have

$$(2\pi)^{\frac{1}{2}}\,\hat{\phi}(x) = -\frac{1}{x^2}\int_{-\infty}^{\infty} \phi''(t)\,e^{ixt}\,dt \quad (x \neq 0).)$$

Now if $\psi = F^*MF\phi$, $\hat{\psi} = F\psi = MF\phi = M\hat{\phi}$ so that we have, for almost all $s \in \mathbb{R}$, $\hat{\psi}(s) = -\operatorname{isgn}(s)\hat{\phi}(s)$. Then, for $\xi \in \mathbb{R}$

$$\int_0^{\xi} \psi(x)\,dx = \int_0^{\xi} dx \frac{1}{(2\pi)^{\frac{1}{2}}}\int_{-\infty}^{\infty}\hat{\psi}(s)\,e^{-isx}\,ds$$

$$= \frac{1}{(2\pi)^{\frac{1}{2}}}\int_{-\infty}^{\infty}\hat{\psi}(s)\frac{e^{-is\xi}-1}{-is}\,ds$$

$$= \int_{-\infty}^{\infty}\hat{\phi}(s)\,\overline{\theta(s)}\,ds \qquad (9.13)$$

where $\theta(s) = (e^{is\xi}-1)/(|s|(2\pi)^{\frac{1}{2}})(s \neq 0)$. (The reversal of the order of integration above is justified by the knowledge that $\int_{-\infty}^{\infty}|\hat{\psi}| < \infty$.) The right hand side of (9.13) is of the form $(F\phi, \theta)$, which equals $(\phi, F^*\theta)$. Now

$$(F^*\theta)(x) = \frac{1}{2\pi}\int_{-\infty}^{\infty}\frac{e^{is\xi}-1}{|s|}e^{-ixs}\,ds$$

$$= \frac{1}{\pi}\int_0^{\infty}\frac{\cos(x-\xi)s - \cos xs}{s}\,ds$$

$$= \frac{1}{\pi}\lim_{a \to 0+}\left\{\int_a^{\infty}\frac{\cos(x-\xi)s}{s}\,ds - \int_a^{\infty}\frac{\cos xs}{s}\,ds\right\}$$

$$= \frac{1}{\pi}\lim_{a \to 0+}\left\{\int_{|x-\xi|a}^{|x|a}\frac{1}{u}\,du + \int_{|x-\xi|a}^{|x|a}\frac{\cos u - 1}{u}\,du\right\}$$

$$= \frac{1}{\pi}\log\left|\frac{x}{x-\xi}\right| \quad (x \neq \xi).$$

Then, from (9.13)

$$\int_0^{\xi}\psi(x)\,dx = (F\phi, \theta) = (\phi, F^*\theta) = \frac{1}{\pi}\int_{-\infty}^{\infty}\phi(x)\log\left|\frac{x}{x-\xi}\right|\,dx.$$

Because ϕ is zero outside some interval of finite length, (9.12) now shows us that

$$\int_0^\xi \psi(x)\,dx = \frac{1}{\pi}\int_0^\xi dx \int_{-\infty}^\infty \frac{\phi(t)}{t-x}\,dt. \tag{9.14}$$

Now (9.14) is true for all $\xi \in \mathbb{R}$ whence for almost all $x \in \mathbb{R}$

$$\psi(x) = \frac{1}{\pi}\int_{-\infty}^\infty \frac{\phi(t)}{t-x}\,dt. \tag{9.15}$$

(Under our assumptions about ϕ, it is not difficult to show that both sides of (9.15) depend continuously on x and are thus equal for all x.) We can summarise these results as follows.

Lemma 9.3

If $\phi: \mathbb{R} \to \mathbb{R}$ is twice differentiable and is zero outside some interval of finite length then, for almost all $x \in \mathbb{R}$,

$$\int_{-\infty}^\infty \frac{\phi(t)}{t-x}\,dt = \pi(F^*MF\phi)(x)$$

where F denotes the Fourier transform and M the operator defined by $(M\phi)(x) = -i\,\mathrm{sgn}(x)\phi(x)$ $(x \in \mathbb{R})$. $\qquad\square$

If we now set $(H_0\phi)(x) = \int_{-\infty}^\infty \phi(t)(t-x)^{-1}\,dt$, then we know that H_0 is well-defined on the subspace of $L_2(-\infty, \infty)$ consisting of twice continuously differentiable functions which are zero outside some interval of finite length (depending on the function) and that for such a function ϕ, $\|H_0\phi\| = \pi\|\phi\|$. H_0 therefore extends by Lemma 9.1 to the operator πF^*MF on $L_2(-\infty, \infty)$. We shall interpret $\int_{-\infty}^\infty \phi(t)(t-x)^{-1}\,dx$ to mean the function $H_0\phi \in L_2(-\infty, \infty)$ whenever $\phi \in L_2(-\infty, \infty)$; if ϕ is not sufficiently smooth the principal value integral may not exist in the normal sense. The operator H_0 is easily seen to be invertible and skew-symmetric, since $M^* = -M = M^{-1}$, and the spectrum of H_0 consists of the two points $\pm i\pi$, both eigenvalues.

In fact, noticing that $M^2 = -I$, we see immediately that $H_0^2 = -\pi^2 I$, which we can use to find a simple expression for the solution of the equation $\mu\phi - H_0\phi = f$. Since $H_0^2 = -\pi^2 I$, $(\mu I - H_0)(\mu I + H_0) = (\mu^2 + \pi^2)I$ and so $(\mu I - H_0)^{-1} = (\mu^2 + \pi^2)^{-1}(\mu I + H_0)$ (for $\mu \neq \pm i\pi$) whence the unique solution of the equation $\mu\phi - H_0\phi = f$ is $(\mu^2 + \pi^2)^{-1}(\mu f + H_0 f)$ for $\mu \neq \pm i\pi$. In the case where $\mu = \pm i\pi$ the operator $\mu I - H_0$ is singular (in more than one sense of the word!) and additional conditions must be imposed on f for a solution ϕ to exist and when a solution does exist it will not be unique. (See Problem 9.8.)

In this chapter our main interest will not be H_0 but the analogous operator on $L_2(0, 1)$. This may be derived from H_0 by letting $J: L_2(0, 1) \to L_2(-\infty, \infty)$ be the operator given by

$$(J\phi)(x) = \begin{cases} \phi(x) & (x \in (0, 1)), \\ 0 & (x \notin (0, 1)). \end{cases}$$

It is easily checked that $\|J\phi\| = \|\phi\|$ and we now define the operator H by setting $H = J^*H_0 J$ so that H is a bounded linear map from $L_2(0, 1)$ to itself, $\|H\| \leqslant \pi$ and $H^* = -H$. Provided ϕ is sufficiently well-behaved that the principal value integral has a meaning $(H\phi)(x)$ will have the value $\int_0^1 \phi(t)(t-x)^{-1} \, dt$ for almost all $x \in (0, 1)$. We shall call H the *Hilbert transform* and summarise its properties below.

Lemma 9.4

Let H be the operator from $L_2(0, 1)$ to itself defined by $H = J^*F^*MFJ$ where $J: L_2(0, 1) \to L_2(-\infty, \infty)$ is defined by $(J\phi)(x) = \phi(x)$ $(x \in (0, 1))$ and $(J\phi)(x) = 0$ $(x \notin (0, 1))$, F is the Fourier transform and $(M\phi)(x) = -i\pi\operatorname{sgn}(x)\phi(x)$. Then if $\phi: (0, 1) \to \mathbb{R}$ is twice differentiable and vanishes outside some interval $[a, b]$ with $0 < a < b < 1$,

$$(H\phi)(x) = \int_0^1 \frac{\phi(t) \, dt}{t - x},$$

for almost all $x \in (0, 1)$. The operator H is bounded, $\|H\| \leqslant \pi$ and $H^* = -H$. ☐

Example 9.5

For each non-zero real constant μ and each $f \in L_2(0, 1)$ the integral equation

$$\mu\phi(x) = \int_0^1 \frac{\phi(t) \, dt}{t - x} + f(x) \quad (0 < x < 1)$$

possesses a unique solution $\phi \in L_2(0, 1)$; the integral is to be interpreted as $H\phi$ in the case where ϕ is not sufficiently smooth that it exists in the usual sense.

The integral equation may be reformulated as the operator equation $\mu\phi = H\phi + f$ in $L_2(0, 1)$. Since $H^* = -H$, we see that iH is self-adjoint (treating $L_2(0, 1)$ as the space of complex-valued functions) and therefore $\sigma(H) \subseteq \{i\alpha : \alpha \in \mathbb{R}\}$. From this it is obvious that if μ is real and non-zero $\mu \notin \sigma(H)$ so $\mu I - H$ is invertible. We can even estimate $\|(\mu I - H)^{-1}\|$ simply, for

$$|\mu|\,\|\phi\|^2 = |\text{Re}((\mu I - H)\phi, \phi)| \leqslant |((\mu I - H)\phi, \phi)| \leqslant \|(\mu I - H)\phi\|\,\|\phi\|$$

for all $\phi \in H$, whence $\|(\mu I - H)^{-1}\| \leqslant |\mu|^{-1}$. $\qquad\qquad\square$

9.5 Equations with Cauchy singular kernels

We are now in a position to solve the prototype integral equation with a Cauchy singular kernel, namely

$$\mu\phi(x) = f(x) + \int_0^1 \frac{\phi(t)\,dt}{t - x} \quad (0 < x < 1). \tag{9.16}$$

In the last section we showed that the Hilbert transform H, defined by $(H\phi)(x) = \int_0^1 (t - x)^{-1}\phi(t)\,dt$, is a bounded operator on $L_2(0, 1)$ and we can therefore consider (9.16) via the equation $\mu\phi = f + H\phi$ in $L_2(0, 1)$.

9.5.1 The second kind equation

Central to our method of solution is $\phi_\gamma \in L_2(0, 1)$ defined by

$$\phi_\gamma(x) = x^{-\gamma}(1 - x)^\gamma \quad (0 < x < 1) \tag{9.17}$$

where $0 < |\gamma| < \frac{1}{2}$, and we need to evaluate $H\phi_\gamma$. Let $0 < \gamma < \frac{1}{2}$ and $0 < x < 1$ and note that

$$(H\phi_\gamma)(x) = (1 - x)\int_0^1 \frac{(1 - t)^{\gamma - 1}\,dt}{t^\gamma(t - x)} - \int_0^1 (1 - t)^{\gamma - 1}t^{-\gamma}\,dt$$

$$= x^{-\gamma}(1 - x)^\gamma \int_0^\infty \frac{u^{\gamma - 1}}{1 - u}\,du - B(\gamma, 1 - \gamma),$$

on setting $t = x(x + u - xu)^{-1}$ in the first integral. Now $\int_0^\infty u^{\gamma - 1}(1 - u)^{-1}\,du = \pi \cot(\pi\gamma)$ for $\gamma \in (0, \frac{1}{2})$ (see Appendix C) and therefore

$$(H\phi_\gamma)(x) = \pi \cot(\pi\gamma)\phi_\gamma(x) - \pi \operatorname{cosec}(\pi\gamma) \quad (0 < x < 1, 0 < \gamma < \tfrac{1}{2}).$$

Now let U be the unitary operator defined on $L_2(0, 1)$ by $(U\phi)(x) = \phi(1 - x)$ and notice that $U\phi_\gamma = \phi_{-\gamma}$. Also $UHU = -H = H^*$ and therefore $HU = -UH$. Suppose that $-\frac{1}{2} < \gamma < 0$ and that $0 < x < 1$, then

$$(UH\phi_\gamma)(x) = -(HU\phi_\gamma)(x) = -(H\phi_{-\gamma})(x) = \pi \cot(\pi\gamma)\phi_{-\gamma}(x) - \pi \operatorname{cosec}(\pi\gamma),$$

by our previous working. Therefore

$$(H\phi_\gamma)(x) = \pi \cot(\pi\gamma)(U\phi_{-\gamma})(x) - \pi \operatorname{cosec}(\pi\gamma)$$

$$= \pi \cot(\pi\gamma)\phi_\gamma(x) - \pi \operatorname{cosec}(\pi\gamma) \quad (0 < x < 1, -\tfrac{1}{2} < \gamma < 0),$$

and we conclude that

$$(H\phi_\gamma)(x) = \pi \cot(\pi\gamma)\phi_\gamma(x) - \pi \operatorname{cosec}(\pi\gamma) \quad (0 < x < 1, 0 < |\gamma| < \tfrac{1}{2}). \quad (9.18)$$

We can immediately solve a particular example of $\mu\phi = f + H\phi$, that in which $f(x) = 1$ $(0 \leqslant x \leqslant 1)$. For suppose that μ is a given, non-zero real number. Then there is a unique value γ_0 such that $\pi \cot(\pi\gamma_0) = \mu$ and $0 < |\gamma_0| < \tfrac{1}{2}$ and from (9.18) we deduce that $\mu\phi = f + H\phi$, where $f = 1$, is satisfied by $\phi = \pi^{-1} \sin(\pi\gamma_0)\phi_{\gamma_0} \in L_2(0, 1)$.

The identity (9.18) actually plays an important part in the derivation of a solution of $\mu\phi = f + H\phi$ for more general f, when used in conjunction with a 'simplifying operator', and the operator U is also a useful tool when dealing with the full equation. The next lemma and the succeeding theorem together establish the central result in our solution method.

Lemma 9.5

Let $0 < \gamma < \tfrac{1}{2}$ and

$$k(x, t) = \frac{\partial}{\partial t} \int_0^x \frac{\log|t - s|}{s^\gamma(x - s)^{1-\gamma}} \, ds \quad (0 < x, t < 1, x \neq t).$$

Then

$$k(x, t) = \begin{cases} \pi \operatorname{cosec}(\pi\gamma) \ t^{-\gamma}(t - x)^{\gamma - 1} & (t > x) \\ -\pi \cot(\pi\gamma) \ t^{-\gamma}(x - t)^{\gamma - 1} & (x > t). \end{cases}$$

Proof

The case $t > x$ presents no problem. It is easy to check that differentiation under the integral sign is permissible and therefore that

$$k(x, t) = \int_0^x \frac{ds}{s^\gamma(x - s)^{1-\gamma}(t - s)}$$

$$= t^{-\gamma}(t - x)^{\gamma - 1} \int_0^1 u^{-\gamma}(1 - u)^{\gamma - 1} \, du$$

$$= \pi \operatorname{cosec}(\pi\gamma) t^{-\gamma}(t - x)^{\gamma - 1} \quad (0 < x < t < 1),$$

on substituting $s = uxt(ux + t - x)^{-1}$.

For $x > t$ the evaluation of k is more awkward. Differentiation under the integral sign results in a Cauchy principal value integral and we have not derived conditions under which this process is valid. We therefore employ a less direct approach which makes use of results which are already available. Let

$$\ell(x, t) = \int_0^x \frac{\log|1 - t/s|}{s^\gamma (x - s)^{1 - \gamma}} \, ds \quad (0 < t < x < 1)$$

and note that, since the integration order may be interchanged,

$$\int_0^x \ell(u, t) \, du = \frac{1}{\gamma} \int_0^x \frac{(x - s)^\gamma}{s^\gamma} \log|1 - t/s| \, ds$$

$$= \frac{x}{\gamma} \int_0^1 \phi_\gamma(u) \log|1 - \xi/u| \, du,$$

where the variable change $s = xu$ has been used to give the second version of the right hand side, we have put $\xi = t/x < 1$, and the notation (9.17) has been employed. By (9.12) it follows that

$$\int_0^x \ell(u, t) \, du = -\frac{x}{\gamma} \int_0^\xi (H\phi_\gamma)(u) \, du$$

$$= -\frac{x}{\gamma} \int_0^\xi \{\pi \cot(\pi\gamma)\phi_\gamma(u) - \pi \operatorname{cosec}(\pi\gamma)\} \, du$$

$$= -\frac{1}{\gamma} \int_0^t \{\pi \cot(\pi\gamma)s^{-\gamma}(x - s)^\gamma - \pi \operatorname{cosec}(\pi\gamma)\} \, ds \quad (t < x < 1),$$

using (9.18) and reverting to the integration variable $s = xu$. Therefore

$$\ell(x, t) = -\pi \cot(\pi\gamma) \int_0^t s^{-\gamma}(x - s)^{\gamma - 1} \, ds.$$

Finally, note that

$$k(x, t) = \frac{\partial}{\partial t} \ell(x, t) = -\pi \cot(\pi\gamma) t^{-\gamma}(x - t)^{\gamma - 1} \quad (0 < t < x < 1). \qquad \square$$

Theorem 9.6

Let the operators H, T and S on $L_2(0, 1)$ be defined by

$$(H\phi)(x) = \int_0^1 \frac{\phi(t) \, dt}{t - x}, \quad (T\phi)(x) = \int_0^x \frac{\phi(t) \, dt}{t^\gamma (x - t)^{1 - \gamma}},$$

$$(S\phi)(x) = \int_0^x \frac{\phi(t) \, dt}{x^\gamma (x - t)^{1 - \gamma}},$$

where $0 < \gamma < \frac{1}{2}$. Then

$$TH = -\pi \cot(\pi\gamma) T + \pi \operatorname{cosec}(\pi\gamma) S^*.$$

Proof

First note that H, T and S are bounded operators on $L_2(0, 1)$. This follows from the last section in the case of H and from Lemma 6.7 for T and S.

Let E denote the subspace of $L_2(0, 1)$ consisting of functions which are differentiable, have a continuous derivative and vanish at the ends of the interval $[0, 1]$. Let $\phi \in E$. Then, integrating by parts, we have, for $\epsilon > 0$,

$$\int_0^{x-\epsilon} \frac{\phi(t)\,dt}{t-x} + \int_{x+\epsilon}^1 \frac{\phi(t)\,dt}{t-x} =$$

$$\{\phi(x-\epsilon) - \phi(x+\epsilon)\} \log \epsilon - \int_0^{x-\epsilon} \log(x-t)\phi'(t)\,dt - \int_{x+\epsilon}^1 \log(t-x)\phi'(t)\,dt.$$

From the definition of a Cauchy principal value and the assumed continuity of ϕ it follows that $H\phi = -L\phi'$ for $\phi \in E$, where $(L\phi)(x) = \int_0^1 \log|x-t|\phi(t)\,dt$ $(0 \leqslant x \leqslant 1)$. Thus $TH\phi = -TL\phi'$ for $\phi \in E$ and so

$$(TH\phi)(x) = -\int_0^x \frac{ds}{s^\gamma(x-s)^{1-\gamma}} \int_0^1 \log|t-s|\phi'(t)\,dt$$

$$= -\int_0^1 \phi'(t)\,dt \int_0^x \frac{\log|t-s|}{s^\gamma(x-s)^{1-\gamma}}\,ds$$

for $\phi \in E$. It is easy to check that the reversal of integration order carried out here is valid. Further, the function defined by the inner integral in the last expression is a continuous function of t in $[0, 1]$ for each $x \in [0, 1]$. Therefore writing the integral over t as the sum of the integral from 0 to x and an integral from x to 1, integrating by parts and invoking the continuity of ϕ again, we obtain

$$(TH\phi)(x) = \int_0^x \phi(t) \frac{\partial}{\partial t} \int_0^x \frac{\log|t-s|}{s^\gamma(x-s)^{1-\gamma}}\,ds\,dt$$

$$+ \int_x^1 \phi(t) \frac{\partial}{\partial t} \int_0^x \frac{\log|t-s|}{s^\gamma(x-s)^{1-\gamma}}\,ds\,dt$$

$$= -\pi \cot(\pi\gamma)(T\phi)(x) + \pi \operatorname{cosec}(\pi\gamma)(S^*\phi)(x),$$

for $\phi \in E$, on using the result in Lemma 9.5.

Finally, since E is a dense subspace of $L_2(0, 1)$, Lemma 9.1 may be applied, and the stated result is proved. $\qquad\square$

Remark

At first glance the proof of the theorem may seem to be rather circuitous, particularly when the supporting lemma is taken into account. An alternative approach would be to show that reversal of the integration order is permissible in $TH\phi$. This tactic, which certainly appears more direct, is actually a good deal more laborious to implement than the one we have used. A proof of the validity of the interchange of integration order in $TH\phi$ is less than straightforward, even for functions ϕ belonging to a carefully constructed subspace of $L_2(0, 1)$. Some idea of the complication which arises may be surmised by referring to the derivation of the relatively simple result (9.12).

The role of T as a 'simplifying operator' for $\mu\phi = f + H\phi$ becomes clear on noting that

$$T(\mu I - H) = (\mu + \pi \cot(\pi\gamma))T - \pi \csc(\pi\gamma)S^* \quad (0 < \gamma < \tfrac{1}{2}), \quad (9.19)$$

by Theorem 9.6.

Suppose that there is a $\phi \in L_2(0, 1)$ which satisfies $\mu\phi = f + H\phi$ where μ is given and $\mu < 0$. Choosing $\gamma \in (0, \tfrac{1}{2})$ such that $\mu = -\pi \cot(\pi\gamma)$ we see from (9.19) that ϕ satisfies $-\pi \csc(\pi\gamma)S^*\phi = Tf$ and, by the technique developed in §9.2, that it therefore satisfies

$$-\pi^2 \csc^2(\pi\gamma) \int_x^1 t^{-\gamma}\phi(t)\, dt = \int_x^1 \frac{(Tf)(t)\, dt}{(t-x)^\gamma}. \quad (9.20)$$

Since $\phi \in L_2(0, 1)$ by hypothesis, both sides of (9.20) are differentiable almost everywhere in $[0, 1]$ and ϕ is given by

$$\phi(x) = x^\gamma \frac{\sin^2(\pi\gamma)}{\pi^2} \frac{d}{dx} \int_x^1 \frac{ds}{(s-x)^\gamma} \int_0^s \frac{t^{-\gamma}f(t)\, dt}{(s-t)^{1-\gamma}} \quad (9.21)$$

for almost all $x \in [0, 1]$.

We showed in Example 9.5 that H is a skew-adjoint operator on $L_2(0, 1)$ and therefore its spectrum is purely imaginary. Hence for all non-zero real μ the operator $\mu I - H$ is invertible and for every $f \in L_2(0, 1)$ the equation $\mu\phi = f + H\phi$ has the unique solution $\phi = (\mu I - H)^{-1}f$ in $L_2(0, 1)$. It follows that for all $\mu < 0$ and $f \in L_2(0, 1)$ the function ϕ given by (9.21) is the unique solution of $\mu\phi = f + H\phi$ in $L_2(0, 1)$.

It is easy to deduce the unique solution of $\mu\phi = f + H\phi$ in $L_2(0, 1)$ for $\mu > 0$, by bringing into play again the operator U where $(U\phi)(x) = \phi(1 - x)$. We have already noted that $UHU = -H$ and hence that $HU = -UH$, since

$U^2 = I$. Let $\mu\phi = f + H\phi$ where $\mu > 0$. Then $\mu U\phi = Uf + UH\phi$, so that

$$(-\mu)U\phi = (-Uf) + HU\phi,$$

and for each $f \in L_2(0, 1)$ there is a unique solution $U\phi \in L_2(0,1)$ given by (9.21). Consequently

$$(U\phi)(x) = -x^\gamma \frac{\sin^2(\pi\gamma)}{\pi^2} \frac{d}{dx} \int_x^1 \frac{ds}{(s-x)^\gamma} \int_0^s \frac{t^{-\gamma}(Uf)(t)}{(s-t)^{1-\gamma}} dt$$

almost everywhere in $[0, 1]$ where $-\mu = -\pi\cot(\pi\gamma)$ and $0 < \gamma < \frac{1}{2}$. Changing the variables in an obvious way we deduce that the unique solution of $\mu\phi = f + H\phi$ in $L_2(0, 1)$, with $\mu > 0$, is given by

$$\phi(x) = (1-x)^\gamma \frac{\sin^2(\pi\gamma)}{\pi^2} \frac{d}{dx} \int_0^x \frac{ds}{(x-s)^\gamma} \int_s^1 \frac{(1-t)^{-\gamma}f(t)}{(t-s)^{1-\gamma}} dt$$

for almost all $x \in [0, 1]$, where $\mu = \pi\cot(\pi\gamma)$ and $0 < \gamma < \frac{1}{2}$.

We can summarise our results as follows.

Theorem 9.7

Consider the equation $\mu\phi = f + H\phi$, where

$$(H\phi)(x) = \int_0^1 \frac{\phi(t)\,dt}{t-x} \quad (0 < x < 1),$$

and where μ is real and non-zero. Let γ be defined by $|\mu| = \pi\cot(\pi\gamma)$ and $0 < \gamma < \frac{1}{2}$. Then for each $f \in L_2(0, 1)$ there is a unique solution $\phi \in L_2(0, 1)$ given almost everywhere in $[0, 1]$ by

$$\phi(x) = x^\gamma \frac{\sin^2(\pi\gamma)}{\pi^2} \frac{d}{dx} \int_x^1 \frac{ds}{(s-x)^\gamma} \int_0^s \frac{t^{-\gamma}f(t)}{(s-t)^{1-\gamma}} dt \quad (\mu < 0),$$

$$\phi(x) = (1-x)^\gamma \frac{\sin^2(\pi\gamma)}{\pi^2} \frac{d}{dx} \int_0^x \frac{ds}{(x-s)^\gamma} \int_s^1 \frac{(1-t)^{-\gamma}f(t)}{(t-s)^{1-\gamma}} dt \quad (\mu > 0). \quad \square$$

Example 9.6

Solve $\mu\phi(x) = x + \int_0^1 \phi(t)(t-x)^{-1}\,dt$ $(0 < x < 1)$ where $\mu < 0$.

Let $\gamma \in (0, \frac{1}{2})$ satisfy $\mu = -\pi\cot(\pi\gamma)$ and let $f(x) = x$ $(0 \leqslant x \leqslant 1)$. Now

$$\int_0^s \frac{t^{-\gamma}f(t)}{(s-t)^{1-\gamma}} dt = s \int_0^1 u^{1-\gamma}(1-u)^{\gamma-1}\,du = sB(2-\gamma, \gamma)$$

on setting $t=su$. Since $B(2-\gamma,\gamma)=(1-\gamma)\pi\,\mathrm{cosec}(\pi\gamma)$ we therefore have

$$\frac{d}{dx}\int_x^1\frac{ds}{(s-x)^\gamma}\int_0^s\frac{t^{-\gamma}f(t)}{(s-t)^{1-\gamma}}\,dt=(1-\gamma)\pi\,\mathrm{cosec}(\pi\gamma)\frac{d}{dx}\int_x^1\frac{s\,ds}{(s-x)^\gamma}$$

$$=(1-\gamma)\pi\,\mathrm{cosec}(\pi\gamma)\frac{d}{dx}\int_0^{1-x}(x+u)u^{-\gamma}\,du$$

$$=\pi\,\mathrm{cosec}(\pi\gamma)(1-x)^{-\gamma}(\gamma-x).$$

According to Theorem 9.7 then, the unique solution of the given equation in $L_2(0,1)$ is given by $\phi(x)=\pi^{-1}\sin(\pi\gamma)x^\gamma(1-x)^{-\gamma}(\gamma-x)$ $(0<x<1)$.   □

Example 9.7

Solve $\mu\phi(x)=x+\int_0^1\phi(t)(t-x)^{-1}\,dt$ $(0<x<1)$ where $\mu>0$.
  Let γ satisfy $\mu=\pi\cot(\pi\gamma)$ and $0<\gamma<\frac{1}{2}$, and let $f(x)=x$ $(0\leqslant x\leqslant1)$.
Changing the integration variable by means of $t=1-(1-s)u$ we have

$$\int_s^1\frac{(1-t)^{-\gamma}f(t)}{(t-s)^{1-\gamma}}\,dt=\int_0^1\{1-(1-s)u\}u^{-\gamma}(1-u)^{\gamma-1}\,du$$

$$=B(\gamma,1-\gamma)-(1-s)B(\gamma,2-\gamma)$$

$$=\pi\,\mathrm{cosec}(\pi\gamma)\{1-(1-s)(1-\gamma)\}.$$

Therefore

$$\frac{d}{dx}\int_0^x\frac{ds}{(x-s)^\gamma}\int_s^1\frac{(1-t)^{-\gamma}f(t)}{(t-s)^{1-\gamma}}\,dt=\pi\,\mathrm{cosec}(\pi\gamma)\frac{d}{dx}\int_0^x\frac{\{1-(1-s)(1-\gamma)\}}{(x-s)^\gamma}\,ds$$

$$=\pi\,\mathrm{cosec}(\pi\gamma)\frac{d}{dx}\int_0^x\{\gamma+(1-\gamma)(x-u)\}u^{-\gamma}\,du$$

$$=\pi\,\mathrm{cosec}(\pi\gamma)x^{-\gamma}(\gamma+x).$$

By Theorem 9.7 the given equation has a unique solution in $L_2(0,1)$ given by

$$\phi(x)=\pi^{-1}\sin(\pi\gamma)x^{-\gamma}(1-x)^\gamma(\gamma+x)\quad(0<x<1).$$ □

Remark

The solution in this last example could of course have been deduced from the solution of Example 9.6 by means of the operator U which we used earlier.

It is possible to widen the class of solutions of $\mu\phi=f+H\phi$ without departing from our strategy of working in $L_2(0,1)$, because of the particular structure of the operator H. If we put $\psi(x) = x\phi(x)$, elementary manipulation gives

$$x(H\phi)(x) = (H\psi)(x) - C \tag{9.22}$$

where $C = \int_0^1 \phi(t) \, dt$.

Suppose now that ϕ satisfies the homogeneous equation

$$\mu\phi(x) = \int_0^1 \frac{\phi(t) \, dt}{t - x} \quad (0 < x < 1). \tag{9.23}$$

We know of course by Theorem 9.7 that if $\mu \neq 0$ the associated equation $\mu\phi = H\phi$ in $L_2(0, 1)$ has only the trivial solution. However (9.22) shows that the function ψ, where $\psi(x) = x\phi(x)$, satisfies $\mu\psi = H\psi - C$, which does have a solution in $L_2(0, 1)$. Moreover we can deduce this solution directly using (9.18), which shows that if $\mu < 0$ and if $\gamma \in (0, \frac{1}{2})$ satisfies $\mu = -\pi \cot(\pi\gamma)$ then $\psi = C\pi^{-1} \sin(\pi\gamma)\phi_{-\gamma}$ satisfies $\mu\psi = H\psi - C$. The corresponding function ϕ is given by $\phi(x) = C\pi^{-1} \sin(\pi\gamma)x^{\gamma-1}(1-x)^{-\gamma}$ $(0 < x < 1)$ and is such that $\int_0^1 \phi(x) \, dx = C$. Therefore C is indeterminate, as we would expect, and (9.18) may be used to check that the derived ϕ does indeed satisfy (9.23). Since (9.23) implies that

$$-\mu\phi(1-x) = \int_0^1 \frac{\phi(1-t) \, dt}{t - x} \quad (0 < x < 1)$$

we deduce that for $\mu > 0$ (9.23) is satisfied by the function ϕ defined by $\phi(x) = C\pi^{-1} \sin(\pi\gamma)x^{-\gamma}(1-x)^{\gamma-1}$ $(0 < x < 1)$, where $\mu = \pi \cot(\pi\gamma)$, $0 < \gamma < \frac{1}{2}$ and C is an arbitrary constant.

We have therefore identified solutions of (9.23) which are continuous in $(0, 1)$ but which grow so rapidly near one of the end points that they do not belong to $L_2(0, 1)$, being merely members of $L_1(0, 1)$. The question arises as to whether further work in this direction would produce yet more solutions and we need to consider what sort of function is acceptable as a 'solution' of (9.23). For practical purposes, we shall wish to consider only solutions $\phi \in L_1(0, 1)$ which are continuous in $(0, 1)$ and which satisfy the conditions $x\phi(x) \to 0$ as $x \to 0+$ and $(1-x)\phi(x) \to 0$ as $x \to 1-$. For such a function $x(1-x)\phi(x)$ defines a function in $L_2(0, 1)$, so we need not consider functions ψ for which $x^2(1-x)^2\psi(x)$ is an L_2-function. Expanding our class of admissible solutions to include those with singularities at both ends for which $x(1-x)\phi(x)$ is in $L_2(0, 1)$ does not, in fact, yield solutions additional to those we have found already. (See Problem 9.11.)

Theorem 9.8

Let μ be a real, non-zero constant and let $\gamma \in (0, \frac{1}{2})$ satisfy $|\mu| = \pi \cot(\pi\gamma)$. The only solution of

$$\mu\phi(x) = \int_0^1 \frac{\phi(t)}{t-x} dt \quad (0<x<1)$$

which is continuous in $(0, 1)$, integrable in $[0, 1]$ and for which $x\phi(x)\to0$ as $x\to0+$ and $(1-x)\phi(x)\to0$ as $x\to1-$ is given by

$$\phi(x) = Cx^{\gamma-1}(1-x)^{-\gamma} \quad (0<x<1, \mu<0),$$
$$\phi(x) = Cx^{-\gamma}(1-x)^{\gamma-1} \quad (0<x<1, \mu>0). \qquad \square$$

Example 9.8

Determine solutions of the integral equations

$$\pm\pi\phi(x) = x + \int_0^1 \frac{\phi(t)}{t-x} dt \quad (0<x<1)$$

which are integrable in $[0, 1]$.

With the minus sign option we deduce from Example 9.6 (by setting $\gamma=\frac{1}{4}$) that the function $\phi \in L_2(0, 1)$, defined by $\phi(x) = 2^{\frac{1}{2}}x^{\frac{1}{4}}(1-x)^{-\frac{1}{4}}(1-4x)/(8\pi)$ $(0\leqslant x<1)$, satisfies the equation. According to Theorem 9.8 the homogeneous version of the equation is satisfied by the integrable function ϕ where $\phi(x) = x^{-\frac{3}{4}}(1-x)^{-\frac{1}{4}}$ $(0<x<1)$. Therefore a solution of the equation which is integrable in $[0, 1]$ is given by

$$\phi(x) = \frac{2^{\frac{1}{2}}}{8\pi} x^{\frac{1}{4}}(1-x)^{-\frac{1}{4}}(1-4x) + Cx^{-\frac{3}{4}}(1-x)^{-\frac{1}{4}} \quad (0<x<1),$$

for any constant C.

With the plus sign option, the homogeneous equation is satisfied by $\phi(x) = x^{-\frac{1}{4}}(1-x)^{-\frac{3}{4}}(0<x<1)$. Using Example 9.7 to deduce the solution of the equation in $L_2(0, 1)$ we find that the equation is satisfied by the integrable function ϕ where

$$\phi(x) = \frac{2^{\frac{1}{2}}}{8\pi} x^{-\frac{1}{4}}(1-x)^{\frac{1}{4}}(1+4x) + Cx^{-\frac{1}{4}}(1-x)^{-\frac{3}{4}} \quad (0<x<1),$$

where C is constant. $\qquad \square$

We now return to the solution of $\mu\phi = f + H\phi$ in $L_2(0, 1)$. The formulae we have derived for this solution, given in Theorem 9.7, are generally more convenient than any variants of them, from the point of view of evaluating ϕ for a particular f. Reversal of the integration order is permissible in each of the versions of ϕ, but the resulting inner integrals cannot be expressed in terms of elementary functions. However we can derive an alternative, relatively simple form for ϕ which gives some insight into the structure of the solution of $\mu\phi = f + H\phi$, even though it is not especially useful from a practical point of view as it involves a Cauchy principal value integral.

We recall from Theorem 9.6 the operator identity $TH = -\pi\cot(\pi\gamma)T + \pi\operatorname{cosec}(\pi\gamma)S^*$ $(0 < \gamma < \frac{1}{2})$ in $L_2(0, 1)$. Since $T\psi = g$ implies that

$$\pi\operatorname{cosec}(\pi\gamma)\int_0^x \frac{\psi(t)}{t^\gamma}\,dt = \int_0^x \frac{g(t)\,dt}{(x-t)^\gamma},$$

we deduce from the identity that

$$\int_0^x \frac{(H\psi)(t)}{t^\gamma}\,dt = -\pi\cot(\pi\gamma)\int_0^x \frac{\psi(t)\,dt}{t^\gamma} + \int_0^x \frac{ds}{(x-s)^\gamma}\int_s^1 \frac{\psi(t)\,dt}{t^\gamma(t-s)^{1-\gamma}},$$

for all $\psi \in L_2(0, 1)$ and $\gamma \in (0, \frac{1}{2})$. Replacing the variables x, s and t by $1-x$, $1-s$ and $1-t$ respectively, we see that

$$\int_x^1 \frac{ds}{(s-x)^\gamma}\int_0^s \frac{(U\psi)(t)\,dt}{(1-t)^\gamma(s-t)^{1-\gamma}} = \int_x^1 \{(UH\psi)(t) + \pi\cot(\pi\gamma)(U\psi)(t)\}\frac{dt}{(1-t)^\gamma} \tag{9.24}$$

for all $\psi \in L_2(0, 1)$ and $\gamma \in (0, \frac{1}{2})$. We have again used the operator U defined by $(U\phi)(x) = \phi(1-x)$.

Now let $\mu\phi = f + H\phi$ in $L_2(0, 1)$ where $\mu < 0$ and let $\gamma \in (0, \frac{1}{2})$ satisfy $\mu = -\pi\cot(\pi\gamma)$. As we have shown, there is a unique solution ϕ for each f and it satisfies

$$-\pi^2\operatorname{cosec}^2(\pi\gamma)\int_x^1 t^{-\gamma}\phi(t)\,dt = \int_x^1 \frac{(Tf)(t)\,dt}{(t-x)^\gamma} \quad (0 < x < 1). \tag{9.20}$$

Differentiation of this last equality led to the expression for ϕ given in Theorem 9.7. We now proceed along a different route, choosing $\psi(x) = x^\gamma(1-x)^{-\gamma}f(1-x)$, for then by (9.24)

$$\int_x^1 \frac{(Tf)(t)\,dt}{(t-x)^\gamma} = \int_x^1 \frac{ds}{(s-x)^\gamma}\int_0^s \frac{(U\psi)(t)\,dt}{(1-t)^\gamma(s-t)^{1-\gamma}}$$

$$= \int_x^1 \{(UH\psi)(t) + \pi\cot(\pi\gamma)(U\psi)(t)\}\frac{dt}{(1-t)^\gamma}, \tag{9.25}$$

provided that this $\psi \in L_2(0, 1)$. If this proviso holds, (9.20) and (9.25) together show that the solution of $\mu\phi = f + H\phi$ in $L_2(0, 1)$ satisfies

$$-\pi^2 \operatorname{cosec}^2(\pi\gamma) \int_x^1 t^{-\gamma}\phi(t)\,dt = \int_x^1 \{(UH\psi)(t) + \pi \cot(\pi\gamma)(U\psi)(t)\}\,\frac{dt}{(1-t)^\gamma}$$

$$(0 < x < 1),$$

and therefore

$$\phi(x) = -\frac{x^\gamma}{(1-x)^\gamma}\,\frac{\sin^2(\pi\gamma)}{\pi^2}\,\{(UH\psi)(x) + \pi \cot(\pi\gamma)(U\psi)(x)\}$$

almost everywhere in $[0, 1]$. Using $UH = -HU$ and $\mu = -\pi \cot(\pi\gamma)$, and putting $(U\psi)(x) = (1-x)^\gamma x^{-\gamma}f(x)$, we conclude that

$$\phi(x) = (\pi^2 + \mu^2)^{-1}\left\{\mu f(x) + \frac{x^\gamma}{(1-x)^\gamma}\int_0^1 \frac{(1-t)^\gamma}{t^\gamma}\,\frac{f(t)\,dt}{t-x}\right\} \tag{9.26}$$

for almost all $x \in [0, 1]$. This ϕ is the unique solution of $\mu\phi = f + H\phi$ belonging to $L_2(0, 1)$, as long as the auxiliary function ψ we have used also lies in $L_2(0, 1)$. Let E denote the subspace of $L_2(0, 1)$ consisting of functions $f \in L_2(0, 1)$ with the property that the functions whose values are $x^{-\gamma}f(x)$ are also members of $L_2(0, 1)$. Since $f \in E$ implies $\psi \in L_2(0, 1)$, the derivation of (9.26) is secure for such f.

That the condition $f \in E$ is necessary to produce (9.26) is evident from the fact that the operator H is required to act on functions with the values of $(1-x)^\gamma x^{-\gamma}f(x)$. We can now take a different viewpoint however and, motivated by (9.26), introduce the new operator R_γ on $L_2(0, 1)$, where for $0 \leqslant |\gamma| < \frac{1}{2}$,

$$(R_\gamma\phi)(x) = \frac{x^\gamma}{(1-x)^\gamma}\int_0^1 \frac{(1-t)^\gamma}{t^\gamma}\,\frac{\phi(t)\,dt}{t-x} \qquad (0 \leqslant x \leqslant 1). \tag{9.27}$$

In terms of R_γ, we have shown that if $f \in E$ and $\mu < 0$ the unique solution $\phi \in L_2(0, 1)$ of $\mu\phi = f + H\phi$ is given by $(\pi^2 + \mu^2)\phi = \mu f + R_\gamma f$, γ being selected so that $\mu = -\pi \cot(\pi\gamma)$. Eliminating ϕ between the equation and its solution, we find that $\mu(R_\gamma - H)f = \pi^2 f + HR_\gamma f$ for $f \in E$. Now E is a dense subspace of $L_2(0, 1)$, H is a bounded operator on $L_2(0, 1)$ and so is R_γ (see Problem 9.13). Therefore by Lemma 9.1 the operator identity $\mu(R_\gamma - H) = \pi^2 I + HR_\gamma$ holds in $L_2(0, 1)$ and it follows that $\phi = (\pi^2 + \mu^2)^{-1}(\mu f + R_\gamma f)$ is the unique solution of $\mu\phi = f + H\phi$ in $L_2(0, 1)$, for all $f \in L_2(0, 1)$, provided that $\mu < 0$.

To extend this result to the case with $\mu > 0$ we use the now familiar device of writing $\mu\phi = f + H\phi$ as $-\mu U\phi = -Uf + HU\phi$. Then $U\phi = (\mu^2 + \pi^2)^{-1}$ $(\mu Uf - R_\gamma Uf)$ where γ is fixed by $\mu = \pi \cot(\pi\gamma)$ and $0 < \gamma < \frac{1}{2}$. Now $UR_\gamma U =$

R_γ^* and $R_\gamma^* = -R_{-\gamma}$, so that $R_\gamma U = -UR_{-\gamma}$, and therefore the solution of $\mu\phi = f + H\phi$ in $L_2(0, 1)$ for $\mu > 0$ is given by

$$\phi(x) = (\pi^2 + \mu^2)^{-1} \left\{ \mu f(x) + \frac{(1-x)^\gamma}{x^\gamma} \int_0^1 \frac{t^\gamma}{(1-t)^\gamma} \frac{f(t)\, dt}{t-x} \right\} \tag{9.28}$$

almost everywhere in $[0, 1]$.

We can combine (9.26) and (9.28) in the following way.

Theorem 9.9

Let μ be a non-zero real constant and let

$$(H\phi)(x) = \int_0^1 \frac{\phi(t)}{t-x}\, dt \quad (0 < x < 1).$$

Then $\mu\phi = f + H\phi$ has a unique solution in $L_2(0, 1)$ given almost everywhere in $[0, 1]$ by

$$(\pi^2 + \mu^2)\phi(x) = \mu f(x) + \frac{(1-x)^\gamma}{x^\gamma} \int_0^1 \frac{t^\gamma}{(1-t)^\gamma} \frac{f(t)\, dt}{t-x},$$

where γ satisfies $0 < |\gamma| < \frac{1}{2}$ and $\mu = \pi \cot(\pi\gamma)$. ☐

Note the reappearance in this final form of the solution of $\mu\phi = f + H\phi$ of the function ϕ_γ defined in (9.17) which we used to get our solution method under way. Note also that the solution may be written as $\phi = (\pi^2 + \mu^2)^{-1} (\mu f + R_{-\gamma} f)$, where R_γ is defined in (9.27).

To conclude this section, we observe that if ϕ satisfies the equation $\mu\phi = f + (H + K)\phi$ in $L_2(0, 1)$, where K is a compact operator, then by solving $\mu\phi = F + H\phi$ with $F = f + K\phi$, it follows that $(\pi^2 + \mu^2)\phi = (\mu I + R_{-\gamma})f + (\mu K + R_{-\gamma}K)\phi$. Now by Theorem 3.5 $\mu K + R_{-\gamma}K$ is a compact operator on $L_2(0, 1)$ and the theory we have developed for such operators may be deployed for the revised equation. This idea is pursued further in Problem 9.21.

9.5.2 The first kind equation

The integral equation

$$\int_0^1 \frac{\phi(t)\, dt}{t-x} = f(x) \quad (0 < x < 1) \tag{9.29}$$

is sometimes described as the aerofoil equation because of its association with the calculation of the flow around a slender aerofoil. It effectively corresponds to the second kind equation (9.16) with $\mu = 0$, a value excluded from the investigation in the last section. We therefore have to make a fresh start with (9.29), but this is not a difficulty; the absence of ϕ outside the integral means that there is a good deal of scope for choosing a 'simplifying operator' for a first kind equation such as (9.29).

We can, for example, simply integrate (9.29) to give

$$\int_0^x ds \int_0^1 \frac{\phi(t)\, dt}{t-s} = \int_0^x f(t)\, dt \quad (0<x<1), \qquad (9.30)$$

which reduces to

$$-\int_0^1 \log 2|x-t|\phi(t)\, dt = \int_0^x f(t)\, dt - C \quad (0<x<1) \qquad (9.31)$$

on using (9.12). Here we have written $C=\int_0^1 \log(2t)\phi(t)\, dt$. Note however that (9.31) is identically satisfied as $x\to 0+$, so that C is indeterminate and we can hereafter regard it as an arbitrary constant. The factor of 2 in the logarithm in (9.31) has been included in anticipation of a modification of that equation. In §1.3.3 we rather tentatively associated an integral equation with a logarithmically singular kernel and one with a Cauchy singular kernel. We are now making this relationship secure.

Since (9.12) extends to give an operator identity in $L_2(0, 1)$, by virtue of Lemma 9.1, we can reinterpret our derivation of (9.31) from (9.29). If ϕ satisfies $H\phi = f$ in $L_2(0, 1)$, H being the Hilbert transform, then it also satisfies (9.31), regarded as defining an equation in $L_2(0, 1)$. This transformation is of no value of course unless we can make some headway with (9.31).

One way forward is to change the variables by $t=\cos^2(\tfrac{1}{2}\theta)$ and $x=\cos^2(\tfrac{1}{2}\sigma)$, where $0<\theta<\pi$ and $0<\sigma<\pi$. If we also write $\phi(\cos^2(\tfrac{1}{2}\theta))\sin\theta = 2\hat\phi(\theta)$, straightforward manipulation converts (9.31) into the equation

$$-\int_0^\pi \log|\cos\theta - \cos\sigma|\hat\phi(\theta)\, d\theta = g(\sigma) \quad (0<\sigma<\pi), \qquad (9.32)$$

where

$$g(\sigma) = \frac{1}{2}\int_\sigma^\pi f(\cos^2(\tfrac{1}{2}\theta))\sin\theta\, d\theta - C.$$

Now the operator K defined on $L_2(0, \pi)$ by

$$(K\hat\phi)(\sigma) = -\int_0^\pi \log|\cos\theta - \cos\sigma|\hat\phi(\theta)\, d\theta \quad (0<\sigma<\pi) \qquad (9.33)$$

is compact and self-adjoint, and we know that its eigenvalue sequence (μ_n) is given by

$$\mu_0 = \pi \log 2, \quad \mu_n = \pi/n \quad (n \in \mathbb{N})$$

(see Example 6.10). Therefore K is a positive operator. The corresponding complete set of eigenvectors is given by

$$\phi_0(x) = \frac{1}{\pi^{\frac{1}{2}}}, \quad \phi_n(x) = \left(\frac{2}{\pi}\right)^{\frac{1}{2}} \cos(nx) \quad (n \in \mathbb{N}).$$

Using this information we can seek a solution of $K\hat\phi = g$ in $L_2(0, \pi)$ by spectral methods. According to the theory developed in Example 4.6, $K\hat\phi = g$ has a solution in $L_2(0, \pi)$ if and only if $\Sigma |\mu_n^{-1}(g, \phi_n)|^2$ is finite. If this condition is satisfied for a particular g the solution of $K\hat\phi = g$, which is given by $\hat\phi = \Sigma \mu_n^{-1}(g, \phi_n)\phi_n$, is certainly unique since $K\hat\phi = 0$ implies $\hat\phi = 0$. Notice that even if (9.32) does have a solution in $L_2(0, \pi)$, it will not necessarily give a solution of (9.31) which is in $L_2(0, 1)$. The point here is that $\hat\phi$ is defined from ϕ by a mapping which is injective but not surjective.

Suppose, however, that, for a particular $f \in L_2(0, 1)$, the mechanism described does produce a solution $\hat\phi$ of $K\hat\phi = g$ which derives from a $\phi \in L_2(0, 1)$. Then this ϕ is the unique solution of (9.31), and therefore of (9.30) also, in $L_2(0, 1)$. Appealing to the fact that if $\int_0^x \psi_1(t)\, dt = \int_0^x \psi_2(t)\, dt$ for all $x \in (0, 1)$ then $\psi_1(x) = \psi_2(x)$ almost everywhere in $[0, 1]$, we deduce that the $\phi \in L_2(0, 1)$ which satisfies (9.30) also satisfies (9.29), at least almost everywhere in $[0, 1]$. On the other hand if there is a solution ϕ of (9.29) in $L_2(0, 1)$, this ϕ also satisfies (9.31) and transforms to a $\hat\phi$ which is certainly in $L_2(0, \pi)$. We can thus be sure of locating the (unique) solution $\phi \in L_2(0, 1)$ of (9.29), if there is one, by considering (9.32).

It is useful at this point to work through a specific example.

Example 9.9

Solve the integral equation

$$\int_0^1 \frac{\phi(t)\, dt}{t - x} = 1 - 2x \quad (0 < x < 1).$$

We first convert the equation into the form (9.31) giving

$$-\int_0^1 \log 2|x - t|\phi(t)\, dt = x - x^2 - C \quad (0 < x < 1) \tag{9.34}$$

where C is an arbitrary constant. Now let $t = \cos^2(\frac{1}{2}\theta)$, $x = \cos^2(\frac{1}{2}\sigma)$ and $\phi(\cos^2(\frac{1}{2}\theta)) \sin \theta = 2\hat\phi(\theta)$ so that $K\hat\phi = g$ in the notation of (9.33), where

$$g(\sigma) = (1 - \cos 2\sigma)/8 - C \quad (0 < \sigma < \pi).$$

Using the information given after (9.33) about the eigenvalues and eigenvectors of K, we find that for the present g we have $(g, \phi_0) = \pi^{\frac{1}{2}}(\frac{1}{8} - C)$ and $(g, \phi_2) = -\pi^{\frac{1}{2}}/(8\sqrt{2})$, all other inner products of the form (g, ϕ_n) being zero. Therefore there is a solution $\hat{\phi} \in L_2(0, \pi)$ of $K\hat{\phi} = g$ in this case, it is unique (for each fixed value of C, that is) and it is given by

$$\hat{\phi}(\sigma) = \frac{(\frac{1}{8} - C)}{\pi \log 2} - \frac{1}{4\pi} \cos(2\sigma) \quad (0 \leqslant \sigma \leqslant \pi).$$

Reverting to the original variables, the function ϕ which results has the values

$$\phi(x) = \left\{ \frac{(\frac{1}{8} - C)}{\pi \log 2} - \frac{1}{4\pi} \right\} x^{-\frac{1}{2}}(1 - x)^{-\frac{1}{2}} + \frac{2}{\pi} x^{\frac{1}{2}}(1 - x)^{\frac{1}{2}} \quad (0 < x < 1).$$

This ϕ is in $L_2(0, 1)$ if and only if we choose $C = (1 - 2 \log 2)/8$ so that $\phi(x) = (2/\pi) x^{\frac{1}{2}}(1 - x)^{\frac{1}{2}}$ $(0 \leqslant x \leqslant 1)$. (One can confirm, incidentally, that for this last ϕ we have $\int_0^1 \phi(x) \log(2x) \, dx = (1 - 2 \log 2)/8$.)

Therefore equation (9.34) has the unique solution $\phi \in L_2(0, 1)$, for the particular choice of C indicated, where

$$\phi(x) = \frac{2}{\pi} x^{\frac{1}{2}}(1 - x)^{\frac{1}{2}},$$

and by our earlier arguments this is also the unique solution in $L_2(0, 1)$ of the given equation. $\qquad \square$

We found solutions of the homogeneous second kind equation $\mu\phi = H\phi$ which were not in $L_2(0, 1)$ and the last example raises the question of whether the function ϕ_0, where

$$\phi_0(x) = x^{-\frac{1}{2}}(1 - x)^{-\frac{1}{2}} \quad (0 < x < 1) \tag{9.35}$$

satisfies $\int_0^1 (t - x)^{-1} \phi_0(t) \, dt = 0$ $(0 < x < 1)$. The most immediate way of examining whether ϕ_0 has the suspected property is by direct calculation. The substitution $t = x(x + u - ux)^{-1}$ gives

$$\int_0^1 \frac{dt}{t^{\frac{1}{2}}(1 - t)^{\frac{1}{2}}(t - x)} = \frac{1}{x^{\frac{1}{2}}(1 - x)^{\frac{1}{2}}} \int_0^\infty \frac{du}{u^{\frac{1}{2}}(1 - u)} \quad (0 < x < 1)$$

and since $\int_0^\infty u^{-\frac{1}{2}}(1 - u)^{-1} \, du = 0$ (see Appendix C), we conclude that ϕ_0 does indeed satisfy $\int_0^1 (t - x)^{-1} \phi_0(t) \, dt = 0$ $(0 < x < 1)$.

The integrable solution of the integral equation in the last example is therefore given by

$$\phi(x) = \frac{2}{\pi} x^{\frac{1}{2}}(1-x)^{\frac{1}{2}} + Cx^{-\frac{1}{2}}(1-x)^{-\frac{1}{2}} \quad (0 < x < 1)$$

where C is any constant. (*cf.* Problem 9.19.)

At the beginning of this section it was suggested that there is more than one way of dealing with the equation $H\phi = f$ and we now substantiate this remark by considering a fresh approach. This involves a new 'simplifying operator' which leads to a familiar equation, as the following lemma reveals.

Lemma 9.10

Let the operators R and H be defined on $L_2(0, 1)$ by

$$(R\phi)(x) = \int_0^x t^{\frac{1}{2}} \phi(t)\, dt, \quad (H\phi)(x) = \int_0^1 \frac{\phi(t)\, dt}{t - x} \quad (0 < x < 1).$$

Then

$$(RH\phi)(x) = \int_0^1 \left\{ -2x^{\frac{1}{2}} + t^{\frac{1}{2}} \log \left| \frac{x^{\frac{1}{2}} + t^{\frac{1}{2}}}{x^{\frac{1}{2}} - t^{\frac{1}{2}}} \right| \right\} \phi(t)\, dt \quad (0 < x < 1).$$

Proof

Let E denote the subspace of $L_2(0, 1)$ consisting of functions which have a continuous derivative in $[0, 1]$. Suppose that $x \in (0, 1)$ and let $\phi \in E$. Then

$$(RH\phi)(x) = \int_0^x s^{\frac{1}{2}}\, ds \int_0^1 \frac{\phi(t)}{t - s}\, dt$$

$$= \int_0^x \frac{ds}{2s^{\frac{1}{2}}} \int_0^1 \left\{ -2 + \frac{t^{\frac{1}{2}}}{t^{\frac{1}{2}} + s^{\frac{1}{2}}} + \frac{t^{\frac{1}{2}}}{t^{\frac{1}{2}} - s^{\frac{1}{2}}} \right\} \phi(t)\, dt.$$

Now put $s = \xi^2$ and $t = \eta^2$, where $0 < \xi < 1$ and $0 \leqslant \eta \leqslant 1$, to give

$$(RH\phi)(x) = \int_0^{x^{\frac{1}{2}}} d\xi \int_0^1 \left\{ -2 + \frac{\eta}{\eta + \xi} + \frac{\eta}{\eta - \xi} \right\} 2\eta\phi(\eta^2)\, d\eta$$

$$= \int_0^1 \left\{ -2x^{\frac{1}{2}} + \eta \log\left(1 + \frac{x^{\frac{1}{2}}}{\eta} \right) - \eta \log\left| 1 - \frac{x^{\frac{1}{2}}}{\eta} \right| \right\} 2\eta\phi(\eta^2)\, d\eta,$$

where we have used (9.12) to deal with the final term in the integrand, the switch in integration order being straightforward for the remaining two terms. Restoring $t = \eta^2$, we arrive at

$$(RH\phi)(x) = \int_0^1 \left\{ -2x^{\frac{1}{2}} + t^{\frac{1}{2}} \log \left| \frac{x^{\frac{1}{2}} + t^{\frac{1}{2}}}{x^{\frac{1}{2}} - t^{\frac{1}{2}}} \right| \right\} \phi(t) \, dt \quad (0 < x < 1)$$

for $\phi \in E$. Since R, H and RH are bounded operators on $L_2(0, 1)$ and since E is a dense subspace of $L_2(0, 1)$, the result extends to functions $\phi \in L_2(0, 1)$, by Lemma 9.1. $\qquad\square$

Now let ϕ satisfy $H\phi = f$ in $L_2(0, 1)$. Then $RH\phi = Rf$, which we can write as

$$\int_0^1 \log \left| \frac{x^{\frac{1}{2}} + t^{\frac{1}{2}}}{x^{\frac{1}{2}} - t^{\frac{1}{2}}} \right| t^{\frac{1}{2}} \phi(t) \, dt = 2Dx^{\frac{1}{2}} + \int_0^x t^{\frac{1}{2}} f(t) \, dt \quad (0 < x < 1), \quad (9.36)$$

where $D = \int_0^1 \phi(t) \, dt$. Note that if, for a given $f \in L_2(0, 1)$, there is a $\phi \in L_2(0, 1)$ such that $(RH\phi) = (Rf)(x)$ for $0 < x < 1$, then $(H\phi)(x) = f(x)$ almost everywhere in $[0, 1]$. We can therefore solve $H\phi = f$ in $L_2(0, 1)$ by finding the solutions of (9.36) in the same space.

The main virtue of (9.36) is that we can solve it explicitly in closed form using the TT^* factorisation derived in Example 6.7. We have already deduced a formula for the solution of an equation of the form (9.36) in Example 9.1, but it is convenient to deal with (9.36) directly.

Using the TT^* form we have referred to, (9.36) can be written as

$$\int_0^x \frac{ds}{(x-s)^{\frac{1}{2}}} \int_s^1 \frac{t^{\frac{1}{2}} \phi(t)}{(t-s)^{\frac{1}{2}}} \, dt = 2Dx^{\frac{1}{2}} + \int_0^x t^{\frac{1}{2}} f(t) \, dt \quad (0 < x < 1), \quad (9.37)$$

and we can extricate ϕ by using the methods developed in §9.2. All of the interchanges of integration order in what follows are permissible as long as the functions with the values of $x^{\frac{1}{2}} \phi(x)$ and $x^{\frac{1}{2}} f(x)$ are members of $L_2(0, 1)$. Note first that (9.37) implies

$$\pi \int_0^x ds \int_s^1 \frac{t^{\frac{1}{2}} \phi(t)}{(t-s)^{\frac{1}{2}}} \, dt = \int_0^x \frac{ds}{(x-s)^{\frac{1}{2}}} \left\{ 2Ds^{\frac{1}{2}} + \int_0^s t^{\frac{1}{2}} f(t) \, dt \right\}$$

$$= D\pi x + 2 \int_0^x (x-t)^{\frac{1}{2}} t^{\frac{1}{2}} f(t) \, dt \quad (0 < x < 1)$$

which may be written as

$$\int_0^x ds \int_s^1 \frac{t^{\frac{1}{2}} \phi(t)}{(t-s)^{\frac{1}{2}}} \, dt = \int_0^x ds \left\{ D + \frac{1}{\pi} \int_0^s \frac{t^{\frac{1}{2}} f(t)}{(s-t)^{\frac{1}{2}}} \, dt \right\} \quad (0 < x < 1).$$

It follows that

$$\int_x^1 \frac{t^{\frac{1}{2}}\phi(t)}{(t-x)^{\frac{1}{2}}} dt = D + \frac{1}{\pi} \int_0^x \frac{t^{\frac{1}{2}}f(t)}{(x-t)^{\frac{1}{2}}} dt$$

almost everywhere in $[0, 1]$, whence

$$\pi \int_x^1 t^{\frac{1}{2}}\phi(t)\, dt = \int_x^1 \frac{ds}{(s-x)^{\frac{1}{2}}}\left\{ D + \frac{1}{\pi} \int_0^s \frac{t^{\frac{1}{2}}f(t)}{(s-t)^{\frac{1}{2}}} dt \right\} \quad (0<x<1)$$

which after a minor rearrangement gives

$$\int_x^1 t^{\frac{1}{2}}\phi(t)\, dt = \frac{1}{\pi^2} \int_x^1 \frac{ds}{(s-x)^{\frac{1}{2}}} \int_0^s \frac{t^{\frac{1}{2}}f(t)}{(s-t)^{\frac{1}{2}}} dt + \frac{2D}{\pi}(1-x)^{\frac{1}{2}} \quad (0<x<1). \quad (9.38)$$

Now recall that we are trying to find a solution of (9.36) in $L_2(0, 1)$. That there is at most one such solution follows from the fact that the operator defined by the left hand side of (9.36) (considered as applied to $x^{\frac{1}{2}}\phi(x)$) is positive, as we showed in Example 6.7. Thus if the function defined by

$$\phi(x) = -\frac{x^{-\frac{1}{2}}}{\pi^2} \frac{d}{dx} \int_x^1 \frac{ds}{(s-x)^{\frac{1}{2}}} \int_0^s \frac{t^{\frac{1}{2}}f(t)}{(s-t)^{\frac{1}{2}}} dt + \frac{Dx^{-\frac{1}{2}}}{\pi}(1-x)^{-\frac{1}{2}} \quad (9.39)$$

exists for all but finitely many x and is in $L_2(0, 1)$, the differentiations being possible at all but finitely many points and the intermediate integrals defining continuous functions (with value 0 at $x=1$ and $s=0$), for a particular $f \in L_2(0, 1)$ and a particular value of D, then the solution process employed is certainly valid and this ϕ is the unique solution of (9.36), and therefore of $H\phi = f$, in $L_2(0, 1)$. That D is an arbitrary constant, as just implied, follows from the fact that $\int_0^1 (t-x)^{-1}\phi_0(t)\, dt = 0\,(0<x<1)$ in the notation (9.35), as we showed earlier. Thus a solution of $H\phi = f$ in $L_2(0, 1)$, if there is one, can be amplified by the addition of any constant multiple of the integrable function ϕ_0.

Alternative representations of the solution of $H\phi = f$ can be derived from (9.38) and (9.39). Using the result in Problem 6.5, for instance, we see that (9.38) can be written as

$$\int_x^1 t^{\frac{1}{2}}\phi(t)\, dt = \frac{1}{\pi^2} \int_0^1 \log\left| \frac{(1-x)^{\frac{1}{2}}+(1-t)^{\frac{1}{2}}}{(1-x)^{\frac{1}{2}}-(1-t)^{\frac{1}{2}}} \right| t^{\frac{1}{2}}f(t)\, dt + \frac{2D}{\pi}(1-x)^{\frac{1}{2}}$$

$$(0<x<1). \quad (9.40)$$

Differentiation of this equality gives another version of the solution of (9.29). We can also use (9.40) to derive a form of the solution involving a Cauchy principal value integral. For

$$\int_0^1 \log\left|\frac{(1-x)^{\frac{1}{2}}+(1-t)^{\frac{1}{2}}}{(1-x)^{\frac{1}{2}}-(1-t)^{\frac{1}{2}}}\right| t^{\frac{1}{2}}f(t)\, dt$$

$$= \int_0^1 \log\left|\frac{(1-x)^{\frac{1}{2}}+t^{\frac{1}{2}}}{(1-x)^{\frac{1}{2}}-t^{\frac{1}{2}}}\right| (1-t)^{\frac{1}{2}}f(1-t)\, dt$$

$$= \int_0^{1-x} s^{\frac{1}{2}}\, ds \int_0^1 \frac{(1-t)^{\frac{1}{2}}f(1-t)\, dt}{t^{\frac{1}{2}}(t-s)} + 2(1-x)^{\frac{1}{2}} \int_0^1 \frac{(1-t)^{\frac{1}{2}}f(1-t)\, dt}{t^{\frac{1}{2}}}$$

$$= -\int_x^1 (1-s)^{\frac{1}{2}}\, ds \int_0^1 \frac{t^{\frac{1}{2}}f(t)\, dt}{(1-t)^{\frac{1}{2}}(t-s)} + 2(1-x)^{\frac{1}{2}} \int_0^1 \frac{t^{\frac{1}{2}}f(t)\, dt}{(1-t)^{\frac{1}{2}}},$$

by Lemma 9.10, provided the function $(1-x)^{-\frac{1}{2}}f(x)$ lies in $L_2(0, 1)$. If this proviso holds, (9.40) can be written as

$$\int_x^1 t^{\frac{1}{2}}\phi(t)\, dt = -\frac{1}{\pi^2}\int_x^1 (1-s)^{\frac{1}{2}}\, ds \int_0^1 \frac{t^{\frac{1}{2}}f(t)\, dt}{(1-t)^{\frac{1}{2}}(t-s)} + D'\int_x^1 \frac{ds}{(1-s)^{\frac{1}{2}}} \quad (0<x<1),$$

where D' is a constant, and therefore

$$\phi(x) = -\frac{1}{\pi^2}\frac{(1-x)^{\frac{1}{2}}}{x^{\frac{1}{2}}}\int_0^1 \frac{t^{\frac{1}{2}}}{(1-t)^{\frac{1}{2}}}\frac{f(t)\, dt}{t-x} + \frac{D'}{x^{\frac{1}{2}}(1-x)^{\frac{1}{2}}} \tag{9.41}$$

almost everywhere in $[0, 1]$.

The first term on the right hand side of (9.41) is the counterpart for $H\phi = f$ of that given in Theorem 9.9 for the second kind equation. However we have shown only that (9.41) provides the unique solution of $H\phi = f$ if it defines a function in $L_2(0, 1)$ and if the function with values $(1-x)^{-\frac{1}{2}}f(x)$ belongs to $L_2(0, 1)$. We cannot use Lemma 9.1 to extend the formula (9.41) to $f \in L_2(0, 1)$ because the transformation defined by

$$(R_{\frac{1}{2}}\phi)(x) = \frac{(1-x)^{\frac{1}{2}}}{x^{\frac{1}{2}}}\int_0^1 \frac{t^{\frac{1}{2}}}{(1-t)^{\frac{1}{2}}}\frac{\phi(t)\, dt}{t-x}$$

is not a bounded operator on $L_2(0, 1)$ (see Problem 9.13).

Example 9.10

Solve the integral equation

$$\int_0^1 \frac{\phi(t)\, dt}{t-x} = 1 - 2x \quad (0<x<1).$$

We solved this equation in Example 9.9 by spectral methods and we now use the formula (9.39).

Note first that, putting $t = su$,

$$\int_0^s \frac{t^{\frac{1}{2}}(1-2t)}{(s-t)^{\frac{1}{2}}} \, dt = s \int_0^1 u^{\frac{1}{2}}(1-u)^{-\frac{1}{2}} \, du - 2s^2 \int_0^1 u^{\frac{3}{2}}(1-u)^{-\frac{1}{2}} \, du$$

$$= B(\tfrac{3}{2}, \tfrac{1}{2})s - 2B(\tfrac{5}{2}, \tfrac{1}{2})s^2 = \pi(2s - 3s^2)/4.$$

Further, with $s = x + v$,

$$\int_x^1 \frac{(s-3s^2/2)}{(s-x)^{\frac{1}{2}}} \, ds = \int_0^{1-x} \left\{ (x - 3x^2/2)v^{-\frac{1}{2}} + (1 - 3x)v^{\frac{1}{2}} - 3v^{\frac{3}{2}}/2 \right\} dv$$

$$= (1-x)^{\frac{1}{2}}(1 + 8x - 24x^2)/15.$$

This function is continuous on $[0, 1]$ and differentiable on $[0, 1)$ and therefore, using (9.39), we have

$$\phi(x) = -\frac{x^{-\frac{1}{2}}}{2\pi} \frac{d}{dx} \left\{ (1-x)^{\frac{1}{2}}(1 + 8x - 24x^2)/15 \right\} + \frac{D}{\pi} x^{-\frac{1}{2}}(1-x)^{-\frac{1}{2}}$$

$$= \frac{2}{\pi} x^{\frac{1}{2}}(1-x)^{\frac{1}{2}} + \frac{1}{\pi}(D - \tfrac{1}{4})x^{-\frac{1}{2}}(1-x)^{-\frac{1}{2}} \quad (0 < x < 1).$$

This ϕ is in $L_2(0, 1)$ if and only if $D = \tfrac{1}{4}$, in which case we have the unique solution in that space. More generally we have an integrable solution with D arbitrary. ☐

Example 9.11

Solve the integral equation

$$\int_0^1 \frac{\phi(t) \, dt}{t - x} = 1 - x \quad (0 < x < 1).$$

Since $f(x) = 1 - x$ here the function with values $(1-x)^{-\frac{1}{2}}f(x)$ is certainly in $L_2(0, 1)$ and (9.41) can therefore be used. Now

$$\int_0^1 \frac{t^{\frac{1}{2}}(1-t)^{\frac{1}{2}}}{t-x} \, dt = x(1-x) \int_0^1 \frac{t^{-\frac{1}{2}}(1-t)^{-\frac{1}{2}}}{t-x} \, dt + \int_0^1 (1-t)^{\frac{1}{2}} t^{-\frac{1}{2}} \, dt$$

$$- x \int_0^1 t^{-\frac{1}{2}}(1-t)^{-\frac{1}{2}} \, dt$$

$$= B(\tfrac{3}{2}, \tfrac{1}{2}) - xB(\tfrac{1}{2}, \tfrac{1}{2}) = \pi/2 - \pi x,$$

the first integral on the right hand side vanishing, as we have shown previously following (9.35). The formula (9.41) now gives

$$\phi(x) = -\frac{1}{2\pi} \frac{(1-x)^{\frac{1}{2}}}{x^{\frac{1}{2}}} (1 - 2x) + \frac{E}{x^{\frac{1}{2}}(1-x)^{\frac{1}{2}}} \quad (0 < x < 1),$$

which is not in $L_2(0, 1)$ for any value of E. This ϕ does satisfy the given integral equation, however, as can be confirmed by a direct calculation.

We conclude that the equation has no solution in $L_2(0, 1)$, and that it has a solution which is integrable in $[0, 1]$ but which is not unique. Note that there is an integrable solution which is bounded at $x = 1$, corresponding to the choice $E = 0$, and there is an integrable solution which is bounded at $x = 0$, given by setting $E = 1/(2\pi)$. □

Remark

This example shows that H is not invertible on $L_2(0, 1)$.

Example 9.12

Solve the integral equation

$$\int_0^1 \left\{ \frac{1}{t-x} + \frac{1}{t+x} \right\} \phi(t)\, dt = 2x^2 - 1 \quad (0 < x < 1).$$

Equations such as this, in which the integral consists of a principal value part and a 'regular' part, are common in practice (see §1.3.3). There are a number of strategies available, one being to deduce an equation for ϕ with a weakly singular kernel by applying a suitable operator. In particular,

$$\int_0^x ds \int_0^1 \left\{ \frac{1}{t-s} + \frac{1}{t+s} \right\} \phi(t)\, dt = \int_0^x (2t^2 - 1)\, dt \quad (0 < x < 1),$$

which reduces to

$$\int_0^1 \log \left| \frac{t+x}{t-x} \right| \phi(t)\, dt = \tfrac{2}{3}x^3 - x \quad (0 < x < 1)$$

on using (9.12).

By Problem 6.7 this last equation may be written as $TT^*\phi = f$ where

$$(T\phi)(x) = \int_0^x \left(\frac{2t}{x^2 - t^2} \right)^{\frac{1}{2}} \phi(t)\, dt \quad (0 \leqslant x \leqslant 1)$$

and $f(x) = 2x^3/3 - x \ (0 \leqslant x \leqslant 1)$. Now $TT^*\phi = f$ has at most one solution in $L_2(0, 1)$ and, by the arguments we have used earlier in this section, such a solution is also the unique solution of the equation in $L_2(0, 1)$ given at the start of this example.

We can solve $TT^*\phi = f$ as two successive Volterra equations, employing the method described in §9.2. From

$$\int_0^x \left(\frac{2t}{x^2 - t^2} \right)^{\frac{1}{2}} (T^*\phi)(t)\, dt = \tfrac{2}{3}x^3 - x$$

we deduce that

$$\pi \int_0^x (2t)^{\frac{1}{2}} (T^*\phi)(t)\, dt = \int_0^x (\tfrac{2}{3}t^3 - t)\, \frac{2t\, dt}{(x^2 - t^2)^{\frac{1}{2}}}$$

$$= \tfrac{2}{3}x^4 B(\tfrac{5}{2}, \tfrac{1}{2}) - x^2 B(\tfrac{3}{2}, \tfrac{1}{2})$$

$$= \tfrac{1}{4}\pi x^4 - \tfrac{1}{2}\pi x^2, \quad (0 < x < 1)$$

where the substitution $t = xu^{\frac{1}{2}}$ has been made to evaluate the integral. It follows that $(2x)^{\frac{1}{2}}(T^*\phi)(x) = x^3 - x$ $(0 < x < 1)$ which we may write as

$$\int_x^1 \frac{2\phi(t)\, dt}{(t^2 - x^2)^{\frac{1}{2}}} = x^2 - 1.$$

Therefore

$$\pi \int_x^1 \phi(t)\, dt = \int_x^1 \frac{(t^2 - 1)t}{(t^2 - x^2)^{\frac{1}{2}}}\, dt = -\tfrac{2}{3}(1 - x^2)^{\frac{3}{2}},$$

in this case using the variable change $t^2 = u + x^2$ to simplify the integral. We conclude that $\phi(x) = -2\pi^{-1} x(1 - x^2)^{\frac{1}{2}}$ $(0 \leqslant x \leqslant 1)$ gives the unique solution of the equation in $L_2(0, 1)$. □

Example 9.13

Solve the integral equation

$$\int_0^1 \left\{ \frac{1}{t - x} - \frac{1}{t + x} \right\} \phi(t)\, dt = x(\pi x - 2) \quad (0 < x < 1).$$

This equation is similar to the one considered in the last example, but we use a different approach here. Let $t^2 = \eta$ and $x^2 = \xi$ where $0 \leqslant \eta \leqslant 1$ and $0 < \xi < 1$ and put $\phi(\eta^{\frac{1}{2}}) = \eta^{\frac{1}{2}}\bar{\phi}(\eta)$. Then $\bar{\phi}$ satisfies

$$\int_0^1 \frac{\bar{\phi}(\eta)\, d\eta}{\eta - \xi} = \pi \xi^{\frac{1}{2}} - 2 \quad (0 < \xi < 1).$$

We now seek a solution $\bar{\phi} \in L_2(0, 1)$ by using the formula (9.39). Putting $t = su$ we have

$$\int_0^s \frac{(\pi t - 2t^{\frac{1}{2}})}{(s - t)^{\frac{1}{2}}}\, dt = \int_0^1 \frac{(\pi s^{\frac{3}{2}} u - 2su^{\frac{1}{2}})}{(1 - u)^{\frac{1}{2}}}\, du$$

$$= \pi s^{\frac{3}{2}} B(2, \tfrac{1}{2}) - 2s B(\tfrac{3}{2}, \tfrac{1}{2})$$

$$= \tfrac{4}{3}\pi s^{\frac{3}{2}} - \pi s \quad (s > 0),$$

so that we also need to evaluate

$$I(\xi) = \frac{d}{d\xi} \int_\xi^1 \{\tfrac{4}{3}s^{\frac{3}{2}} - s\} \frac{ds}{(s - \xi)^{\frac{1}{2}}} \quad (0 < \xi < 1).$$

Substituting $s = \xi \sec^2 \theta$ $(0 \leqslant \theta < \tfrac{1}{2}\pi)$ gives

$$I(\xi) = \frac{d}{d\xi} \left\{ \frac{8\xi^2}{3} \int_0^{\theta_0} \sec^5 \theta \, d\theta - 2\xi^{\frac{3}{2}} \int_0^{\theta_0} \sec^4 \theta \, d\theta \right\}$$

where $\sec \theta_0 = \xi^{-\frac{1}{2}}$. Elementary integration now leads to

$$I(\xi) = \frac{d}{d\xi} \left\{ \xi^2 \log \left(\frac{1 + (1-\xi)^{\frac{1}{2}}}{\xi^{\frac{1}{2}}} \right) - \tfrac{1}{3}\xi(1-\xi)^{\frac{1}{2}} \right\}$$

$$= 2\xi \log \left(\frac{1 + (1-\xi)^{\frac{1}{2}}}{\xi^{\frac{1}{2}}} \right) - \tfrac{1}{3}(1-\xi)^{-\frac{1}{2}}.$$

Therefore, according to (9.39) we have

$$\phi(\xi) = \frac{1}{\pi}(D + \tfrac{1}{3})\xi^{-\frac{1}{2}}(1-\xi)^{-\frac{1}{2}} - \frac{2\xi^{\frac{1}{2}}}{\pi} \log \left(\frac{1 + (1-\xi)^{\frac{1}{2}}}{\xi^{\frac{1}{2}}} \right) \quad (0 < \xi < 1),$$

where D is a arbitrary constant. Since $\phi(x) = x\hat{\phi}(x^2)$ the corresponding solution of the given equation is defined by

$$\phi(x) = \frac{1}{\pi}(D + \tfrac{1}{3})(1 - x^2)^{-\frac{1}{2}} - \frac{2x^2}{\pi} \log \left(\frac{1 + (1 - x^2)^{\frac{1}{2}}}{x} \right) \quad (0 < x < 1),$$

and choosing $D = -\tfrac{1}{3}$ we see that the unique solution $\phi \in L_2(0, 1)$ is such that

$$\phi(x) = -\frac{2x^2}{\pi} \log \left(\frac{1 + (1 - x^2)^{\frac{1}{2}}}{x} \right) \quad (0 < x \leqslant 1). \qquad \square$$

9.6 Fourier transform methods

Hitherto in this book we have restricted attention almost entirely to integral equations on an interval of finite length and in many parts of the theory this restriction has been essential to the methods employed. In this section we shall depart from this position and consider integral equations (of a fairly restricted class) where the independent variable is allowed to range over the whole real line or the positive half-line. We shall not establish a general theory analogous to the contents of earlier chapters for such 'singular' integral equations but merely notice at the outset that we may expect some significant differences. One such difference is that most of the integral operators we encounter will not be compact, in contrast to the situation on an interval of finite length where operators arising from 'well-behaved' kernels are compact (non-compact operators arising only from kernels which are unbounded when considered as functions). A further point worth noting is that of the class of functions to which we demand that a solution belongs: on the interval $[a, b]$ every continuous function belongs to $L_2(a, b)$, the space in which we have chosen to define our operators, but it is not the case that every continuous function on $(-\infty, \infty)$ belongs to $L_2(-\infty, \infty)$ and a general treatment would have to consider issues such as this.

Example 9.14

Consider the equation

$$\phi(x) - \int_{-\infty}^{\infty} e^{-a|x-t|}\phi(t)\,dt = e^{-|x|} \quad (x \in \mathbb{R}) \tag{9.42}$$

where $a > 2$.

Let $F\phi$ denote the Fourier transform of ϕ, so that $(F\phi)(x) = 1/(2\pi)^{\frac{1}{2}} \int_{-\infty}^{\infty} \phi(t)\,e^{itx}\,dt$ and notice that

$$\frac{1}{(2\pi)^{\frac{1}{2}}} \int_{-\infty}^{\infty} e^{-a|t|+ixt}\,dt = \left(\frac{2}{\pi}\right)^{\frac{1}{2}} \frac{a}{a^2+x^2} \tag{9.43}$$

and

$$\frac{1}{(2\pi)^{\frac{1}{2}}} \int_{-\infty}^{\infty} e^{ixs}\,ds \int_{-\infty}^{\infty} e^{-a|s-t|}\phi(t)\,dt = \frac{1}{(2\pi)^{\frac{1}{2}}} \int_{-\infty}^{\infty} \phi(t)\,e^{ixt}\,dt$$

$$\times \int_{-\infty}^{\infty} e^{ix(s-t)-a|s-t|}\,ds$$

$$= (F\phi)(x)\frac{2a}{a^2+x^2},$$

the reversal of the order of integration being justifiable by Fubini's Theorem if, for example, $\int_{-\infty}^{\infty} |\phi(t)|\,dt < \infty$. Subject to this proviso we see, by taking Fourier transforms of both sides of (9.42), that

$$(F\phi)(x) = \left(\frac{2}{\pi}\right)^{\frac{1}{2}} \frac{(a^2+x^2)}{(1+x^2)(x^2+a^2-2a)}$$

$$= \left(\frac{2}{\pi}\right)^{\frac{1}{2}} \left\{\frac{a^2-1}{b^2-1}\cdot\frac{1}{1+x^2} + \frac{b^2-a^2}{b^2-1}\cdot\frac{1}{x^2+b^2}\right\} \quad (x \in \mathbb{R})$$

where $b = (a^2 - 2a)^{\frac{1}{2}}$ and we presume $b \neq 1$. From (9.43) we can recognise the right hand side of the foregoing equation as a Fourier transform, whence

$$\phi(x) = \frac{a^2-1}{b^2-1}e^{-|x|} + \frac{b^2-a^2}{b^2-1}\cdot\frac{1}{b}e^{-b|x|} \quad (x \in \mathbb{R}).$$

Noticing that, for this ϕ, $\int_{-\infty}^{\infty}|\phi| < \infty$ provides retrospective justification of our manipulation and we can easily check that ϕ is a solution of (9.42).

$\square$

It is an obvious question as to the extent to which the ideas used in Example 9.14 can be generalised. To remain within the framework of the operator theory we have already used we shall seek solutions $\phi \in L_2(-\infty, \infty)$ of the integral equations arising. We already know from Lemma 9.2 that the Fourier transform operator is a unitary map from $L_2(-\infty, \infty)$ to itself. If $k \in L_1(-\infty, \infty)$ and we define $(K\phi)(x) = \int_{-\infty}^{\infty} k(x-t)\phi(t)\,dt$ then an application of Schwarz's inequality shows that

$$\int_{-\infty}^{\infty}|K\phi(x)|^2\,dx \leqslant \int_{-\infty}^{\infty}dx \int_{-\infty}^{\infty}|k(x-t)||\phi(t)|^2\,dt \int_{-\infty}^{\infty}|k(x-t)|\,dt$$

$$\leqslant C\int_{-\infty}^{\infty}dt \int_{-\infty}^{\infty}|k(x-t)||\phi(t)|^2\,dx \leqslant C^2\|\phi\|^2,$$

where $C = \int_{-\infty}^{\infty}|k(t)|\,dt$, provided $\phi \in L_2(-\infty, \infty)$, whence K maps $L_2(-\infty, \infty)$ to itself and $\|K\| \leqslant C$. To calculate the Fourier transform of $K\phi$ we notice that

$$\frac{1}{(2\pi)^{\frac{1}{2}}}\int_{-\infty}^{\infty}(K\phi)(x)\,e^{ixs}\,ds = \frac{1}{(2\pi)^{\frac{1}{2}}}\int_{-\infty}^{\infty}\phi(t)\,e^{ixt}\,dt \int_{-\infty}^{\infty}k(s-t)\,e^{ix(s-t)}\,ds$$

$$= (F\phi)(x)\,\hat{k}(x)$$

where $\hat{k}(x) = \int_{-\infty}^{\infty}k(t)\,e^{ixt}\,dt$, the reversal of the order of integration being valid if $\phi \in L_1(-\infty, \infty)$. Therefore if we denote by M the bounded operator on $L_2(-\infty, \infty)$ given by $(M\phi)(x) = \hat{k}(x)\phi(x)$ (bounded since $|\hat{k}(x)| \leqslant \int_{-\infty}^{\infty}|k(t)|\,dt$ $(x \in \mathbb{R})$) then $FK\phi = MF\phi$ for all ϕ belonging to the dense subspace $L_1(-\infty, \infty) \cap L_2(-\infty, \infty)$ of $L_2(-\infty, \infty)$. By Lemma 9.1, then, $FK = MF$ or $K = F^*MF$. The spectrum of K is therefore identical to that of M and since it is an easy matter to show that $\hat{k}$ is continuous we see that $\sigma(K) = \overline{\{\hat{k}(x): x \in \mathbb{R}\}}$. Indeed, we can now calculate $\|M\|$ and hence $\|K\|$ exactly giving $\|K\| = \sup\{|\hat{k}(x)|: x \in \mathbb{R}\}$. Now the Riemann–Lebesgue lemma shows that $\hat{k}(x) \to 0$ as $x \to \pm\infty$ so $0 \in \sigma(K)$ and, apart from the case where $k = 0$ almost everywhere, $\sigma(K)$ is uncountable so that K is certainly not compact. K is self-adjoint if and only if M is, that is, if and only if $\hat{k}$ is real-valued.

We now summarise what we have established in the form of a lemma.

Lemma 9.11

Suppose that $k \in L_1(-\infty, \infty)$. Then the formula $(K\phi)(x) = \int_{-\infty}^{\infty} k(x-t)\phi(t)\,dt$ defines a bounded operator K from $L_2(-\infty, \infty)$ to itself whose spectrum is $\overline{\{\hat{k}(x): x \in \mathbb{R}\}}$ where $\hat{k}(x) = \int_{-\infty}^{\infty} k(t)\,e^{ixt}\,dt$ and which is self-adjoint if and only if $\hat{k}$ is real-valued. $\qquad\square$

The properties of the operator K which we have just proved are essentially the observation that the Fourier transform of the 'convolution' $\int_{-\infty}^{\infty} k(x-t)\phi(t)\,dt$ is $(2\pi)^{\frac{1}{2}}$ times the product of the Fourier transforms of k and ϕ.

Let us now return to the integral equation

$$\phi(x) - \lambda \int_{-\infty}^{\infty} k(x-t)\phi(t)\,dt = f(x) \quad (x \in \mathbb{R})$$

where $k \in L_1(-\infty, \infty)$ and $f \in L_2(-\infty, \infty)$ are given. This is equivalent to $F\phi - \lambda M F\phi = Ff$ where M is, as before, the operator of multiplication by $\hat{k}$. Then, provided λ^{-1} does not belong to the closure of $\{\hat{k}(x): x \in \mathbb{R}\}$ we see that

$$(F\phi)(x) = \frac{(Ff)(x)}{1 - \lambda\hat{k}(x)} \quad (x \in \mathbb{R})$$

and, applying the inverse transform, we see that ϕ satisfies the original equation if and only if

$$\phi(x) = \frac{1}{(2\pi)^{\frac{1}{2}}} \int_{-\infty}^{\infty} \frac{e^{-itx}(Ff)(t)}{1 - \lambda\hat{k}(t)}\,dt$$

for almost all $x \in \mathbb{R}$.

We now turn to the study of the equation $(I - \lambda K)\phi = f$ in the case where $I - \lambda K$ is not invertible, that is, where λ^{-1} belongs to the closure of $\{\hat{k}(x): x \in \mathbb{R}\}$. If it happens that there is a set E of non-zero measure for which $x \in E \Rightarrow \hat{k}(x) = \mu_0$ then μ_0 is an eigenvalue of K and every function whose Fourier transform is zero on $\mathbb{R}\backslash E$ is an eigenfunction; that there are non-trivial functions with this property is easily checked. Notice that the space of eigenvectors has infinite dimension. If $\{x: \hat{k}(x) = \mu_0\}$ has measure zero but μ_0 belongs to the closure of $\{\hat{k}(x): x \in \mathbb{R}\}$ then there is a sequence (x_n) in $\mathbb{R}$ for which $\hat{k}(x_n) \to \mu_0$ as $n \to \infty$. Using this and the continuity of $\hat{k}$ we can produce functions $\phi_n \in L_2(-\infty, \infty)$ for which $(I - \lambda K)\phi_n \to 0$ but $\|\phi_n\| = 1$ for each $n \in \mathbb{N}$, showing that $I - \lambda K$ cannot have a bounded inverse. In this case, even though $(I - \lambda K)\phi = 0$ has no non-trivial solution, the operator $I-\lambda K$ is not invertible, and $(I-\lambda K)\phi_n \to f$ as $n\to\infty$ does not imply that (ϕ_n) converges, let alone that it converges to a solution of $(I-\lambda K)\phi = f$. If $I-\lambda K$ is not invertible then $(I-\lambda K)\phi = f$ will have a solution if and only if the function $(Ff)(x)/(1-\lambda \hat{k}(x))$ belongs to $L_2(-\infty, \infty)$, which, effectively, imposes conditions on the behaviour of Ff near points x at which $\lambda \hat{k}(x)=1$.

Example 9.15

Consider the equation

$$\phi(x)-\lambda\int_{-\infty}^{\infty}\frac{\phi(t)\,dt}{1+(x-t)^2}=\begin{cases}1 & (|x|\leqslant 1)\\0 & (|x|>1).\end{cases} \tag{9.44}$$

Let $k(x)=1/(1+x^2)$ so that $\hat{k}(x)=\pi e^{-|x|}$, and, noticing that the Fourier transform of the function whose value is 1 on $[-1, 1]$ and zero elsewhere is $(2/\pi)^{\frac{1}{2}}(\sin x)/x$ we see that (9.44) is equivalent to

$$(1-\lambda\pi e^{-|x|})(F\phi)(x)=\left(\frac{2}{\pi}\right)^{\frac{1}{2}}\frac{\sin x}{x}\quad(x\in\mathbb{R}).$$

Since $1/(1-\lambda\pi e^{-|x|})$ is a bounded function unless $\lambda\in[\pi^{-1}, \infty)$ we see that for $\lambda\notin[\pi^{-1}, \infty]$ (9.44) has a unique solution ϕ given by

$$\phi(x)=\frac{1}{\pi}\int_{-\infty}^{\infty}\frac{e^{-ixt}}{1-\lambda\pi e^{-|t|}}\frac{\sin t}{t}\,dt\quad(x\in\mathbb{R}).$$

If $\lambda\geqslant\pi^{-1}$ then $1-\lambda\pi e^{-|x|}=0$ for $x=\pm\log\lambda\pi$, and it is easily checked that the function ψ whose value at x is

$$\psi(x)=\left(\frac{2}{\pi}\right)^{\frac{1}{2}}\frac{\sin x}{x}\frac{1}{1-\lambda\pi e^{-|x|}}\quad(x\in\mathbb{R}\setminus\{\pm\log\lambda\pi\})$$

does not belong to $L_2(-\infty, \infty)$ if $\sin(\log\lambda\pi)\neq 0$ (for then the function behaves like $1/|x-x_0|$ as $x\to x_0=\pm\log\lambda\pi$). On the other hand if $\sin(\log\lambda\pi)=0$ and $\lambda\pi\neq 1$, the function ψ tends to a limit as $x\to\pm\log\lambda\pi$ and $\psi\in L_2(-\infty, \infty)$. Therefore (9.44) does have a solution $\phi\in L_2(-\infty, \infty)$, which is unique, if $\lambda=\pi^{-1}e^{n\pi}$ $(n=\pm 1, \pm 2, \ldots)$. (For $\lambda\pi=1$ the point at which $1-\lambda\pi e^{-|x|}=0$ is $x=0$ and in this case $|\psi(x)|\sim(2/\pi)^{\frac{1}{2}}|x^{-1}|$ as $x\to 0$.) $\square$

9.7 An example

We conclude with an example which brings together Fourier transforms and integral equations with unbounded kernels in a way which is frequently encountered in practice.

Let the continuous function $u: [0, \infty) \times [0, \infty) \to \mathbb{R}$ be defined by the boundary value problem

$$u_{xx} + u_{yy} = 0 \qquad (x > 0, y > 0),$$
$$u(x, 0) = 0 \qquad (x \geqslant 0),$$
$$u(0, y) = \beta \sinh(\alpha y) \qquad (0 \leqslant y \leqslant 1),$$
$$u_x(0, y) = e^{-\alpha y} \qquad (y > 1),$$
$$u \to 0 \text{ as } x^2 + y^2 \to \infty.$$

In addition, the first derivatives of u are required to be bounded for $x \geqslant 0$, $y \geqslant 0$. The real parameter α is positive and may be regarded as assigned. The value of the real quantity β is not given in advance and is to be determined in the solution process; we shall see that the conditions imposed on u make this possible.

To dispel the impression that this situation is somewhat contrived, it should be added that the problem originates in water wave theory and is concerned with the scattering of a train of small amplitude surface waves by a submerged barrier. The quantity β is related to the amplitude of the scattered waves and is of particular importance.

We begin by converting the boundary value problem into an integral equation, and this is achieved by using the Fourier sine transform. The classical sine transform is defined by

$$(F_s \phi)(x) = \left(\frac{2}{\pi}\right)^{\frac{1}{2}} \int_0^\infty \sin(xt)\phi(t)\, dt \quad (x \geqslant 0)$$

and extends to a bounded unitary operator on $L_2(0, \infty)$, such that $F_s F_s^* = I$. The proof of this assertion is very similar to that given for the Fourier transform itself and can easily be deduced from the results in §9.4.

Now define the transform

$$U(x, s) = \left(\frac{2}{\pi}\right)^{\frac{1}{2}} \int_0^\infty \sin(sy)u(x, y)\, dy \quad (x \geqslant 0, s \geqslant 0)$$

and note that the differential equation satisfied by u and the boundary condition on $y = 0$ together imply that $U_{xx} - s^2 U = 0$. We must choose $U(x, s) = A(s)\, e^{-sx}$ so that the resulting expression for u is bounded as $x \to \infty$. Thus, using $F_s F_s^* = I$ we have

$$u(x, y) = \left(\frac{2}{\pi}\right)^{\frac{1}{2}} \int_0^\infty \sin(sy)A(s)\, e^{-sx}\, ds \quad (x \geqslant 0, y \geqslant 0)$$

and it remains to satisfy the boundary conditions on $x = 0$.

Now $U_x(0, s) = -sA(s) = (2/\pi)^{\frac{1}{2}} \int_0^\infty \sin(st) u_x(0, t) \, dt \ (s \geqslant 0)$ and so

$$u(x, y) = -\frac{2}{\pi} \int_0^\infty \frac{\sin(sy) e^{-sx}}{s} \, ds \int_0^\infty \sin(st) u_x(0, t) \, dt \quad (x \geqslant 0, y \geqslant 0) \quad (9.45)$$

from which it follows that u is determined once we have found $u_x(0, y)$ for $y \in [0, 1]$. It is actually easier to deal with the function ϕ, where $\phi(y) = u_x(0, y) - e^{-\alpha y} \ (0 \leqslant y \leqslant 1)$, rather than u_x itself. Since $\int_0^\infty \sin(st) e^{-\alpha t} \, dt = s(s^2 + \alpha^2)^{-1}$ and $u_x(0, y) = e^{-\alpha y} \ (y > 1)$, (9.45) gives

$$u(x, y) = -\frac{2}{\pi} \int_0^\infty \frac{\sin(sy) e^{-sx}}{s} \, ds \left\{ \int_0^1 \sin(st) \phi(t) \, dt + s(s^2 + \alpha^2)^{-1} \right\}$$

$$= -\frac{2}{\pi} \int_0^1 \phi(t) \, dt \int_0^\infty \frac{\sin(sy) \sin(st) e^{-sx}}{s} \, ds - \frac{2}{\pi} \int_0^\infty \frac{\sin(sy) e^{-sx}}{s^2 + \alpha^2} \, ds$$

$$(x \geqslant 0, y \geqslant 0).$$

Using the identity

$$2 \int_0^\infty s^{-1} \sin(sy) \sin(st) \, ds = \log \left| \frac{y+t}{y-t} \right| \quad (y > 0, t > 0, y \neq t)$$

and the outstanding boundary condition we deduce that ϕ satisfies the integral equation

$$\int_0^1 \log \left| \frac{y+t}{y-t} \right| \phi(t) \, dt = f(y) \quad (0 \leqslant y \leqslant 1), \quad (9.46)$$

where

$$f(y) = -\pi \beta \sinh(\alpha y) - 2 \int_0^\infty (s^2 + \alpha^2)^{-1} \sin(sy) \, ds \quad (0 \leqslant y \leqslant 1). \quad (9.47)$$

By the terms of the boundary value problem we seek a solution $\phi(y)$ of (9.46) which is bounded for $0 \leqslant y \leqslant 1$. Noting also that $f \in L_2(0, 1)$, we can therefore consider (9.46) via the corresponding equation in $L_2(0, 1)$. Of the several solution processes available for the resulting operator equation, the most direct is provided by Problem 6.7 which shows that it can be written in the form $TT^*\phi = f$ where T is the bounded operator on $L_2(0, 1)$ defined by

$$(T\phi)(y) = \int_0^y \left(\frac{2t}{y^2 - t^2} \right)^{\frac{1}{2}} \phi(t) \, dt \quad (0 \leqslant y \leqslant 1).$$

We can implement the method of §9.3, setting $T\psi = f$ and $\psi = T^*\phi$. It is then easy to show that

$$\psi(y) = \frac{1}{\pi}(2y)^{-\frac{1}{2}} \frac{d}{dy} \int_0^y \frac{2tf(t)\,dt}{(y^2-t^2)^{\frac{1}{2}}}, \quad \phi(y) = -\frac{1}{\pi}\frac{d}{dy}\int_y^1 \left(\frac{2t}{t^2-y^2}\right)^{\frac{1}{2}} \psi(t)\,dt$$
$$(0 \leqslant y \leqslant 1), \quad (9.48)$$

subject to the provisos associated with this solution process, which we discussed in the earlier section. In particular we are certainly seeking a solution $\phi \in L_2(0, 1)$ of $TT^*\phi = f$; that there is at most one such solution follows from the observation that $T^*\phi = 0 \Rightarrow \phi = 0$.

To evaluate ψ we first note that $f(0) = 0$ and that f is continuous on $[0, 1]$ and differentiable on $(0, 1]$. In fact

$$f'(y) = -\pi\alpha\beta\cosh(\alpha y) - 2\int_0^\infty s(s^2+\alpha^2)^{-1}\cos(sy)\,ds \quad (0 < y < 1)$$

and we see that $f' \in L_2(0, 1)$ (the cosine transform $(F_c\phi)(x) = (2/\pi)^{\frac{1}{2}} \int_0^\infty \cos(xt)\phi(t)\,dt$ defines a bounded operator on $L_2(0, \infty)$). With this information in hand it is not difficult to verify that we can write ψ in the form

$$\psi(y) = \frac{1}{\pi}(2y)^{\frac{1}{2}} \int_0^y \frac{f'(t)\,dt}{(y^2-t^2)^{\frac{1}{2}}} \quad (0 \leqslant y \leqslant 1). \quad (9.49)$$

At this point we have to evaluate the integrals arising in this last expression, and we find ourselves on familiar ground. Replacing $\cosh(xt)$ by its Taylor series and using the beta function, there is no difficulty in showing that

$$\int_0^1 (1-t^2)^{-\frac{1}{2}}\cosh(xt)\,dt = \frac{\pi}{2}\sum_{n=0}^\infty \left(\frac{x^2}{4}\right)^n (n!)^{-2}.$$

The sum is in fact equal to the modified Bessel function $I_0(x)$, and therefore

$$\int_0^y \frac{\cosh(\alpha t)\,dt}{(y^2-t^2)^{\frac{1}{2}}} = \frac{\pi}{2}I_0(\alpha y).$$

We also observe that

$$\int_0^y \frac{dt}{(y^2-t^2)^{\frac{1}{2}}}\int_0^\infty \frac{s\cos(st)}{(s^2+\alpha^2)}\,ds = \int_0^\infty u\cos(\alpha yu)\,du \int_0^1 \frac{ds}{(1-s^2)^{\frac{1}{2}}(u^2+s^2)}$$
$$= \frac{\pi}{2}\int_0^\infty \frac{\cos(\alpha yu)}{(u^2+1)^{\frac{1}{2}}}\,du = \frac{\pi}{2}K_0(\alpha y) \quad (y > 0)$$

which follows on making some variable changes, evaluating a straightforward integral and using the cosine transform

$$(2/\pi)^{\frac{1}{2}} \int_0^\infty (t^2+1)^{-\frac{1}{2}} \cos(xt)\, dt = (2/\pi)^{\frac{1}{2}} K_0(x) \quad (x>0).$$

K_0 is a second solution of the modified Bessel equation of order zero and is mentioned in Problem 3.16, where we have indicated how both I_0 and K_0 may be approximated to any desired accuracy. One can deduce from Problem 3.16 that $K_0(x) = -\log x + k_0(x)$ where k_0 is continuous and has a continuous derivative in $[0, 1]$ such that $k_0'(0)=0$.

Returning to (9.49) we now see that

$$\psi(y) = -\tfrac{1}{2}(2y)^{\frac{1}{2}}\{\pi\alpha\beta I_0(\alpha y)+2K_0(\alpha y)\} \quad (0\leqslant y \leqslant 1),$$

so that $\psi \in L_2(0, 1)$ and, by (9.48),

$$\phi(y) = \frac{1}{\pi}\frac{d}{dy}\int_y^1 \frac{t\{\pi\alpha\beta I_0(\alpha t)+2K_0(\alpha t)\}}{(t^2-y^2)^{\frac{1}{2}}}\, dt \quad (0\leqslant y \leqslant 1). \tag{9.50}$$

The final step is to ensure that this solution is bounded, a requirement which determines the value of β. From earlier remarks we see that the Bessel functions I_0 and K_0 are such that

$$\pi\alpha\beta I_0(\alpha t)+2K_0(\alpha t) = -2\log t + h(t)$$

where h is continuous and has a continuous derivative in $[0, 1]$ such that $h'(0)=0$. It is not difficult to show that

$$\frac{d}{dy}\int_y^1 \frac{t\log t\, dt}{(t^2-y^2)^{\frac{1}{2}}} = \frac{d}{dy}\left\{y\tan^{-1}\!\left(\frac{(1-y^2)^{\frac{1}{2}}}{y}\right)-(1-y^2)^{\frac{1}{2}}\right\}$$

$$= \tan^{-1}\!\left(\frac{(1-y^2)^{\frac{1}{2}}}{y}\right)$$

where $\tan^{-1}:\mathbb{R}\to(-\pi/2, \pi/2)$. Further,

$$\frac{d}{dy}\int_y^1 \frac{t\, h(t)\, dt}{(t^2-y^2)^{\frac{1}{2}}} = \frac{d}{dy}\left\{(1-y^2)^{\frac{1}{2}}h(1)-\int_y^1 (t^2-y^2)^{\frac{1}{2}}h'(t)\, dt\right\}.$$

We infer that for the function ϕ given by (9.50) to be bounded in $[0, 1]$ it is necessary that $h(1)=0$. This condition is also sufficient, because

$$\phi(y) = -\frac{2}{\pi}\tan^{-1}\!\left(\frac{(1-y^2)^{\frac{1}{2}}}{y}\right)+\frac{1}{\pi}y\int_y^1 \frac{h'(t)\, dt}{(t^2-y^2)^{\frac{1}{2}}} \quad (0\leqslant y \leqslant 1) \tag{9.51}$$

is certainly bounded in $[0, 1]$, the second term by virtue of the properties of h'.

It is easily checked that (9.51) can be written as

$$\phi(y) = \frac{1}{\pi} y \int_y^1 \frac{g'(t)\,dt}{(t^2 - y^2)^{\frac{1}{2}}} \quad (0 \leqslant y \leqslant 1), \qquad (9.52)$$

where $g(t) = \pi\alpha\beta I_0(\alpha t) + 2K_0(\alpha t)$. Since $h(1) = 0 \Rightarrow g(1) = 0$ the required value of β is given by $\beta = -2K_0(\alpha)/\pi\alpha I_0(\alpha)$ and the solution of the boundary value problem is complete.

One may derive alternative forms for ϕ. For instance, (9.48) and (9.49) can be combined to give

$$\phi(y) = -\frac{1}{\pi^2} \frac{d}{dy} \int_0^1 f'(t)\,dt \int_{\max(y,t)}^1 \frac{2s\,ds}{(s^2 - y^2)^{\frac{1}{2}}(s^2 - t^2)^{\frac{1}{2}}}$$

$$= -\frac{1}{\pi^2} \frac{d}{dy} \int_0^1 \log \left| \frac{(1 - y^2)^{\frac{1}{2}} + (1 - t^2)^{\frac{1}{2}}}{(1 - y^2)^{\frac{1}{2}} - (1 - t^2)^{\frac{1}{2}}} \right| f'(t)\,dt \quad (0 \leqslant y \leqslant 1). \qquad (9.53)$$

This version of ϕ can also be deduced if a different tactic is used at the outset and (9.46) is solved by spectral methods. Reference to Problem 6.4 shows that this is possible following a change of variables. The version (9.53) of ϕ is less useful than (9.52) from the point of view of computing its values. As we have remarked in similar situations previously, the most practically convenient versions of solutions of singular integral equations are usually those expressed in terms of successive indefinite integrals, the prototype being the solution of Abel's equation (1.1).

Problems

9.1 Let the operators L and M be defined on $L_2(0, 1)$ by

$$(L\phi)(x) = \int_0^x \ell(x - t)\phi(t)\,dt, \quad (M\phi)(x) = \int_0^x m(x - t)\phi(t)\,dt \quad (0 \leqslant x \leqslant 1).$$

Show that the kernel of ML is

$$k(X) = \int_0^1 m(X(1 - u))\ell(Xu)X\,du$$

where $X = x - t > 0$.

To solve the equation $L\phi = f$ one may seek a 'simplifying operator' M such that k is a function which allows $ML\phi = Mf$ to be solved easily. The following examples illustrate the method.

(i)   Given that

$$\int_0^{\pi/2} \cosh(X^{\frac{1}{2}}\cos\theta)\cos(X^{\frac{1}{2}}\sin\theta)\,d\theta = \tfrac{1}{2}\pi \quad (X\geqslant 0),$$

$$\int_0^{\pi/2} \cos(X\cos^2\theta)\sin(X\sin^2\theta)\,d\theta = \tfrac{1}{4}\pi\sin X \quad (X\geqslant 0),$$

solve the integral equations

$$\int_0^x \frac{\cos\{(x-t)^{\frac{1}{2}}\}}{(x-t)^{\frac{1}{2}}}\phi(t)\,dt = f(x), \quad \int_0^x \frac{\cosh\{(x-t)^{\frac{1}{2}}\}}{(x-t)^{\frac{1}{2}}}\phi(t)\,dt = f(x),$$

$$\int_0^x \frac{\cos(x-t)}{(x-t)^{\frac{1}{2}}}\phi(t)\,dt = f(x), \quad \int_0^x \frac{\sin(x-t)}{(x-t)^{\frac{1}{2}}}\phi(t)\,dt = f(x),$$

where $0\leqslant x\leqslant 1$.

(ii)   Solve

$$\int_0^x J_0(x-t)\phi(t)\,dt = f(x) \quad (0\leqslant x\leqslant 1)$$

for suitable functions f, where J_0 is the zero order Bessel function and it is given that

$$X\int_0^1 J_0(X-Xu)J_0(Xu)\,du = \sin X \quad (X\geqslant 0).$$

Show also that the second kind equation

$$\phi(x) = f(x) + \lambda\int_0^x J_0(x-t)\phi(t)\,dt \quad (0\leqslant x\leqslant 1)$$

can be solved by a similar technique.

(The Maclaurin series for J_0 which can be derived as indicated in Problem 3.16 may be regarded as the definition of the Bessel function.)

9.2   Determine the integrable solution of

$$\int_0^1 \log\left|\frac{t+x}{t-x}\right|\phi(t)\,dt = x(1-x) \quad (0\leqslant x\leqslant 1).$$

9.3   Use Problem 6.9 to show that the operator T defined on $L_2(0,1)$ by

$$(T\phi)(x) = 2\int_0^x \frac{xt^{\frac{1}{2}(v-1)}\phi(t)}{(x^2-t^2)^{\frac{1}{2}(v+1)}}\,dt \quad (0\leqslant x\leqslant 1)$$

where $0<v<1$, is such that

$$(TT^*\phi)(x) = B(\tfrac{1}{2}v,\tfrac{1}{2}(1-v))\int_0^1 \{|x-t|^{-v}+(x+t)^{-v}\}\phi(t)\,dt \quad (0\leqslant x\leqslant 1).$$

Hence show that the unique solution of the equation

$$\int_0^1 \{|x-t|^{-v} + (x+t)^{-v}\}\phi(t)\,dt = 1 \quad (0 \leqslant x \leqslant 1)$$

where $0 < v < 1$, in $L_2(0, 1)$ is defined by

$$\phi(x) = \pi^{-1}\cos(\tfrac{1}{2}\pi v)(1-x^2)^{-\frac{1}{2}(1-v)} \quad (0 \leqslant x < 1).$$

9.4   Let $(K\phi)(x) = \int_0^1 |x-t|^{-v}\phi(t)\,dt$ $(0 \leqslant x \leqslant 1)$, where $0 < v < 1$ and let $f(x) = x$ $(0 \leqslant x \leqslant 1)$.

(i)   Using the factorisation of K given in Example 9.2 show that the unique solution of $K\phi = f$ in $L_2(0, 1)$ is given by

$$\phi(x) = (\pi v)^{-1}\cos(\tfrac{1}{2}\pi v)\{x(1-x)\}^{\frac{1}{2}(v-1)}\{x + \tfrac{1}{2}(v-1)\} \quad (0 < x < 1).$$

(ii)   Let S be the bounded operator on $L_2(0, 1)$ defined by $(S\phi)(x) = \int_0^x t^{\frac{1}{2}(v-1)}(x-t)^{\frac{1}{2}(v-1)}\phi(t)\,dt$ where $0 < v < 1$. Show that $(ST\phi)(x) = \pi\sec(\tfrac{1}{2}\pi v)\int_0^x t^{\frac{1}{2}(v-1)}\phi(t)\,dt$ $(0 \leqslant x \leqslant 1)$, where T is the operator used in Example 9.2, and deduce that

$$vB(v, \tfrac{1}{2}-\tfrac{1}{2}v)(SKS^*\phi)(x) = \pi^2\sec^2(\tfrac{1}{2}\pi v)\int_0^1 \min(x^v, t^v)\phi(t)\,dt$$
$$(0 \leqslant x \leqslant 1).$$

Now put $\phi = S^*\psi$ in $K\phi = f$ and deduce that ψ satisfies the integral equation

$$\int_0^1 \min(x^v, t^v)\psi(t)\,dt = (2\pi)^{-1}\cos(\tfrac{1}{2}\pi v)x^{v+1} \quad (0 \leqslant x \leqslant 1).$$

Differentiate to obtain ψ and deduce that

$$\phi(x) = -(\pi v)^{-1}\cos(\tfrac{1}{2}\pi v)x^{\frac{1}{2}(v-1)}(1-x)^{\frac{1}{2}(1+v)} \quad (0 < x \leqslant 1).$$

This 'solution' is not the same as that obtained in part (i) which, by the work in the text, is the unique solution. The ϕ we have found is wrong, despite the plausibility of the steps given above. This illustrates the need for care in using these techniques, particularly in distinguishing between conditions which are necessary and those which are sufficient, and the rather cavalier progress above has led us astray. The equation for ψ is necessarily satisfied by any function ψ for which $SKS^*\psi = Sf$, and in turn an equation for ψ may be deduced, necessarily satisfied by ψ. To proceed further we need to check that the ψ we have found *does* satisfy the equation, which turns out to be false. The essential difficulty here is that not every function is of the form $S^*\psi$ and the solution of the equation in hand turns out not to be of the required form.

9.5   In Example 9.3 we considered the equation

$$\int_0^1 \left\{ \frac{1}{|x-t|^v} - \frac{1}{(x+t)^v} \right\} \phi(t)\, dt = f(x) \quad (0<x<1) \tag{9.54}$$

$f \in L_2(0,1)$ and $v \in (0,1)$ being given. By introducing the operator T where $(T\phi)(x) = 2\int_0^x (t/(x^2-t^2))^{\frac{1}{2}(v+1)} \phi(t)\, dt$ and successively solving $T\psi = g$ and $T^*\phi = \psi$ (with $g = B(\frac{1}{2}v, \frac{1}{2}-\frac{1}{2}v)f$) we found a solution to the original equation. This problem aims to explore the validity of this solution.

Show first that if (9.54) has a solution $\phi \in L_2(0,1)$ then $\psi = T^*\phi \in L_2(0,1)$ and ψ and ϕ are equal almost everywhere to the formulae given. Therefore, if there is a solution $\phi \in L_2(0,1)$ it is unique and given by (9.8).

Suppose now that $G(x) = \int_0^x tg(t)(x^2-t^2)^{-\frac{1}{2}(1-v)}\, dt$ is continuous on $[0,1)$ with $G(0)=0$, is differentiable at all but finitely many points of $(0,1)$ and that its derivative belongs to $L_1(0,1)$; prove that, with ψ given by the quoted formula $T\psi = g$. Show further that if $\psi(x)$ exists for almost all x and is such that $H(x) = \int_x^1 t^{\frac{1}{2}(1-v)}\psi(t)(t^2-x^2)^{-\frac{1}{2}(1-v)}\, dt$ exists, is continuous on $(0,1]$ with $H(1)=0$, and is differentiable for all but finitely many x, and the resulting derivative is in $L_1(0,1)$ then ϕ satisfies (9.54). Notice that we have checked that the function ϕ given by

$$\phi(x) = -\frac{B(\frac{1}{2}v, \frac{1}{2}-\frac{1}{2}v)}{\pi^2 \sec^2(\frac{1}{2}\pi v)} \frac{d}{dx} \int_x^1 \frac{s^{-v}\, ds}{(s^2-x^2)^{\frac{1}{2}(1-v)}} \frac{d}{ds} \int_0^s \frac{tf(t)\, dt}{(s^2-t^2)^{\frac{1}{2}(1-v)}}$$

satisfies the equation (9.54) provided the two functions to be differentiated are indeed differentiable at all but finitely many points, that they are continuous at all internal points and at the appropriate end points (that is, the integral with respect to t is continuous at $s=0$, and the other continuous at $x=1$) and that the resulting functions are well enough behaved to be in $L_1(0,1)$. If we seek $\phi \in L_2(0,1)$ then, of course, that is an additional demand; $\phi \in L_2(0,1)$ forces $\psi = T^*\phi$ to be in $L_2(0,1)$ also.

9.6   (i)   Let $0<v<1$ and

$$(K_v\phi)(x) = \int_0^x \frac{\phi(t)\, dt}{(x-t)^v} \quad (0\leqslant x \leqslant 1).$$

Show that $K_{1-v}(K_v + K_v^*)K_{1-v}^*$ has the kernel $\pi v^{-1} \operatorname{cosec}(\pi v)(t^v + x^v - |x-t|^v)$ $(0\leqslant x,t\leqslant 1)$. Using the result established in Example 6.7 deduce that the operator L defined on $L_2(0,1)$ by

$$(L\phi)(x) = \int_0^1 |x-t|^v \phi(t)\, dt \quad (0\leqslant x \leqslant 1)$$

has only one positive eigenvalue and infinitely many negative eigenvalues.

(ii) Let ϕ satisfy the integral equation

$$\int_0^1 |x-t|^v \phi(t)\,dt = 1 \quad (0 \leqslant x \leqslant 1),$$

where $0 < v < 1$. By using the relationship between L and $K_{1-v}(K_v + K_v^*)K_{1-v}^*$ arising in (i) and proceeding as in Example 9.4, show that the corresponding equation in $L_2(0, 1)$ has no solution. Show also that the integral equation is satisfied by the integrable function ϕ where

$$\phi(x) = \pi^{-1}\cos(\tfrac{1}{2}\pi v)\{x(1-x)\}^{-\frac{1}{2}(1+v)} \quad (0 < x < 1).$$

The result

$$\int_x^1 \frac{\{t(1-t)\}^{-\frac{1}{2}(1+v)}}{(t-x)^{1-v}}\,dt = B(\tfrac{1}{2}-\tfrac{1}{2}v, v)\{x(1-x)\}^{-\frac{1}{2}(1-v)} \quad (0 < x < 1),$$

which follows on setting $t = x(x+s-sx)^{-1}$, will prove useful. (Part (i) generalises a result given in the case $v = \tfrac{1}{2}$ in Problem 7.17; bounds for the negative eigenvalues of L can be obtained by extending the procedure set out in that problem. Part (ii) generalises the problem in Example 9.4.)

9.7   Let the operator K on $L_2(0, \pi)$ be defined by

$$(K\phi)(\theta) = \int_0^\pi \log\left|\frac{\sin\tfrac{1}{2}(\theta+\sigma)}{\sin\tfrac{1}{2}(\theta-\sigma)}\right|\phi(\sigma)\,d\sigma \quad (0 \leqslant \theta \leqslant \pi)$$

and let $f(\theta) = \sin\theta$ $(0 \leqslant \theta \leqslant \pi)$. Solve $K\phi = f$ (i) by using the factorisation $K = SS^*$ given in Problem 6.8 and (ii) by spectral methods, using the Fourier series

$$\sum_{n=1}^\infty n^{-1}\cos(n\theta) = -\tfrac{1}{2}\log\{2(1-\cos\theta)\} \quad (0 < \theta < 2\pi).$$

(The Fourier series is derived in Appendix C.)

9.8   Let H_0 denote the Hilbert transform on $L_2(-\infty, \infty)$. Show that if $\phi \in L_2(-\infty, \infty)$ and $i\pi\phi = H_0\phi$ then $(F\phi)(x) = 0$ for almost all $x \in (0, \infty)$. Show further that the equation $i\pi\phi = f + H_0\phi$, where $f \in L_2(-\infty, \infty)$ is given, has a solution if and only if $(Ff)(x) = 0$ for almost all $x \in (-\infty, 0)$ and that if f satisfies this condition $\phi = (2i\pi)^{-1}f + \psi$ where ψ is any function in $L_2(-\infty, \infty)$ for which $(F\psi)(x) = 0$ for almost all $x \in (0, \infty)$.

9.9   Let $\phi_\beta(x) = x^{-\beta}(1-x)^\beta$ $(0 < x < 1, 0 < |\beta| < \tfrac{1}{2})$ and let H denote the Hilbert transform. Show that the unique solution of $\mu\phi = \phi_{-\beta} + H\phi$ in $L_2(0, 1)$, where μ is real and non-zero, is given by

$$\pi\sin(\pi\beta)\{\cot(\pi\beta) + \cot(\pi\gamma)\}\phi = \sin(\pi\beta)\phi_{-\beta} + \sin(\pi\gamma)\phi_\gamma$$

where $\mu = \pi\cot(\pi\gamma)$ and $0 < |\gamma| < \tfrac{1}{2}$.

9.10 By converting the integral equation $\int_0^1 (t-x)^{-1}\phi(t)\,dt=0$ $(0<x<1)$ to one with a logarithmically singular kernel and using spectral methods (see Example 9.9), prove that $H\phi=0$, where H is the Hilbert transform, has only the trivial solution in $L_2(0,1)$.

9.11 Let ϕ satisfy $\mu\phi(x)=\int_0^1 (t-x)^{-1}\phi(t)\,dt$ $(0<x<1)$ and let $\chi(x)=x(1-x)\phi(x)$ $(0<x<1)$. Show that $\mu\chi=f+H\chi$ where H is the Hilbert transform and $f(x)=C_1-(1-x)C_2$ $(0\leqslant x\leqslant1)$, C_1 and C_2 being undetermined constants. By solving for χ in $L_2(0,1)$ deduce that the only solution ϕ which is continuous in $(0,1)$, integrable in $[0,1]$ and satisfies $x\phi(x)\to0$ as $x\to0+$ and $(1-x)\phi(x)\to0$ as $x\to1-$ is as given in Theorem 9.8.

9.12 Let $f(x)=x(1-x)$ $(0\leqslant x\leqslant1)$ and let H denote the Hilbert transform.

(i)   Solve $\mu\phi=f+H\phi$ in $L_2(0,1)$ in the cases $\mu=\pm\pi$.
(ii) Show that $H\phi=f$ has no solution in $L_2(0,1)$.
(iii) Find solutions of

$$\int_0^1 \frac{\phi(t)\,dt}{t-x}=x(1-x) \quad (0\leqslant x\leqslant1)$$

which are integrable in $[0,1]$ and (a) bounded at $x=0$ and (b) bounded at $x=1$.

9.13 Let $-\tfrac12<\gamma<\tfrac12$ and define T_γ by

$$(T_\gamma\phi)(x)=\frac{1}{x}\int_0^\infty \frac{(t/x)^\gamma-1}{(t/x)-1}\phi(t)\,dt \quad (x>0).$$

Use the method of Problem 3.7 (extended to $L_2(0,\infty)$) to show that T_γ is a bounded operator from $L_2(0,\infty)$ to itself and that $\|T_\gamma\|\leqslant\int_0^\infty (t^\gamma-1)t^{-\frac34}/(t-1)\,dt$. (The value of the integral is $|\pi\tan\pi\gamma|$.)

Now set $(U\phi)(x)=\phi(x/(1-x))/(1-x)$ $(0<x<1)$ and show that $U:$ $L_2(0,\infty)\to L_2(0,1)$ is a unitary operator with $(U^{-1}\phi)(x)=\phi(x/(1+x))/(1+x)$ $(x>0)$. Show that

$$R_\gamma=UT_{-\gamma}U^{-1}+H$$

where R_γ is the operator defined by (9.27). Deduce that R_γ is a bounded operator for $|\gamma|<\tfrac12$.

The corresponding operator R_γ with $\gamma=\tfrac12$ does not map $L_2(0,1)$ to itself. This may be seen by setting $\phi(t)=t^{\frac14}(1-t)^{\frac14}$ and observing that the formula for $R_{\frac12}\phi$ produces a function which is not square-integrable.

9.14 In §9.5.1 it is shown that the equation $\mu\phi=f+H\phi$, where H is the Hilbert transform and μ is a real, non-zero parameter, has the unique solution given by $\phi=(\pi^2+\mu^2)^{-1}(\mu f+R_{-\gamma}f)$ in $L_2(0,1)$; $\mu=\pi\cot(\pi\gamma)$, $0<|\gamma|<\tfrac12$ and R_γ is the operator defined by (9.27). Show that $R_{-\gamma}H=HR_{-\gamma}=\mu(R_{-\gamma}-H)-\pi^2 I$.

By using (9.18) to evaluate the inner integral, show that

$$\int_0^1 \left(\frac{t}{1-t}\right)^\gamma \phi(t)\,dt \int_0^1 \left(\frac{1-s}{s}\right)^\gamma \frac{ds}{(s-x)(t-s)} = \mu\{(R_{-\gamma}\phi)(x) - (H\phi)(x)\}$$

for $\phi \in L_2(0,1)$ and for almost all $x \in [0,1]$. Deduce that

$$\int_0^1 \left(\frac{1-s}{s}\right)^\gamma \frac{ds}{s-x} \int_0^1 \left(\frac{t}{1-t}\right)^\gamma \frac{\phi(t)\,dt}{t-s}$$

$$- \int_0^1 \left(\frac{t}{1-t}\right)^\gamma \phi(t)\,dt \int_0^1 \left(\frac{1-s}{s}\right)^\gamma \frac{ds}{(s-x)(t-s)} = -\pi^2 \phi(x)$$

for $0 < |\gamma| < \tfrac{1}{2}$, $\phi \in L_2(0,1)$ and almost all $x \in [0,1]$.

(This is a form of the *Poincaré–Bertrand formula* for changing the integration order in two successive Cauchy principal value integrals. This formula can be used as a starting point in a study of the equation $\mu\phi = f + H\phi$ and indeed this is the traditional approach to Cauchy singular integral equations, expounded, for example, by N. I. Muskhelishvili in his text *Singular Integral Equations* (Noordhoff, 1953). We have deduced the formula, having solved $\mu\phi = f + H\phi$ by other means; its direct proof requires sophisticated real and complex analysis which our approach has avoided.)

9.15 Let ℓ be defined by

$$\ell(x,t) = \int_0^x \frac{s^{-\frac{1}{2}}(x-s)^{-\frac{1}{2}}}{t-s}\,ds \quad (0 < x, t < 1,\ x \neq t).$$

By putting $s = uxt(ux + t - x)^{-1}$ show that $\ell(x,t) = \pi t^{-\frac{1}{2}}(t-x)^{-\frac{1}{2}}$ $(0 < x < t < 1)$; by using $\int_0^\infty u^{-\frac{1}{2}}(1-u)^{-1}\,du = 0$ (see Appendix C) show that $\ell(x,t) = 0$ $(0 < t < x < 1)$.

Now let T be the bounded operator on $L_2(0,1)$ defined by

$$(T\phi)(x) = \int_0^x t^{\frac{1}{2}}(x-t)^{-\frac{1}{2}}\phi(t)\,dt$$

and let H denote the Hilbert transform. Use the method of Theorem 9.6 to prove that

$$(TH\phi)(x) = \pi \int_x^1 t^{\frac{1}{2}}(t-x)^{-\frac{1}{2}}\phi(t)\,dt - \pi \int_0^1 \phi(t)\,dt \quad (0 \leq x \leq 1)$$

for $\phi \in L_2(0,1)$.

Using T as a 'simplifying operator' show that if $H\phi = f$ has a solution $\phi \in L_2(0,1)$ then it is given by the formula (9.39).

9.16 Let T, S and P be the bounded operators on $L_2(0,1)$ defined by

$$(T\phi)(x) = \int_0^x \frac{t^\gamma \phi(t)\,dt}{(x-t)^\gamma}, \quad (S\phi)(x) = \int_0^x \frac{x^\gamma \phi(t)\,dt}{(x-t)^\gamma}, \quad (P\phi)(x) = \int_0^1 \phi(t)\,dt,$$

where $0<\gamma<1$. Show that

$$TH=\pi\cot(\pi\gamma)T+\pi\operatorname{cosec}(\pi\gamma)(S^*-P),$$

H being the Hilbert transform.

Now let $\mu\phi=f+H\phi$ in $L_2(0,1)$, where μ is real. By using T as a simplifying operator show that

$$\phi(x)=x^{-\gamma}\frac{\sin^2(\pi\gamma)}{\pi^2}\frac{d}{dx}\int_x^1\frac{ds}{(s-x)^{1-\gamma}}\int_0^s\frac{t^\gamma f(t)\,dt}{(s-t)^\gamma}+Cx^{-\gamma}(1-x)^{\gamma-1},\quad(9.55)$$

where $\mu=\pi\cot(\pi\gamma)$ and $0<\gamma<1$, C being a constant; there is a unique value of C for which the right hand side of (9.55) belongs to $L_2(0,1)$.

Use the operator identity further to deduce that, if μ is real and non-zero, the unique solution of $\mu\phi=f+H\phi$ in $L_2(0,1)$ is given by

$$(\pi^2+\mu^2)\phi(x)=\mu f(x)+\frac{(1-x)^\gamma}{x^\gamma}\int_0^1\frac{t^\gamma}{(1-t)^\gamma}\frac{f(t)\,dt}{t-x}+\frac{D}{x^\gamma(1-x)^{1-\gamma}},$$

where $\mu=\pi\cot(\pi\gamma)$ and $\gamma\in(0,\frac12)\cup(\frac12,1)$, for some particular choice of D, provided that $t^{-\gamma}f(1-t)\in L_2(0,1)$ if $\gamma\geq\frac12$.

(This problem extends to the second kind equation, the solution method given in the previous problem for $H\phi=f$. The fact that the single formula (9.55) applies for all real μ (including $\mu=0$) suggests that the present solution process is more satisfactory than that given in §9.5. Note however that Theorem 9.7 provides exact formulae for the unique solution $\phi\in L_2(0,1)$ in the case $\mu\neq0$, the analysis preceding it establishing the existence of such a solution. The first term on the right hand side of (9.55) is not in $L_2(0,1)$ for all $f\in L_2(0,1)$ and all $\mu\neq0$. Note also that Theorem 9.8 shows that if a wider class of solutions than those in $L_2(0,1)$ is accepted, the constant C may be chosen arbitrarily. For $\phi\in L_2(0,1)$ the value of C is determined, even though our process has not found it explicitly. Although (9.55) does not provide a convenient base from which to establish the existence and uniqueness of a solution of $\mu\phi=f+H\phi$ in $L_2(0,1)$, for μ real and non-zero, it is a useful formula. The condition on f is required for $\gamma\geq\frac12$ since the resulting operator R_γ (in the notation of (9.27)) is not a bounded linear map on $L_2(0,1)$.)

9.17 This problem considers the weakly singular Fredholm equation

$$a\int_0^x\frac{\phi(t)\,dt}{(x-t)^\gamma}+b\int_x^1\frac{\phi(t)\,dt}{(t-x)^\gamma}=f(x)\quad(0<x<1),\quad(9.56)$$

where a and b are real, non-zero constants and $\gamma\in(0,1)$.

(i)   Let the operators T, S and P be defined on $L_2(0,1)$ by

$$(T\phi)(x)=\int_0^x\frac{t^\gamma\phi(t)\,dt}{(x-t)^\gamma},\ (S\phi)(x)=\int_0^x\frac{x^\gamma\phi(t)\,dt}{(x-t)^\gamma},\ (P\phi)(x)=\int_0^1\phi(t)\,dt$$

where $0<\gamma<1$. Note that $TH=\pi\cot(\pi\gamma)T+\pi\operatorname{cosec}(\pi\gamma)(S^*-P)$, by the previous problem. Show that ϕ satisfies (9.56) interpreted as an equation in $L_2(0, 1)$, if and only if it satisfies

$$(\mu I - H)T^*\phi = g,$$

where $\mu = \pi\{(b/a)\operatorname{cosec}(\pi\gamma) - \cot(\pi\gamma)\}$ and $g(x) = \pi \operatorname{cosec}(\pi\gamma)\{a^{-1}x^{\gamma}f(x) - (P\phi)(x)\}$ $(0 \leqslant x \leqslant 1)$. Show also that if $\psi \in L_2(0, 1)$, where $\phi(x) = x^{\gamma}\psi(x)$ $(0 \leqslant x \leqslant 1)$, then ψ satisfies

$$T(\nu I - H)\psi = h,$$

where $\nu = \pi\{\cot(\pi\gamma) - (a/b)\operatorname{cosec}(\pi\gamma)\}$ and $h = -\pi \operatorname{cosec}(\pi\gamma)\{b^{-1}f - P\psi\}$ if and only if ϕ satisfies (9.56).

(ii) Now consider the homogeneous version of (9.56), with $a = -b = 1$, via the equation $(\mu I - H)T^*\phi = -\pi \operatorname{cosec}(\pi\gamma)P\phi$. Use Theorem 9.9 to determine $T^*\phi$ and, with the aid of the integral given in Problem 9.6(ii), deduce that the integrable function $\phi(x) = \{x(1-x)\}^{\frac{1}{2}\gamma - 1}$ $(0 < x < 1)$ satisfies

$$\int_0^x \frac{\phi(t)\,dt}{(x-t)^{\gamma}} - \int_x^1 \frac{\phi(t)\,dt}{(t-x)^{\gamma}} = 0 \quad (0 < x < 1).$$

(Using either of its reformulations in terms of the operators H and T, the solution of (9.56) can be determined in a variety of forms. It can be shown that the homogeneous version of (9.56) has a solution which is continuous in $(0, 1)$ and integrable over $[0, 1]$ if and only if $ab < 0$.)

9.18 Let $T \in B(L_2(0, 1))$ be defined by

$$(T\phi)(x) = \int_0^x \frac{\phi(t)\,dt}{t^{\gamma}(x-t)^{1-\gamma}}$$

where $0 < \gamma < \frac{1}{2}$. Solve $\mu\phi = f + H\phi$ by putting $\phi = T^*\psi$ and making use of the operator identity given in Theorem 9.6. In particular, show that for $\mu > 0$ and $\mu\phi = f + H\phi$ then

$$\phi(x) = \frac{\sin^2(\pi\gamma)}{\pi^2} \int_x^1 \frac{ds}{x^{\gamma}(s-x)^{1-\gamma}} \frac{d}{ds} \int_0^s \frac{t^{\gamma}f(t)\,dt}{(s-t)^{\gamma}}$$

for almost all x in $[0, 1]$ where $\mu = \pi \cot(\pi\gamma)$ and $0 < \gamma < \frac{1}{2}$. This formula for ϕ is certainly correct if $\phi = T^*\psi$ for some $\psi \in L_2(0, 1)$; by showing that this condition is equivalent to $f \in \mathcal{R}(S)$ with S as defined in Theorem 9.6, deduce that the formula is valid for f belonging to a dense subspace of $L_2(0, 1)$. (This restriction cannot be wholly removed since f must be such that $d/ds\int_0^s t^{\gamma}f(t)/(s-t)^{\gamma}\,dt$ makes sense.) Show also that the difference between this expression for ϕ and

$$x^{-\gamma}\frac{\sin^2(\pi\gamma)}{\pi^2}\frac{d}{dx}\int_x^1 \frac{ds}{(s-x)^{1-\gamma}}\int_0^s \frac{t^{\gamma}f(t)\,dt}{(s-t)^{\gamma}}$$

is proportional to $x^{-\gamma}(1-x)^{1-\gamma}$ for $0 < x < 1$. (*cf.* (9.55).)

(This problem illustrates, amongst other things, a simplifying operator used 'internally' rather than 'externally'. Different solution formulae result from the two approaches and in a particular application one may be more useful than the other.)

9.19 (i)   Put $u = x(t-s)\{s(t-x)\}^{-1}$ to show that

$$\int_x^t \frac{ds}{s(s-x)^{\frac{1}{2}}(t-s)^{\frac{1}{2}}} = \pi(xt)^{-\frac{1}{2}} \quad (0 < x < t).$$

(ii)   Put $u = (sxt)^{-\frac{1}{2}}\{t^{\frac{1}{2}}(s-x)^{\frac{1}{2}} + x^{\frac{1}{2}}(s-t)^{\frac{1}{2}}\}$ to show that

$$\int_{\max(x,t)}^1 \frac{ds}{s(s-x)^{\frac{1}{2}}(s-t)^{\frac{1}{2}}} = (xt)^{-\frac{1}{2}} \log\left|\frac{t^{\frac{1}{2}}(1-x)^{\frac{1}{2}} + x^{\frac{1}{2}}(1-t)^{\frac{1}{2}}}{t^{\frac{1}{2}}(1-x)^{\frac{1}{2}} - x^{\frac{1}{2}}(1-t)^{\frac{1}{2}}}\right|$$

$$(0 < x, t < 1, x \neq t).$$

(iii) Let the bounded operator S on $L_2(0, 1)$ defined by

$$(S\phi)(x) = \int_0^x \frac{t^{\frac{1}{2}}}{x^{\frac{1}{2}}} \frac{\phi(t)\,dt}{(x-t)^{\frac{1}{2}}}$$

be a 'simplifying operator' for $H\phi = f$ in $L_2(0, 1)$, where H denotes the Hilbert transform; deduce an expression for $(SH\phi)(x)$ from the expression for $(TH\phi)(x)$ given in Problem 9.15. Hence show that $H\phi = f$ has a solution ϕ in $L_2(0, 1)$ if and only if ϕ satisfies

$$\pi^2 \int_x^1 \phi(t)\,dt = (S^*Sf)(x) + 2\pi C \tan^{-1}\{x^{-\frac{1}{2}}(1-x)^{\frac{1}{2}}\}, \quad (9.57)$$

where $C = \int_0^1 \phi(t)\,dt$. Deduce that the solution of $H\phi = f$ in $L_2(0, 1)$, if it exists, is given by

$$\phi(x) = -\frac{1}{\pi^2} \frac{d}{dx} \int_0^1 \log\left|\frac{t^{\frac{1}{2}}(1-x)^{\frac{1}{2}} + x^{\frac{1}{2}}(1-t)^{\frac{1}{2}}}{t^{\frac{1}{2}}(1-x)^{\frac{1}{2}} - x^{\frac{1}{2}}(1-t)^{\frac{1}{2}}}\right| f(t)\,dt + \frac{C\pi}{x^{\frac{1}{2}}(1-x)^{\frac{1}{2}}}$$

for some value of C.

9.20 Let ϕ satisfy the *integro-differential equation*

$$r(x)\phi(x) - \int_0^1 \frac{\phi'(t)\,dt}{t-x} = f(x) \quad (0 < x < 1),$$

where r and f are given functions, and let $\phi(0) = \phi(1) = 0$. By considering the corresponding equation $H\phi' = r\phi - f$ in $L_2(0, 1)$ (that is seeking a solution ϕ such that $\phi' \in L_2(0, 1)$) and using part (iii) of Problem 9.19, show that ϕ also satisfies the integral equation

$$\phi(x) = -\frac{1}{\pi^2} \int_0^1 \log\left|\frac{t^{\frac{1}{2}}(1-x)^{\frac{1}{2}} + x^{\frac{1}{2}}(1-t)^{\frac{1}{2}}}{t^{\frac{1}{2}}(1-x)^{\frac{1}{2}} - x^{\frac{1}{2}}(1-t)^{\frac{1}{2}}}\right| \{r(t)\phi(t) - f(t)\}\,dt \quad (0 \leq x \leq 1).$$

Now consider the particular case in which $r(x) = x^{-\frac{1}{2}}(1-x)^{-\frac{1}{2}}b(x)$, where

$b>0$ in $[0, 1]$, and let $\hat{b}(\theta)=\{b(\sin^2(\theta/2))\}^{\frac{1}{2}}$ $(0\leqslant\theta\leqslant\pi)$. Show that the function $\hat{\phi}$, where $\hat{\phi}(\theta)=\hat{b}(\theta)\phi(\sin^2(\theta/2))$ $(0\leqslant\theta\leqslant\pi)$, satisfies the integral equation

$$\hat{\phi}(\theta)=-\frac{1}{\pi^2}\int_0^\pi \log\left|\frac{\sin\frac{1}{2}(\theta+\sigma)}{\sin\frac{1}{2}(\theta-\sigma)}\right|\hat{b}(\theta)\hat{b}(\sigma)\hat{\phi}(\sigma)\,d\sigma + \hat{f}(\sigma) \quad (0\leqslant\sigma\leqslant\pi),$$

where

$$\hat{f}(\theta)=\frac{\hat{b}(\theta)}{2\pi^2}\int_0^\pi \log\left|\frac{\sin\frac{1}{2}(\theta+\sigma)}{\sin\frac{1}{2}(\theta-\sigma)}\right|(\sin\sigma)f(\sin^2(\sigma/2))\,d\sigma.$$

By referring to Problem 6.8, show that the corresponding equation in $L_2(0, \pi)$ can be written in the form $(I+Q^*Q)\hat{\phi}=\hat{f}$, where Q is a bounded operator. Deduce that, for the r considered, if $\hat{f}\in L_2(0, \pi)$ then there is a unique $\hat{\phi}\in L_2(0, \pi)$ which satisfies the integral equation for $\hat{\phi}$. Show further that if f is such that the resulting function ϕ is differentiable and $\phi'\in L_2(0, 1)$ then, in the notation of (9.57),

$$\pi^2\int_x^1 \phi'(t)\,dt=(S^*S(r\phi-f))(x)$$

and hence ϕ satisfies the given integro-differential equation and boundary conditions.

(The integro-differential equation arises in aerofoil theory; ϕ is the circulation around a wing of finite span, $1/r(x)$ representing the chord length at position x (so that $1/r(x)$ vanishes at $x=0$ and at $x=1$ and is symmetric about $x=\frac{1}{2}$). The term f is related to the angle of incidence of the wing.)

9.21 Suppose that k is differentiable and that k' is bounded, and show that the operator L defined by

$$(L\phi)(x)=\int_0^1 \frac{k(t-x)\phi(t)}{t-x}\,dt \quad (0\leqslant x\leqslant 1)$$

is of the form $k(0)H+K$ where H denotes the Hilbert transform and K is compact. Deduce that the Fredholm Alternative holds for the integral equation

$$\mu\phi(x)=(L\phi)(x)+f(x) \quad (0\leqslant x\leqslant 1)$$

where μ is a non-zero real number.

Use this to show that for all non-zero real numbers μ and all $f\in L_2(0, 1)$ the equation

$$\mu\phi(x)=\int_0^1 \frac{1+(t-x)}{t-x}\phi(t)\,dt + f(x) \quad (0\leqslant x\leqslant 1)$$

has a unique solution $\phi\in L_2(0, 1)$.

9.22 (i) The operator K is defined on $L_2(-\infty, \infty)$ by

$$(K\phi)(x) = \int_{-\infty}^{\infty} \operatorname{sech}\{a(x-t)\}\phi(t)\,dt \quad (a>0).$$

Given that $\int_0^\infty \cos(xt)\operatorname{sech}(at)\,dt = (\pi/2a)\operatorname{sech}(\pi x/2a)$ $(x\geqslant 0,\ a>0)$, determine the spectrum of K and obtain a formula for the solution of $\phi = f + \lambda K\phi$ in $L_2(-\infty, \infty)$ if λ^{-1} is not in the spectrum of K.

(ii) Let $(L\phi)(x) = \int_0^\infty (x+t)^{-1}\phi(t)\,dt$ $(0\leqslant x<\infty)$. Prove that L is a bounded operator on $L_2(0, \infty)$. (The method given in Problem 3.7 can be used.) By making the variable changes $t=e^{2s}$ and $y=e^{2x}$ and using (i) determine the spectrum of L and deduce a formula for the solution of $\phi = f + \lambda L\phi$ in $L_2(0, \infty)$, assuming that λ^{-1} is not in the spectrum of L.

9.23 (i) Prove that the Fourier sine and cosine transforms

$$(F_s\phi)(x) = \left(\frac{2}{\pi}\right)^{\frac{1}{2}} \int_0^\infty \sin(xt)\phi(t)\,dt, \quad (F_c\phi)(x) = \left(\frac{2}{\pi}\right)^{\frac{1}{2}} \int_0^\infty \cos(xt)\phi(t)\,dt,$$

$$(x\geqslant 0)$$

define bounded operators on $L_2(0, \infty)$ such that $F_s^2 = F_c^2 = I$.

(ii) Let

$$(W\phi)(x) = \left(\frac{2}{\pi}\right)^{\frac{1}{2}} \int_0^\infty \frac{\{x\cos(xt) - \alpha\sin(xt)\}}{(x^2+\alpha^2)^{\frac{1}{2}}}\phi(t)\,dt \quad (x\geqslant 0)$$

$$(K\phi)(x) = \left(\frac{2}{\pi}\right)^{\frac{1}{2}} \int_0^\infty \frac{te^{-\alpha x}}{t^2+\alpha^2}\phi(t)\,dt, \quad (x\geqslant 0)$$

$$(N\phi)(x) = \int_0^x e^{-\alpha(x-t)}\phi(t)\,dt \quad (x\geqslant 0)$$

and $(M\phi)(x) = (x^2+\alpha^2)^{-\frac{1}{2}}\phi(x)$ $(x\geqslant 0)$, where α is a positive parameter. Show that W, K, N and M are bounded operators on $L_2(0, \infty)$, that $W\phi_0 = 0$ where $\phi_0(x) = (2\alpha)^{\frac{1}{2}}e^{-\alpha x}$ $(x\geqslant 0)$, that $WN = -MF_s$ and that $NF_s = -W^*M + K$. Deduce that $WW^* = I$ and that $W^*W = I - P$, where $P\phi = (\phi, \phi_0)\phi_0$.

(We have illustrated, in §9.7, how F_s may be used to deduce an integral equation from a boundary value problem for $u(x, y)$ with $u=0$ on one boundary. The operator W plays the same role if u satisfies the mixed boundary condition $u_n + \alpha u = 0$, where u_n denotes the derivative of u normal to the boundary.)

Appendix A: Functional analysis

Suppose that X is a vector space over the real or complex numbers. We say $\|\cdot\|$ is a *norm* on X if, for each $\phi \in X$, $\|\phi\|$ is a non-negative number and

(i) $\|\phi\| = 0 \Leftrightarrow \phi = 0$,

(ii) for all $\phi, \psi \in X$, $\|\phi + \psi\| \leqslant \|\phi\| + \|\psi\|$, and

(iii) for all $\phi \in X$ and scalars μ, $\|\mu\phi\| = |\mu| \|\phi\|$.

The norm defines an idea of distance in X, so that we say that $\phi_n \to \phi$ as $n \to \infty$ in X if $\|\phi_n - \phi\| \to 0$ as $n \to \infty$. A sequence (ϕ_n) in X is said to be *bounded* if $\{\|\phi_n\| : n \in \mathbb{N}\}$ is a bounded set, and to be a *Cauchy sequence* if it satisfies the condition:

for all $\epsilon > 0$ there exists N such that for all $m, n \geqslant N$ $\|\phi_m - \phi_n\| < \epsilon$.

It is easy to see that if (ϕ_n) tends to a limit it must be a Cauchy sequence, but the converse is false in general. A normed vector space in which every Cauchy sequence tends to a limit is called a *Banach space*, and is the only type which will concern us in this book. Examples of Banach spaces are:

$C[a, b]$, the set of all continuous complex-valued functions with domain $[a, b]$, is a Banach space with the natural algebraic operations and the norm $\|\phi\| = \sup\{|f(x)| : a \leqslant x \leqslant b\}$.

$L_2(a, b)$, the set of all equivalence classes of complex-valued Lebesgue measurable functions ϕ (see Appendix B) satisfying the condition $\int_a^b |\phi|^2 < \infty$, is also a Banach space with norm $\|\phi\| = \{\int_a^b |\phi|^2\}^{\frac{1}{2}}$.

In both these examples the functions may be chosen to be real-valued instead of complex-valued, producing an analogous Banach space with real scalars.

The elements of the Banach spaces with which we shall be concerned will be functions (or equivalence classes of functions). The advantage of the viewpoint of functional analysis is partly that one regards an element of a Banach space as a vector (or 'point' in the space) which is conceptually simpler than thinking of it as a function. Only when interpreting the results of the functional analysis do we return to consider these vectors as functions in their own right. The two examples we have given have norms which give rise to known forms of convergence of sequences of functions: $\phi_n \to \phi$ in $C[a, b]$ may be expressed in classical terms as $\phi_n \to \phi$ uniformly on $[a, b]$, while $\phi_n \to \phi$ in $L_2(a, b)$ corresponds to $\int_a^b |\phi_n - \phi|^2 \to 0$, that is, $\phi_n \to \phi$ *in the mean* on $[a, b]$.

A *Hilbert space* H is a Banach space which is additionally endowed with an *inner product*. The inner product of ϕ and ψ is denoted by (ϕ, ψ), has scalar values, and satisfies

(i)   for all ϕ, χ and $\psi \in H$ $(\phi + \chi, \psi) = (\phi, \psi) + (\chi, \psi)$,
(ii)   for all $\phi, \psi \in H$ and scalars μ, $(\mu\phi, \psi) = \mu(\phi, \psi)$,
(iii)   for all $\phi, \psi \in H$ $(\phi, \psi) = \overline{(\psi, \phi)}$,
(iv)   for all $\phi \in H$ $(\phi, \phi) = \|\phi\|^2$.

The space $L_2(a, b)$ above is a Hilbert space, the inner product being given by $(\phi, \psi) = \int_a^b \phi(t)\overline{\psi(t)}\, dt$.

If X and Y are Banach spaces over the same field of scalars and $T: X \to Y$ is a linear map, we say T is a *bounded* linear map if $\{\|T\phi\| : \phi \in X, \|\phi\| \leq 1\}$ is bounded; when this is the case we define the *norm* of T, $\|T\|$, by $\|T\| = \sup\{\|T\phi\| : \phi \in X, \|\phi\| \leq 1\}$. We denote the set of all bounded linear maps from X to Y by $B(X, Y)$; for brevity, $B(X, X)$ is denoted by $B(X)$. We shall call a bounded linear map an *operator* (or, for emphasis, a bounded operator); an operator in X will mean an element of $B(X)$. Notice that if $S, T \in B(X)$ then we may form ST, defined by $ST\phi = S(T\phi)$ and then $\|ST\| \leq \|S\|\|T\|$.

For each Banach space X, we denote by I the *identity operator* defined by $I\phi = \phi$ $(\phi \in X)$. If $S, T \in B(X)$ we say S and T *commute* if $ST = TS$.

Theorem A1

If X and Y are Banach spaces then $B(X, Y)$, with the norm just described, is a Banach space. $\square$

We can form the idea of an infinite series in a Banach space X. $\Sigma\phi_n$ is said to converge in X if (s_n) tends to a limit as $n \to \infty$, where $s_n = \phi_1 + \phi_2 + \cdots + \phi_n$. If $s_n \to \phi$ as $n \to \infty$, we say ϕ is the sum of the series $\Sigma\phi_n$. (The convergence here is, of course, in X, that is, $\|s_n - \phi\| \to 0$ as $n \to \infty$.) The series $\Sigma\phi_n$ is said to be *absolutely convergent* if $\Sigma\|\phi_n\|$ converges (as a series of real numbers). The main result is the following.

Theorem A2

If X is a Banach space, then any absolutely convergent series in X converges in X. $\square$

Corollary A3

If $T: X \to X$ is a bounded linear map, $I: X \to X$ is the identity map and $\| T \| < 1$, then ΣT^n converges in the set $B(X)$ of bounded linear maps from X to X. Also if $S = \Sigma_{n=0}^{\infty} T^n$, then $S(I - T) = (I - T)S = I$, so S is the inverse of $I - T$. In particular, $I - T$ has a bounded inverse. □

A subset S of a Banach space X is said to be *closed* if, whenever $\phi_n \in S$ for all $n \in \mathbb{N}$ and $\phi_n \to \phi \in X$ as $n \to \infty$, $\phi \in S$.

Theorem A4

Let E be a vector subspace of a Banach space X. Then if E is a closed set, the vector space E, equipped with the norm inherited from X, is a Banach space. □

Theorem A5

Every finite-dimensional subspace of a Banach space forms a closed set (and, by Theorem A4, is therefore a Banach space in its own right). □

Theorem A6

Suppose that X and Y are Banach spaces and that $T: X \to Y$ is a linear map. If X has finite dimension then T is a bounded linear map. □

The theory in this book makes use of Hilbert space rather than the more general Banach space, although it is worth noticing that if H is a Hilbert space then the set $B(H)$ of all bounded linear maps from H to itself is not normally a Hilbert space. One of the distinguishing features of Hilbert space is the idea of orthogonality which arises from the inner product.

Definition A1

Two elements, ϕ and ψ, of a Hilbert space H are said to be *orthogonal* if $(\phi, \psi) = 0$. A sequence (ϕ_n) in H is said to be orthogonal if $(\phi_m, \phi_n) = 0$ whenever $m \neq n$. An orthogonal sequence (ϕ_n) is called *orthonormal* if it satisfies the additional condition that $\| \phi_n \| = 1$ $(n \in \mathbb{N})$; an equivalent statement that (ϕ_n) be orthonormal is that

$$(\phi_m, \phi_n) = \begin{cases} 1 & (m = n), \\ 0 & (m \neq n). \end{cases}$$ □

Lemma A7 (Bessel's inequality)

Let H be a Hilbert space and (ϕ_n) be an orthonormal sequence in H. Then if

$\phi \in H$

$$\sum_{n=1}^{\infty} |(\phi, \phi_n)|^2 \leqslant \|\phi\|^2. \qquad \square$$

Given an orthonormal sequence (ϕ_n), we may consider expressing ϕ in terms of the sum of its 'components in the direction of ϕ_n', that is, $\Sigma(\phi, \phi_n)\phi_n$. This is called the *Fourier series* of ϕ with respect to the sequence (ϕ_n). To use this we need a further idea.

Definition A2

An orthonormal sequence (ϕ_n) in the Hilbert space H is said to be *complete* if

$$(\phi, \phi_n) = 0 \ (n = 1, 2, \ldots) \Rightarrow \phi = 0. \qquad \square$$

If we set $\phi_n(x) = \begin{cases} 1 & (n=1) \\ 2^{\frac{1}{2}} \cos(\pi n x) & (n \text{ even}, n \geqslant 2) \\ 2^{\frac{1}{2}} \sin(\pi(n-1)x) & (n \text{ odd}, n \geqslant 3) \end{cases} (0 \leqslant x \leqslant 1)$

then (ϕ_n) is a complete orthonormal sequence in $L_2(0, 1)$, by the usual classical theory of Fourier series. Other complete orthonormal sequences are given by $\phi_n(x) = 2^{\frac{1}{2}} \sin n\pi x \ (n = 1, 2, \ldots)$, and by

$$\phi_n(x) = \begin{cases} 1 & (n=0) \\ 2^{\frac{1}{2}} \cos n\pi x & (n = 1, 2, \ldots) \end{cases}$$

where, for convenience, we choose the index set to be $\{0, 1, 2, \ldots\}$ in the last example.

Theorem A8

Let H be a Hilbert space and (ϕ_n) be an orthonormal sequence in H. The following are equivalent:

(i) (ϕ_n) is a complete orthonormal sequence;
(ii) for all $\phi \in H$, $\phi = \Sigma_{n=1}^{\infty} (\phi, \phi_n)\phi_n$;
(iii) for all $\phi \in H$, $\|\phi\|^2 = \Sigma_{n=1}^{\infty} |(\phi, \phi_n)|^2$. $\qquad \square$

That condition (iii) holds for a complete orthonormal sequence (ϕ_n) is called *Parseval's Theorem*. For a complete orthonormal sequence (ϕ_n), let us call the sequence $((\phi, \phi_n): n = 1, 2, \ldots)$ the sequence of *Fourier coefficients* of ϕ with respect to (ϕ_n). Then from A7 and A8 we deduce the following.

Theorem A9 (Riesz–Fischer)

Let (ϕ_n) be a complete orthonormal sequence in $L_2(a, b)$. Then the sequence (ξ_n) of complex numbers is the sequence of Fourier coefficients with respect to (ϕ_n) of some function $\phi \in L_2(a, b)$ if and only if $\Sigma|\xi_n|^2 < \infty$. □

One may construct Hilbert spaces which contain orthonormal sets which are uncountable. It turns out, in fact, that one can characterise a Hilbert space in terms of the cardinality of its complete orthonormal sets, all such sets in the same space having the same cardinality. (For this to make sense, of course, one has to extend the definitions above to orthonormal sets indexed by sets other than $\mathbb{N}$.) Since we know that $L_2(0, 1)$ has a sequence which is complete and orthonormal, and a simple change of variable shows this to hold for $L_2(a, b)$, we see that every complete orthonormal set is countable. This is included in the following theorem.

Theorem A10

Every orthonormal sequence in $L_2(a, b)$ may be extended to a complete orthonormal sequence, in the sense that if (ϕ_n) is orthonormal there is a sequence (ψ_n), containing all the members of (ϕ_n), which is orthonormal and complete. □

It is worth remarking here that the results stated above about the convergence of $\Sigma(\phi, \phi_n)\phi_n$ state that it converges in the Hilbert space concerned. Thus if $\phi \in L_2(a, b)$ and (ϕ_n) is a complete orthonormal sequence in $L_2(a, b)$, and $s_n(x) = \Sigma_{j=1}^{n}(\phi, \phi_j)\phi_j(x)$, then $s_n \to \phi$ in the norm in $L_2(a, b)$, that is, $s_n \to \phi$ in the mean, or, in symbols,

$$\int_a^b |\phi(t) - \sum_{j=1}^{n} (\phi, \phi_j)\phi_j(t)|^2 \, dt \to 0 \text{ as } n \to \infty.$$

Before leaving the topic of orthogonal sequences, let us notice that if (ϕ_n) is a linearly independent sequence of vectors in a Hilbert space H, then we may construct an orthonormal sequence (ψ_n) from it by the *Gram–Schmidt* process as follows:

$$\psi_1 = \phi_1/\|\phi_1\|,$$

$$\psi_{n+1} = \frac{\phi_{n+1} - \Sigma_{j=1}^{n}(\phi_{n+1}, \psi_j)\psi_j}{\|\phi_{n+1} - \Sigma_{j=1}^{n}(\phi_{n+1}, \psi_j)\psi_j\|}.$$

Definition A3

Let H_1 and H_2 be Hilbert spaces and $T \in B(H_1, H_2)$. Then there is a unique bounded linear map $T^*: H_2 \to H_1$ with the property that

$$\text{for all } \phi \in H_1, \; \psi \in H_2 \; (T\phi, \psi) = (\phi, T^*\psi).$$

This map T^* is called the *adjoint* of T. $\qquad\square$

Lemma A11

The adjoint operation $*$ has the following properties. For all bounded linear maps S, T from H_1 to H_2 and all scalars μ, $(S + T)^* = S^* + T^*$, $(\mu T)^* = \bar{\mu} T^*$, $(ST)^* = T^* S^*$, $(T^*)^* = T$, $I^* = I$ (if $H_1 = H_2$), $\|T^* T\| = \|T\|^2$ and $\|T\| = \|T^*\|$. $\qquad\square$

Definition A4

Let H_1 and H_2 be Hilbert spaces and $T: H_1 \to H_2$ be a bounded linear map. Then we denote by $\mathcal{N}(T)$ the *nullspace* of T

$$\mathcal{N}(T) = \{\phi \in H_1 : T\phi = 0\}$$

and by $\mathcal{R}(T)$ the *image* (or *range*) of T

$$\mathcal{R}(T) = \{T\phi : \phi \in H_1\}.$$

If $E \subseteq H$, then by the *orthogonal complement*, $E^\perp$, of E we mean the set $E^\perp = \{\phi \in H : \text{for all } \psi \in E, (\phi, \psi) = 0\}$. $\qquad\square$

Lemma A12

If E is a subset of a Hilbert space then $E^\perp$ is a closed subspace. If E is itself a closed subspace then $(E^\perp)^\perp = E$ and every vector $\phi \in H$ can be decomposed in a unique way into the form $\phi = \phi_1 + \phi_2$ where $\phi_1 \in E$ and $\phi_2 \in E^\perp$. $\qquad\square$

An operator $P \in B(H)$ is called a *projection* (or orthogonal projection) if $P = P^2 = P^*$. The map P defined by $P\phi = \phi_1$, where $\phi = \phi_1 + \phi_2$ ($\phi_1 \in E$, $\phi_2 \in E^\perp$) is a projection, the *orthogonal projection* onto E, when E is a closed subspace of H.

The closure of a subset S of a Banach space X is the set of all $\phi \in X$ for which there is a sequence (ϕ_n) with $\phi_n \in S$ for all n and for which $\phi_n \to \phi$ as $n \to \infty$. The closure of S is denoted by $\bar{S}$, and S is closed if and only if $S = \bar{S}$. A subset S of X is called *dense* if $\bar{S} = X$.

Theorem A13

Let H_1 and H_2 be Hilbert spaces over the same field of scalars and $T \in B(H_1, H_2)$. Then

$$\mathcal{N}(T) = \mathcal{R}(T^*)^\perp \text{ and } \overline{\mathcal{R}(T)} = \mathcal{N}(T^*)^\perp. \qquad \square$$

(For details and proofs of the functional analysis, see *An Introduction to Hilbert Space* by N. Young, Cambridge University Press, 1988.)

Appendix B: Measure theory and integration

Measure theory and (Lebesgue) integration are both very detailed and technical topics, to an extent that can be off-putting to anyone seeking results to use elsewhere. In this appendix we have attempted to suppress detail that is not required for the purposes in hand – for example, we do not need to know the definition of a measurable function, only enough to show that the functions we shall encounter are measurable. We have not stated the results in their greatest generality.

We shall need to use a vector space of functions which forms a Hilbert space and the natural choice of norm associated with this is given by $\|\phi\| = \{\int |\phi(t)|^2 \, dt\}^{\frac{1}{2}}$. This definition does indeed define a norm on the set of continuous functions on $[a, b]$ but the resulting space is not complete and we need to enlarge our class of functions, which in turn requires a suitable integral, the Lebesgue integral.

A necessary condition for a function $\phi: S \to \mathbb{R}$ (where S is, for the moment, considered to be an interval, possibly of infinite length, in $\mathbb{R}$) to be Lebesgue integrable is that it be *measurable*. The definition of measurability need not concern us here; what we do need to know is that all continuous functions, monotonic functions and all step functions are measurable and that the modulus of a measurable function, the sum and product of two measurable functions and a real-valued function which is the pointwise limit of a sequence of measurable functions are all measurable.

If $\phi: S \to [0, \infty)$ is measurable then the integral $\int_S \phi$ is meaningful provided we admit ∞ as a value. We therefore consider the set of all measurable functions ϕ for which $\int_S |\phi| < \infty$, such a function being said to be *Lebesgue integrable*; it is then automatic that $\int_S \phi$ is meaningful. When the function ϕ is integrable in the simpler sense of Riemann integration the two integrals (Riemann and Lebesgue) have the same value.

To form a Banach space we need to consider functions for which $\int_S |\phi| = 0$. Such functions are said to be equal to zero *almost everywhere* (written as $\phi = 0$ (a.e.)). To be more precise about this we require the notion of the (Lebesgue) measure of a set, which is generalisation of the idea of the length of an interval. Not all subsets of $\mathbb{R}$ are measurable, but the empty set, all intervals, and all countable sets are measurable, while a union of countably many measurable sets is measurable. Moreover, letting $\mu(E)$ denote the measure of E, $\mu(E) \geqslant 0$ for all E, $\mu(\emptyset) = 0$, $\mu((a, b)) = b - a$, and if (E_n) is a sequence of pairwise disjoint measurable sets $\mu(\bigcup_{n=1}^{\infty} E_n) = \Sigma_{n=1}^{\infty} \mu(E_n)$. The measurable function ϕ is said to be equal to 0 *almost everywhere* if and only if $\{t: \phi(t) \neq 0\}$ has measure 0. From this we see that if $f_n = 0$ (a.e.) $(n = 1, 2, 3, \ldots)$ then $\Sigma_{n=1}^{\infty} f_n = 0$ (a.e.) since $\{t: \Sigma_{n=1}^{\infty} f_n(t) \neq 0\} \subseteq \bigcup_{n=1}^{\infty} \{t: f_n(t) \neq 0\}$.

To construct the Banach space $L_2(S)$ we proceed as follows. We consider the set of all measurable functions ϕ defined on S (with real or complex values, as desired) which satisfy the property that $\int_S |\phi|^2 < \infty$. Two such functions ϕ and ψ are said to be equivalent if $\phi = \psi$ almost everywhere and $L_2(S)$ consists of the set of all such equivalence classes. The norm of an equivalence class is $\{\int_S |\phi|^2\}^{\frac{1}{2}}$ where ϕ is a function which is a member of the equivalence class, the inner product being given by $\int_S \phi(t)\overline{\psi(t)} \, dt$ where ϕ and ψ are chosen from the two equivalence classes. The algebraic operations are defined by choosing members of the equivalence classes involved, performing the algebraic operation on them and forming the equivalence class which contains the result. Notice that if we choose a member ϕ of one of the equivalence classes which forms $L_2(S)$ then this is only determined almost everywhere in the sense that if we choose a second representative ϕ' of the same class then $\phi = \phi'$ almost everywhere.

Having made this elaborate construction of $L_2(S)$ it is customary to be more casual than is strictly correct and speak of 'an L_2 function' rather than the equivalence class it generates. This will not lead to error provided the function is regarded as only being determined almost everywhere. Where we need to distinguish between a function and the class it generates we shall refer to a *square-integrable* function as one whose equivalence class belongs to $L_2(S)$. This is relevant if we wish to show, for example, that the class contains a continuous function.

A similar definition is made of $L_1(S)$, the set of equivalence classes of measurable functions for which $\int_S |\phi| < \infty$ with the norm defined by $\|\phi\| = \int_S |\phi|$. This is a Banach space (but not a Hilbert space).

In dealing with the integrals involved we need a number of results, the simplest of which is *Schwarz's inequality* that if ϕ and ψ are square-integrable functions on S then $|\int_S \phi(t)\psi(t) \, dt|^2 \leqslant \int_S |\phi(t)|^2 \, dt \int_S |\psi(t)|^2 \, dt$ with equality possible only if one of the functions is equal almost everywhere to a constant multiple of the other.

Notice that, since (a, b) and $[a, b]$ differ by a set of measure zero, $L_2((a, b))$

and $L_2([a, b])$ are equal, so we shall use $L_2(a, b)$ to denote them. Notice also that Schwarz's inequality shows that $L_2(a, b) \subseteq L_1(a, b)$.

Theorem B1

The sets $C[a, b]$ and $C^\infty[a, b]$ of continuous and of infinitely differentiable functions respectively on $[a, b]$ are dense in $L_1(a, b)$ and in $L_2(a, b)$. So are the sets $C_0[a, b]$ and $C_0^\infty[a, b]$ of such functions which take the value 0 at a and b. □

(We shall not labour the point again, but notice that the set which is dense in $L_1(a, b)$ is the set of all *equivalence classes* which contain a member of $C[a, b]$, etc.)

Theorem B2

The set of all C^∞ functions on $\mathbb{R}$ (or on $(0, \infty)$) which are zero outside some interval of finite length (the interval depending on the function) is dense in $L_1(\mathbb{R})$ and $L_2(\mathbb{R})$ (or $L_1(0, \infty)$ and $L_2(0, \infty)$). □

Theorem B3 (Monotone Convergence Theorem)

Let $\phi_n: S \to \mathbb{R}$ be a sequence of functions in $L_1(S)$ such that, for almost all $x \in S$, $(\phi_n(x))$ is an increasing sequence. Then if $\phi_n(x) \to \phi(x)$(a.e.) as $n \to \infty$, $\lim_{n \to \infty} \int_S \phi_n = \int_S \phi$ (where ∞ is acceptable as a value on both sides). □

Theorem B4 (Dominated Convergence Theorem)

Let $\phi_n: S \to \mathbb{R}$ be a sequence of functions in $L_1(S)$ and let $\phi_n(x) \to \phi(x)$ (a.e.) as $n \to \infty$. If there is a function $\psi \in L_1(S)$ with $|\phi_n(x)| \leqslant \psi(x)$ for all n and almost all x, then $\int_S \phi_n \to \int_S \phi$ as $n \to \infty$. □

Theorem B5

Suppose that ϕ_n, $\phi \in L_2(S)$ and that $\|\phi_n - \phi\| \to 0$ as $n \to \infty$. If, in addition, $\phi_n \to \psi$ pointwise (i.e. for all $x \in S$ $\phi_n(x) \to \psi(x)$), then $\phi = \psi$ almost everywhere. □

Theorem B6 (The Fundamental Theorem of Calculus)

(i) Suppose that $\phi \in L_1(a, b)$ and that $\Phi(x) = \int_a^x \phi$. Then Φ is differentiable at almost all points of $[a, b]$ and $\Phi' = \phi$ (a.e.).

(ii) Suppose that ψ is a continuous function on $[a, b]$, that ψ is differentiable at all but finitely many points of $[a, b]$ and that $\psi' \in L_1(a, b)$. Then $\int_a^b \psi' = \psi(b) - \psi(a)$. □

Notice that one immediate corollary of Theorem B6 is that if $\phi \in L_1(a, b)$ and $\int_a^x \phi = 0$ (a.e.) then $\phi = 0$ (a.e.).

Similar definitions apply to integrating functions of several real variables, the main point of interest for us being conditions for the existence of integrals and the interchange of the order of integration in a repeated integral. For a function of two variables to be integrable it must be measurable. We notice only that continuous functions, functions of the form $\phi(x)\psi(y)$ and $\phi(x - y)$ where ϕ and ψ are measurable, and characteristic functions of open and closed sets are all measurable, also that the sum and product of two measurable functions and the pointwise limit of a sequence of measurable functions are all measurable. The principal results are Fubini's and Tonelli's Theorems.

Theorem B7 (Tonelli's Theorem)

Let E and F be intervals in $\mathbb{R}$ and let $\phi: E \times F \to \mathbb{C}$ be measurable. Then

(i)   for almost all $s \in E, t \mapsto \phi(s, t)$ is a measurable function. If this function is integrable for almost all $s \in E$, then $s \mapsto \int_F \phi(s, t) \, dt$ is a measurable function of s.

(ii)   if $\int_E \{\int_F |\phi(s, t)| \, dt\} \, ds$ is finite then $\phi \in L_1(E \times F)$.

(iii) the corresponding results obtained by interchanging the roles of s and t are also true.   □

Theorem B8 (Fubini's Theorem)

Let E and F be intervals in $\mathbb{R}$ and let $\phi: E \times F \to \mathbb{C}$ belong to $L_1(E \times F)$. Then

(i)   for almost all $s \in E, t \mapsto \phi(s, t)$ defines an element of $L_1(F)$.

(ii) $s \mapsto \int_F \phi(s, t) \, dt$ defines an element of $L_1(E)$.

(iii) for almost all $t \in E, s \mapsto \phi(s, t)$ defines an element of $L_1(E)$.

(iv) $t \mapsto \int_E \phi(s, t) \, ds$ defines an element of $L_1(F)$.

(v) $\int_E \{\int_F \phi(s, t) \, dt\} \, ds = \int_F \{\int_E \phi(s, t) \, ds\} \, dt = \iint_{E \times F} \phi(s, t) \, ds \, dt$.   □

The principal practical point is part (v) of Fubini's Theorem, that the values of the two repeated integrals are equal, and Tonelli's Theorem shows that this is true if one of the two integrals $\int_E \{\int_F |\phi(s, t)| \, dt\} \, ds$ and $\int_F \{\int_E |\phi(s, t)| \, ds\} \, dt$ is finite.

A good reference for the measure theory and integration is *An Introduction to Classical Real Analysis* by K. R. Stromberg, Wadsworth International, 1981.

Appendix C: Miscellaneous results

Some Fourier series

Let $f(x) = \log 2(1 - \cos x)$ $(0 < x < 2\pi)$ and write

$$f(x) = \tfrac{1}{2}a_0 + \sum_{n=1}^{\infty} \{a_n \cos(nx) + b_n \sin(nx)\} \quad (0 < x < 2\pi).$$

Then $b_n = \pi^{-1} \int_0^{2\pi} f(x) \sin(nx)\,dx = 0$ $(n \in \mathbb{N})$ since $f(2\pi - x) = f(x)$ $(0 < x < 2\pi)$. For $n \in \mathbb{N}$,

$$a_n = \frac{1}{\pi} \int_0^{2\pi} f(x) \cos(nx)\,dx = \frac{2}{\pi} \int_0^{2\pi} \log(2 \sin\tfrac{1}{2}x) \cos(nx)\,dx$$

$$= \frac{4}{\pi} \int_0^{\pi} \log(2 \sin x) \cos(2nx)\,dx = -\frac{2}{n\pi} \int_0^{\pi} \frac{\cos x \sin(2nx)}{\sin x}\,dx, \quad \text{(C1)}$$

on integrating by parts.

Now let

$$C_n = \int_0^{\pi} \frac{\sin(2n-1)x}{\sin x}\,dx \quad (n \in \mathbb{N})$$

and note that

$$C_{n+1} - C_n = 2 \int_0^{\pi} \cos(2nx)\,dx = 0 \quad (n \in \mathbb{N}).$$

Since $C_1 = \pi$ we therefore have $C_n = \pi$ $(n \in \mathbb{N})$ and from (C1) $a_n = -(C_{n+1} + C_n)/n\pi = -2/n$ $(n \in \mathbb{N})$.

We also require

$$a_0 = \frac{1}{\pi} \int_0^{2\pi} \log 2(1 - \cos x)\,dx = \frac{4}{\pi} \int_0^{\pi} \log(2 \sin x)\,dx. \quad \text{(C2)}$$

Therefore

$$a_0 = \frac{8}{\pi} \int_0^{\pi/2} \log(2 \sin x)\,dx = \frac{8}{\pi} \int_0^{\pi/2} \log(2 \cos x)\,dx,$$

whence

$$2a_0 = \frac{8}{\pi} \int_0^{\pi/2} \log(2 \sin 2x)\, dx = \frac{4}{\pi} \int_0^{\pi} \log(2 \sin x)\, dx$$

and reference to (C2) now shows that $a_0 = 0$.

We have proved that

$$-\tfrac{1}{2} \log 2(1 - \cos x) = \sum_{n=1}^{\infty} n^{-1} \cos(nx) \quad (0 < x < 2\pi).$$

It follows that

$$\log\{2 \sin \tfrac{1}{2}(x+t)\} = - \sum_{n=1}^{\infty} n^{-1} \cos\{n(x+t)\}$$

$$(0 \leqslant x, t \leqslant \pi, \ x+t \neq 0, \ x+t \neq 2\pi) \qquad \text{(C3)}$$

and that

$$\log\{2 \sin \tfrac{1}{2}(x-t)\} = - \sum_{n=1}^{\infty} n^{-1} \cos\{n(x-t)\} \quad (0 \leqslant t < x \leqslant \pi). \qquad \text{(C4)}$$

Adding (C3) and (C4) we find that

$$\log\{2(\cos t - \cos x)\} = - \sum_{n=1}^{\infty} 2n^{-1} \cos(nx) \cos(nt) \quad (0 \leqslant t < x \leqslant \pi)$$

and, by symmetry,

$$\log\{2|\cos x - \cos t|\} = - \sum_{n=1}^{\infty} 2n^{-1} \cos(nx) \cos(nt) \quad (0 \leqslant x, t \leqslant \pi, \ x \neq t).$$

Subtraction of (C4) from (C3) similarly gives

$$\log \left| \frac{\sin \tfrac{1}{2}(x+t)}{\sin \tfrac{1}{2}(x-t)} \right| = \sum_{n=1}^{\infty} 2n^{-1} \sin(nx) \sin(nt) \quad (0 \leqslant x, t \leqslant \pi, \ x \neq t).$$

Note that $\log\{2|\cos x + \cos t|\} = -\sum_{n=1}^{\infty} 2n^{-1}(-1)^n \cos(nx) \cos(nt)$ $(0 \leqslant x, t \leqslant \pi, \ x+t \neq \pi)$ and hence that

$$\log \left| \frac{\cos x + \cos t}{\cos x - \cos t} \right| = \sum_{n=1}^{\infty} 4(2n-1)^{-1} \cos\{(2n-1)x\} \cos\{(2n-1)t\}$$

$$(0 \leqslant x, t \leqslant \pi, \ x \neq t, \ x+t \neq \pi).$$

A further Fourier series used in the text is easily established. First note that

$$\log(1 - a\,e^{ix}) = -\sum_{n=1}^{\infty} n^{-1} a^n\, e^{inx} \quad (0 < a < 1, 0 \leqslant x \leqslant 2\pi)$$

and take real parts to give

$$\log(1 - 2a\cos x + a^2) = -\sum_{n=1}^{\infty} 2n^{-1} a^n \cos(nx) \quad (0 < a < 1, 0 \leqslant x \leqslant 2\pi).$$

The gamma and beta functions

The gamma function $\Gamma: (0, \infty) \to \mathbb{R}$ is defined by

$$\Gamma(x) = \int_0^\infty e^{-t}\, t^{x-1}\, dt \quad (x > 0)$$

and the beta function $B: (0, \infty) \times (0, \infty) \to \mathbb{R}$ is defined by

$$B(x, y) = \int_0^1 t^{x-1}(1 - t)^{y-1}\, dt \quad (x > 0, y > 0).$$

It is easily shown that $\Gamma(1) = 1$ and an integration by parts gives the recurrence formula

$$\Gamma(x + 1) = x\Gamma(x) \quad (x > 0),$$

from which it follows that

$$\Gamma(n) = (n - 1)! \quad (n \in \mathbb{N}).$$

The gamma and beta functions are connected by the relationship

$$B(x, y) = B(y, x) = \Gamma(x)\Gamma(y)/\Gamma(x + y) \quad (x > 0, y > 0). \tag{C5}$$

This formula, and others following, are derived in many texts on analysis or calculus. The reflection formula for gamma functions is

$$\Gamma(x)\Gamma(1 - x) = \pi\,\mathrm{cosec}(\pi x) \quad (0 < x < 1) \tag{C6}$$

and from (C5) we see that

$$B(x, 1 - x) = \pi\,\mathrm{cosec}(\pi x) \quad (0 < x < 1).$$

The relationship

$$\Gamma(2x) = \pi^{-\frac{1}{2}}\, 2^{2x-1}\Gamma(x)\Gamma(x + \tfrac{1}{2}) \quad (x > 0)$$

is called the duplication formula. From this and also from (C5) we see that $\Gamma(\tfrac{1}{2}) = \pi^{\frac{1}{2}}$. Other values of Γ used in the text are $\Gamma(\tfrac{1}{4}) = 3.6256\ldots$ and

$\Gamma(\frac{3}{4})=2^{\frac{1}{2}}\pi/\Gamma(\frac{1}{4})=1.2254\ldots$. If $a>0$ and $b>0$ then $x^{b-a}\Gamma(x+a)/\Gamma(x+b)=1+O(x^{-1})$ as $x\to\infty$.

A further function which is sometimes useful in evaluating integrals is the psi (or digamma) function $\psi:(0,\infty)\to\mathbb{R}$ defined by

$$\psi(x)=\frac{d}{dx}\log_e\Gamma(x)=\Gamma'(x)/\Gamma(x) \quad (x>0), \tag{C7}$$

or by

$$\psi(x)=\int_0^\infty \left\{\frac{e^{-t}}{t}-\frac{e^{-xt}}{1-e^{-t}}\right\} dt \quad (x>0). \tag{C8}$$

A Cauchy principal value integral

Let

$$C_\gamma=\int_0^\infty \frac{u^{\gamma-1}}{1-u} du \quad (0<\gamma<1)$$

where the integral is to be interpreted as a principal value. Setting $u=e^s$ we have

$$C_\gamma=\int_{-\infty}^\infty \frac{e^{\gamma s}}{1-e^s} ds=\lim_{\epsilon\to 0}\left\{\int_{-\infty}^{-\epsilon}\frac{e^{\gamma s}}{1-e^s} ds+\int_\epsilon^\infty\frac{e^{\gamma s}}{1-e^s} ds\right\}$$

$$=\lim_{\epsilon\to 0}\int_\epsilon^\infty\left\{\frac{e^{-\gamma s}}{1-e^{-s}}+\frac{e^{\gamma s}}{1-e^s}\right\} ds.$$

Therefore

$$C_\gamma=\int_0^\infty \frac{\sinh\{(\frac{1}{2}-\gamma)s\}}{\sinh(\frac{1}{2}s)} ds \tag{C9}$$

and in particular

$$C_{\frac{1}{2}}=\int_0^\infty \frac{u^{-\frac{1}{2}} du}{1-u}=0.$$

Note from (C6) and (C7) that for $0<x<1$ $\psi(1-x)-\psi(x)=-d/dx\log_e\{\Gamma(x)\Gamma(1-x)\}=\pi\cot(\pi x)$ and therefore, using (C8)

$$\pi\cot(\pi x)=\int_0^\infty \frac{(e^{-xt}-e^{-(1-x)t})}{1-e^{-t}} dt=\int_0^\infty\frac{\sinh\{(\frac{1}{2}-x)t\}}{\sinh(\frac{1}{2}t)} dt \quad (0<x<1).$$

We conclude that $C_\gamma=\pi\cot(\pi\gamma)$.

The integral in (C9) can be evaluated in other ways, using contour integration, for example.

Notation Index

Symbol	Description	page
$\overline{A}$	closure of A	356
A_m	mth trace	223
$B(X), B(X, Y)$	spaces of operators	352
$B(x, y)$	beta function	364
$C([a, b]), C^\infty([a, b])$	spaces of functions	360
det A	determinant of A	209
$E^\perp$	orthogonal complement of E	356
$\mathscr{F}$	field of scalars ($\mathbb{R}$ or $\mathbb{C}$)	95
I	identity operator	55, 352
$J(p)$	particular functional	248
$J_K(p)$	particular functional	204
$J(p, q)$	particular functional	260
$K^2, K^{1/2}$	power of operator K	10, 167
$\|K\|$	norm of operator K	352
$L_1(S)$	space of Lebesgue integrable functions	359
$L_2(a, b)$	space of square-integrable functions	351, 359
$\mathscr{N}(T)$	nullspace of T	103, 356
$O(h^n)$	order	205
$R_K(\phi)$	Rayleigh quotient	141
$\mathscr{R}(T)$	image (range) of T	103, 356
sgn(x)	sign of x	–
T^*	adjoint of T	356
$\bar{z}$	complex conjugate	–
$\Gamma(x)$	gamma function	364
λ, μ	parameters	4, 96
$\mu_n, \mu_n(K)$	eigenvalues	141
μ_n^+, μ_n^-	eigenvalues	141
$\rho(T)$	spectral radius of T	86
$\sigma(T)$	spectrum of T	96

367

$\phi_n, \phi_n^+, \phi_n^-$	eigenvectors	141
$\phi_\gamma(x)$	particular function	308
$\psi(x)$	digamma function	365
$(a \leqslant x \leqslant b)$	for all x in $[a, b]$	2
$\| \cdot \|$	norm	351, 352
(ϕ, ψ)	inner product (also ordered pair)	352

Index

Abel's equation, 1, 23, 291
Abel's problem, 21
absolute convergence, 352
adjoint equation, 61, 104, 105–6
adjoint operator, 102, 104–5, 356–7
aerofoil equation, 319
Airy's equation, 9, 24
almost everywhere, 66, 67, 114, 358–9
alternative, *see* Fredholm Alternative
approximation
    by finite-rank operator, 146, 245, 277,
        287–8
    by particular classes of function, 274, 286
    by powers of operator, 137, 279, 280
    of compact operator, 211, 212
    of continuous kernel, 245, 277, 288
    of eigenvalues, eigenvectors etc., 203–40
    of solution of inhomogeneous equation,
        241–90
associated Legendre equation, 51–2

Banach space, 351
Banach space, equations in, 63–4
Bessel function, 20, 92, 93, 181, 236, 337, 338,
    340
Bessel's equation, 45, 48, 51, 92, 93, 188, 235
Bessel's inequality, 353
beta function, 364
bivariational principle, 260
boundary condition
    change of, 149, 162
    mixed, 17, 18, 52, 289, 335
    regular/singular, 44
    weak, 45, 52
boundary-value problem, 13, 16, 34, 44, 149,
    150, 155, 182ff, 289, 335
bounded linear map, *see* operator
bounded below, 164, 165
bounded sequence, 351
bounds
    complementary, 214, 215, 224, 227, 232–3,
        261ff, 267, 285
    pointwise, 264
    see also upper bounds and lower bounds

Calculus, Fundamental Theorem of, 360
Cauchy sequence, 351
Cauchy principal value, 304, 306, 309, 365
Cauchy singular, 7, 20
Cauchy singular equation, 308ff
change of boundary conditions, 149, 162
classification of integral equations, 2
closed subset, 103, 353
commutativity, 133, 152, 166, 352
compact operator, 69, 72, 75, 95–138, 135,
    136, 211, 212
compactness, of operator generated by
    L_2-kernel, 72
    of operator generated by weakly singular
        kernel, 76
comparison of operators, 176–8, 227ff
complementary bounds, 214, 215, 224, 227,
    232–3, 261ff, 267, 285
completely continuous, *see* compact operator
completeness, of normed vector space, 351
    of orthonormal sequence, 112, 354
compression of operator, 154, 182, 212
continuity condition on solutions of ODEs,
    7, 8
continuous dependence on data, 138, 195,
    243
continuous kernel, 126, 128–31, 245
convergence, in the mean, 66, 120, 132, 352,
    355
    of sequence of projections, 211–2
    uniform/pointwise, *see* uniform or
        pointwise convergence
conversion of differential equation to integral
    equation, 31ff
convolution, 333
cosine transform, 337–8, 350
coupled equations, 16

degenerate kernel, 57ff, 70, 94, 146, 150
dense set, 301, 356
derivative of eigenvector, 189
diagonalisation, 113
differential equations,
    Airy's equation, 9, 24

369

associated Legendre equation, 51–2
Bessel's equation, 45, 48, 51, 92, 93, 188, 235
Gegenbauer's equation, 52, 230
Hill's equation, 49
Laplace's equation, 16, 18, 30, 197, 335
Legendre's equation, 46–7
Tricomi's equation, 17, 20
  see also ordinary or partial differential equation
digamma function, 365
dimension of nullspace, 99, 106
Dini's Theorem, 126, 136
dominant eigenvalue, 203
Dominated Convergence Theorem, 360
dual bounds, *see* complementary bounds
dual extremum principles, 250, 257
dual integral equation, 19

eigenvalue, 14, 95, 99, 100, 101, 108, 109, 110, 120, 122, 164, 206ff, 214ff
  approximation of, 206ff
  conventions for numbering, 141
  inequalities for, 146ff, 162
  minimax results, 142, 144, 145
eigenvector, 95, 101, 108, 110, 120, 122, 189, 206, 218ff
  approximation of, 217ff
  completeness of sequence of, 112, 191, 193
  conventions for numbering, 141
  derivative of, 189
elliptic partial differential equation, 16, 289
equivalence class in $L_2(a, b)$, choice of representative, 67, 69, 114–5
equivalence of differential equations and integral equations, 31ff, 41
essential supremum, 93
essentially bounded, 93, 155
estimates for eigenvalues, 142ff, 163ff
existence of solution of integral equation, 116, 117, 342
  see also Fredholm Alternative
expansion of kernel, 120, 122, 126, 129
expansion of operator, spectral, 109–10, 118
extension of operator, 301
extremum principle, *see* maximum or minimum principle
extremum principles, dual, 250, 257

finite rank, 69, 70, 94, 102, 153, 353
finite-dimensional space, 353
first kind equation, 3, 4, 116, 117, 119, 259, 291ff, 295ff, 319ff
Fourier cosine transform, 337–8, 350
Fourier series, 354, 355, 362ff
Fourier sine transform, 17, 335, 350
Fourier transform, 1, 181, 301, 302, 306, 307, 330ff

Fredholm Alternative, 61, 69, 70, 78, 104, 105, 114
Fredholm equation, 3
free term, 2
Fubini's Theorem, 74, 105, 361
functional, 205
Fundamental Theorem of Calculus, 360

Galerkin's method, 210, 269ff
Galerkin's method, iterated, 277–9
gamma function, 364
Gegenbauer's equation, 52, 230
Gegenbauer polynomial, 230
Gram–Schmidt process, 355

hermitian kernel, 107, 120, 124, 128–31, 158–61
hermitian matrix, 208
hermitian operator, *see* self-adjoint
Hilbert space, 352
Hilbert transform, 307ff
Hill's equation, 49
homogeneous equation, 5

image (or range) of operator, 103, 356, 357
inclusion principle, 216
inequalities for eigenvalues, 146ff, 162
infimum, attainment of, 144
inhomogeneous equation, 5
initial values/initial-value problem, 7, 33, 84
inner product, 352
integral, *see* Lebesgue
integral, principal value, 304, 306, 308, 309, 365
integro-differential equation, 348
inverse, 69, 165
invertible/invertibility, 69, 87, 96, 97, 133, 135, 160
iterated Galerkin method, 277–9
iterated kernel, 82
iterated trial function, 222, 253

Kantorovich's method, 276, 281
Kellogg's method, 217
kernel, 2
  approximation of, 245, 277, 288
  continuous, 126, 128–31, 245
  degenerate, 57ff, 70, 94, 146, 150
  differentiable, 158, 161, 277, 288
  hermitian, 107, 120, 124, 128–31, 158–61
  in sense of of vectors annihilated,
    see nullspace
  L_2-, 72, 83, 104, 105, 107, 120, 122, 124, 132, 160
  non-negative-valued, 159
  Poisson, 134
  resolvent, 13, 115, 130–2, 194
  restriction of domain, 172

Schmidt, 123–5, 132, 195
Schur, 74, 105, 107, 133, 135, 136, 160, 173
skew-symmetric, 240
symmetric, 14, 108
weakly singular, 6, 74, 76, 173, 174, 291ff
kind (of equation), 3
  see also first kind, second kind

Lagrange multiplier, 205, 206
Laplace's equation, 16, 18, 30, 197, 335
Laplace transform operator, 178
latent root, *see* eigenvalue
Lebesgue Convergence Theorems, 360
Lebesgue integral, 358
Lebesgue measure/measurable, 358, 359
Legendre's equation, 46
linear integral equation, 5
linear map, 63
linear transformation, 55
logarithmically singular kernel, 6, 320
lower bound, 214, 215, 218–9, 220–1, 224–5, 227, 232–3, 234, 255, 259, 261, 263, 267, 285
L_2-kernel, 72, 83, 104, 105, 107, 120, 122, 124, 132, 160

maximum, attainment of, 144
maximum principle, 206, 248, 249, 250, 256–7
mean convergence, 66, 120, 132, 352, 355
measure, 358
measurable function, 358
Mercer's Theorem, 126, 130, 171
minimax results, 142, 144, 145
minimum, attainment of, 144
minimum principle, 206, 250, 256–7, 259
mixed boundary condition, 17, 52, 289, 335
mixed boundary-value problem, 17, 18, 52, 289, 335
Monotone Convergence Theorem, 360

Neumann series, 78ff
non-negative operator, 126, 129, 150, 160, 163ff
non-self-adjoint equation, bounds for, 266–7
non-singular, *see* invertible
norm, 64, 69, 351, 352
norm-preserving operator, 168, 200
normal operator, 199
nullspace, 103, 106, 356, 357

operator, 69, 72ff, 352
operator, simplifying, 294, 309, 323, 339, 345, 348
  Volterra, 83, 291ff
order of integration, reversal of, 105, 345, 361
ordinary differential equation,
  boundary-value problem, 13, 34, 182ff
  conventions about solution, 8

conversion to integral equation, 7ff, 31ff
  equivalence to integral equation, 41
  initial-value problem, 7, 33, 84
orthogonal complement, 356
orthogonal projection, 356
  see also projection
orthogonal vectors, 108, 112, 353
orthogonality with respect to weight function, 192
orthonormal sequence, 99, 110, 111, 112, 353, 354, 355
  complete, 112, 354, 355

Parseval's Theorem, 73, 354
partial differential equation, 16, 30, 112, 289, 335
piecewise constant function, 275
piecewise linear function, 275
Poincaré–Bertrand formula, 345
pointwise bounds, 264
pointwise convergence, 124
Poisson kernel, 134
positive operator, 163–202, 248, 254, 259
  comparison results, 176ff
positive semi-definite, *see* non-negative,
power method, 220
principal value, 304, 306, 309, 365
principal value integral, reversal of
  integration order, 345
principles, *see* maximum principle, minimum
  principle etc.
projection, 153, 154, 160, 211, 212, 356
projection methods, 270
psi function, 365

radius (of spectrum), 86
range (or image) of operator, 103, 356, 357
rank, 69
Rayleigh quotient, 139, 203, 206
Rayleigh–Ritz method, 207, 210, 213, 214, 260, 267, 272
reciprocal principle, 260, 268
recovery of properties of solution, 67, 69, 114–15
regular boundary value problem, 44
residual error, 251, 271
resolvent kernel/operator, 13, 115, 130–2, 194
restriction of kernel to smaller domain, 172
reversal of order of integration, 74, 105, 345, 361
Riesz–Fischer Theorem, 116, 120, 355

scalar product, *see* inner product
Schmidt kernel, 123–5, 132, 195
Schmidt's Theorem, 124
Schur kernel, 74, 105, 107, 133, 135, 136, 160, 173
Schwarz's inequality, 359

Schwarz's method, 220
Schwarz quotient, 221
second kind equation, definition of, 3, 5
second-order accuracy, 207, 246, 261
self-adjoint operator, 107ff
separation of variables, 112
simple eigenvalues, 225
simplifying operator, 294, 309, 323, 339, 345, 348
sine transform, 17, 335, 350
singular boundary point, 45
singular boundary value problem, 44ff
singular integral, *see* principal value integral,
singular integral equation, 3, 6, 7, 20, 308ff, 330ff
singular value, 118, 122
skew-adjoint, 135, 240
spectral radius, 86
Spectral Theorem, 109
spectrum, 96, 97, 108, 109, 133, 135, 280, 332, of compact operator, 98, 101, 280
square-integrable, 359
square root, 166, 196
stationary principle, 206, 260, 266
stationary point, 205–7, 260
stationary value, 204–7, 261, 266
strongly singular (Cauchy singular), 7, 20, 308ff
Sturm–Liouville problem, 34, 150, 155, 161, 182ff
successive substitution, 11

supremum, attainment of, 144
symmetric kernel, 14, 108

test function, 207
third kind equation, 3, 4, 5
Tonelli's Theorem, 361
trace, 120, 129, 223, 225
trial function, 207
Tricomi's equation, 17, 20

uniform convergence, 124, 125, 126, 127, 195
uniqueness of solutions, *see* Fredholm Alternative
unitary operator, 177, 200, 303
upper bound, 214, 215, 219, 222–3, 224–5, 227, 232–3, 256, 261, 263, 267, 285

variation-iteration, 222
variational methods, accuracy of, 207
variational principle, 203ff, 246ff
Volterra equation, 3, 9, 83
Volterra operator, 83, 291ff

weak boundary condition, 45, 52
weakly convergent sequence, 99–100
weakly singular equation/kernel, 6, 74, 76, 173, 174, 291ff
weight function, 192
well-posedness, 138, 195, 243
Wronskian, 32